# Estrogen and the Vessel Wall

**The Endothelial Cell Research Series**

A series of significant reviews of basic and clinical research related to the endothelium.
Edited by ***Gabor M. Rubanyi,*** *Berlex Biosciences, Richmond, California.*

---

**Volume One**

Endothelium-Derived Hyperpolarizing Factor
edited by *Paul M. Vanhoutte*

**Volume Two**

Endothelial Modulation of Cardiac Function
edited by *Malcolm J. Lewis* and *Ajay M. Shah*

**Volume Three**

Estrogen and the Vessel Wall
edited by *Gabor M. Rubanyi* and *Raymond Kauffman*

**Volumes in Preparation**

Modern Visualisation of Endothelium
*J. Polak*

Morphogenesis of Endothelium
*W. Risau*

Pathophysiology and Clinical Applications of Nitric Oxide
*G. M. Rubanyi*

Vascular Endothelium in Human Physiology and Pathophysiology
*P. Vallance*

Mechanical Forces and the Endothelium

---

This book is part of a series. The publisher will accept continuation orders which may be cancelled at any time and which provide for automatic billing and shipping of each title in the series upon publication. Please write for details.

# Estrogen and the Vessel Wall

Edited by

**Gabor M. Rubanyi**
*Berlex Biosciences*
*Richmond*
*USA*

and

**Raymond Kauffman**
*Lilly Research Laboratories*
*Indianapolis*
*USA*

harwood academic publishers

Australia • Canada • China • France • Germany • India • Japan
Luxembourg • Malaysia • The Netherlands • Russia • Singapore
Switzerland • Thailand

Printed in India.

Amsteldijk 166
1st Floor
1079 LH Amsterdam
The Netherlands

---

**British Library Cataloguing in Publication Data**

Estrogen and the vessel wall. – (Endothelial cell research series; 3)
1. Estrogen – Therapeutic use 2. Cardiovascular pharmacology
I. Rubanyi, Gabor M., 1947– II. Kauffman, Raymond
615.7'1

ISBN 90-5702-283-4

# CONTENTS

# PREFACE

A large and growing body of epidemiological data support biological effects of estrogen on the cardiovascular system resulting in protection from cardiovascular morbidity and mortality in women. A substantial portion of this benefit cannot be accounted for by the well-described effects of estrogen on plasma lipids. These observations, together with results obtained in animal models and in cell culture, have led to the hypothesis that estrogen exerts direct effects upon the artery wall which contributes to its beneficial actions on the cardiovascular system.

The past few years witnessed significant progress in our understanding of the molecular and cellular mechanisms of action of estrogens on the various components (e.g., endothelium and smooth muscle) of the blood vessel wall and their interaction with circulating blood cells (e.g., monocytes, leukocytes, and platelets). This book is the first to give a comprehensive overview of the current knowledge of effects of estrogen on the artery wall. State of the art lectures presented at the workshop "Estrogen and the Blood Vessel Wall," a Satellite Symposium of the IX International Vascular Biology Meeting in Seattle, Washington in September 1996 formed the basis for this book. Additional chapters written by leading experts on topics relevant to the theme but not presented at the symposium completed the volume.

The book is divided into five parts. Recent advances in molecular pharmacology, signal transduction and gene expression regulation by estrogen and estrogen receptors are summarized in Part I, including the discovery of the novel transcriptionally active estrogen receptor, ER-$\beta$. Part II gives an overview of the various actions of estrogen on endothelial cells. Effects on LDL oxidation, lipid accumulation in vessel wall, LP(a), hemostasis and leukocytes, all relevant to the vasculoprotective actions of estrogen, are summarized in Part III. Regulation of vascular smooth muscle tone is the topic of Part IV. Effects of estrogen in postmenopausal women with coronary artery disease, and in animal models of atherosclerosis, vasospasm, and vascular injury are discussed in Part V. In addition, results were presented from *in vivo* cardiovascular studies with a selective estrogen receptor modulator, a novel class of compounds with potential for improved tissue selectivity of estrogenic action.

Reflected by the variety of topics covered in this book it is clear that important advances have been made in the past few years in our knowledge of estrogen's effect on the blood vessel wall and the cardiovascular system. Even more exciting, however, is the anticipation of future advances that will undoubtedly occur as investigators build upon the biological foundations summarized in this book. Indeed, it is no overstatement to predict that we are at the threshold of a new era in understanding cardiovascular actions of estrogen. Importantly, these advances will likely yield novel therapeutic approaches to the prevention and treatment of cardiovascular disease in women, and potentially even in men. Thus, it is with great anticipation that future developments in the cardiovascular biology of estrogen are awaited.

The Editors would like to express their gratitude to the leading experts who contributed to this book and to the staff of Harwood Academic Publishers for their very professional, efficient and timely publication of this book.

Gabor M. Rubanyi
Raymond F. Kauffman

# CONTRIBUTORS

**Adams, Michael R.**
Department of Comparative Medicine
Bowman Gray School of Medicine
Medical Center Boulevard
Winston-Salem NC 27157-1040
USA

**Banka, Carole L.**
Department of Immunology
The Scripps Research Institute
10550 N. Torrey Pines Road
La Jolla CA 92037
USA

**Barber, Dustan A.**
Mayo Clinic & Foundation
Medical Sciences Building 4-54
200 First Street, SW
Rochester MN 55905
USA

**Bean, James S.**
Lilly Research Laboratories
Eli Lilly and Company
Lilly Corporate Center
Indianapolis Indiana 46285
USA

**Bender, Jeffrey R.**
Boyer Center for Molecular Medicine
Yale University School of Medicine
295 Congress Avenue, Room 454
New Haven CT 06536-0812
USA

**Bensch, William R.**
Lilly Research Laboratories
Eli Lilly and Company
Lilly Corporate Center
Indianapolis Indiana 46285
USA

**Caulin-Glaser, Teresa**
Division of Cardiovascular Medicine
Boyer Center for Molecular Medicine
Yale University School of Medicine
295 Congress Avenue, Room 454
New Haven CT 06536-0812
USA

**Cid, Maria C.**
Department of Internal Medicine
Hospital Clinic
Provincial Barcelona
Spain

**Clarkson, Thomas B.**
Department of Comparative Medicine
Clinical Research Center
Bowman Gray School of Medicine
Medical Center Boulevard
Winston-Salem NC 27157-1040
USA

**Collins, Peter**
Department of Cardiac Medicine
National Hearth and Lung Institute
Dovehouse Street
University of London
London SW3 6LY

**Couse, John F.**
Receptor Biology Section
National Institute of Environmental Health Sciences
PO Box 12233
Research Triangle Park NC 27613
USA

**Creager, Mark A.**
Department of Medicine, Cardiovascular Division
Brigham and Women's Hospital
Harvard Medical School
75 Francis Street
Boston MA 02115-6195
USA

**Curtis, Sylvia W.**
Receptor Biology Section
Lab of Reproductive & Developmental Toxicology
National Institute of Environmental Health Sciences
PO Box 12233
Research Triangle Park NC 27613
USA

**DiAugustine, Richard**
Laboratory of Biochemical Risk Analysis
National Institute of Environmental Health Sciences
PO Box 12233
Research Triangle Park NC 27613
USA

**Foegh, Marie L.**
Department of Surgery
Building D, Room 398
Georgetown University Medical Center
Washington DC 20007
USA

**Ganz, Peter**
Department of Medicine, Cardiovascular Division
Brigham and Women's Hospital
Harvard Medical School
75 Francis Street
Boston MA 02115-6195
USA

**Geary, Randolph L.**
Division of Surgical Sciences-General
Bowman Gray School of Medicine
Medical Center Boulevard
Winston-Salem NC 27157-1040
USA

**Gerhard, Marie**
Department of Medicine, Cardiovascular Division
Brigham and Women's Hospital
Harvard Medical School
75 Francis Street
Boston MA 02115-6195
USA

**Grant, Derrick S.**
Thomas Jefferson College of Medicine
11th and Walnut Streets
Philadelphia PA 19107
USA

**Gustafsson, Jan-Åke**
Department of Medical Nutrition
Karolinska Institute
NOVUM, S-14186 Huddinge
Sweden

**Hermsmeyer, Kent**
Oregon Regional Primate Research Center
505 N.W. 185th Avenue
Beaverton OR 97006-3499
USA

**Herrington, David M.**
Medicine/Cardiology
Department of Internal Medicine
The Bowman Gray School of Medicine
Medical Center Boulevard
Winston-Salem NC 27157
USA

**Honoré, Erika K.**
Department of Physiology and Medicine
Southwest Foundation for Biomedical Research
7620 N.W. Loop 410
San Antonio TX 78228
USA

**Kauffman, Raymond F.**
Lilly Research Laboratories
Eli Lilly and Company
Lilly Corporate Center
Indianapolis Indiana 46285
USA

**Kauser, Katalin**
Berlex Biosciences
Cardiovascular Research
15049 San Pablo Avenue
Richmond CA 94806-0099
USA

**Kim-Schulze, Seunghee**
Department of Pediatrics
Northwestern School of Medicine
Chicago Illanois
USA

**Kinlay, Scott**
Department of Medicine, Cardiovascular Division
Brigham and Women's Hospital
Harvard Medical School
75 Francis Street
Boston MA 02115-6195
USA

**Kleinman, Hynda K.**
NIH/NIDR/DIR/LDB
30 Convent Drive
Building 30, Room 433
MSC 4370
Bethesda MD 20892-4370
USA

**Korach, Kenneth S.**
Environmental Toxicology Program
National Institute of Environmental Health Sciences
PO Box 12233
Research Triangle Park NC 27613
USA

**Kuiper, George G.J.M.**
Center for Biotechnology
Karolinska Institute
NOVUM, S-14186 Huddinge
Sweden

**Losordo, Douglas**
St Elizabeth's Hospital
Medical Center
736 Cambridge Street
Boston MA 02135
USA

**Lourdes Ponce, M.**
National Institute of Dental Research
NIH, 30 Convent Drive
Bethesda MD 20892-3740
USA

**Malinda, Katherine M.**
National Institute of Dental Research
NIH, 30 Convent Drive
Bethesda MD 20892-3740
USA

**McDonnell, Donald P.**
Department of Pharmacology
Box 3813
Duke University Medical School
Durham North Carolina 27710
USA

**McGowan, Kelly A.**
National Institute of Dental Research
NIH, 30 Convent Drive
Bethesda MD 20892-3740
USA

**Miller, Virginia M.**
Mayo Clinic & Foundation
Medical Sciences Building 4-54
200 First Street, SW
Rochester MN 55905
USA

**Minshall, Richard D.**
Oregon Regional Primate Research Center
505 N.W. 185th Avenue
Beaverton OR 97006-3499
USA

**Miyagawa, Koichi**
Oregon Regional Primate Research Center
505 N.W. 185th Avenue
Beaverton OR 97006-3499
USA

**Nilsson, Stefan**
Karobio AB
NOVUM, S-14186 Huddinge
Sweden

**Norris, John D.**
Department of Pharmacology
Box 3813
Duke University Medical School
Durham North Carolina 27710
USA

**Øian, Pål**
Department of Obstetrics & Gynecology
Institute of Medical Biology
University of Tromsø
N-9037 Tromsø
Norway

**Østerud, Bjarne**
Department of Biochemistry
Institute of Medical Biology
University of Tromsø
N-9037 Tromsø
Norway

**Ramwell, Peter W.**
Department of Surgery
Building D, Room 398
Georgetown University Medical Center
Washington DC 20007
USA

**Register, Thomas C.**
Department of Comparative Medicine
Bowman Gray School of Medicine
Medical Center Boulevard
Winston-Salem NC 27157-1040
USA

**Rubanyi, Gabor M.**
Berlex Biosciences
Cardiovascular Research
15049 San Pablo Avenue
Richmond CA 94806-0099
USA

**Schnaper, H. William**
Department of Pediatrics
Northwestern School of Medicine
Chicago Illanois
USA

**Schwenke, Dawn C.**
Department of Pathology
Clinical Research Center
Bowman Gray School of Medicine
Medical Center Boulevard
Winston-Salem NC 27157-1040
USA

**Selwyn, Andrew P.**
Department of Medicine,
Cardiovascular Division
Brigham and Women's Hospital
Harvard Medical School
75 Francis Street
Boston MA 02115-6195
USA

**Sessa, William C.**
Department of Pharmacology
Boyer Center for Molecular Medicine
Yale University School of Medicine
295 Congress Avenue, Room 4360
New Haven CT 06536-0812
USA

**Sewall, Charles**
Environmental Toxicology Program
National Institute of Environmental
Health Sciences
PO Box 12233
Research Triangle Park NC 27613
USA

**Spyridopoulos, Ioakim**
St Elizabeth's Hospital
Medical Center
736 Cambridge Street
Boston MA 02135
USA

**Vidgoff, Jaclyn**
Oregon Regional Primate Research
Center
505 N.W. 185th Avenue
Beaverton OR 97006-3499
USA

**Wagner, Janice D.**
Department of Comparative Medicine
Clinical Research Center
Bowman Gray School of Medicine
Medical Center Boulevard
Winston-Salem NC 27157-1040
USA

**Washburn, Todd**
Receptor Biology Section
Lab of Reproductive & Developmental
Toxicology
National Institute of Environmental
Health Sciences
PO Box 12233
Research Triangle Park NC 27613
USA

**Williams, J Koudy**
Department of Comparative Medicine
Clinical Research Center
Bowman Gray School of Medicine
Medical Center Boulevard
Winston-Salem NC 27157-1040
USA

# 1 Cloning, Expression and Function of a Novel Estrogen Receptor

**George G.J.M. Kuiper *, Stefan Nilsson and Jan-Åke Gustafsson**

*Center for Biotechnology and Department of Medical Nutrition, Karolinska Institute, Huddinge, Sweden, and KaroBio AB (S.N.) Huddinge, Sweden.*
** Corresponding author: Dr. George Kuiper, Center for Biotechnology, Karolinska Institute, NOVUM, S-14186 Huddinge, Sweden, Tel.: +46 8 6089147, Fax: +46 8 7745538, E-mail: george.kuiper@cbt.ki.se.*

We have cloned a novel member of the nuclear receptor superfamily; estrogen receptor $\beta$ (ER$\beta$). The cDNA of ER$\beta$ was isolated from a rat prostate cDNA library and it encodes a protein of 485 amino acid residues with a calculated molecular weight of 54200. The ER$\beta$ protein is highly homologous to the previously cloned ER$\alpha$ protein, particularly in the DNA-binding domain (95%) and in the C-terminal ligand-binding domain (55%). Expression of ER$\beta$ in rat tissues was investigated by *in situ* hybridisation and RT-PCR; moderate to high expression was found in prostate (secretory epithelial cells), ovary (granulosa cells), lung, bladder, brain, uterus and testis. Saturation ligand-binding analysis of *in vitro* synthesized rat ER$\beta$ protein revealed a single binding component for 16$\alpha$-iodo-3,17$\beta$-estradiol with high affinity (Kd = 0.4 nM). In ligand-competition experiments the binding affinity decreased in the order dienestrol > 4-OH-tamoxifen > diethylstilbestrol > ICI-164384 > *17$\beta$-estradiol* > estrone > estriol > tamoxifen. In co-transfection experiments of Chinese hamster ovary cells with an ER$\beta$ expression vector and an estrogen-regulated reporter gene, maximal stimulation of reporter gene activity was found during incubation with 1 nM of 17$\beta$-estradiol.

The detailed biological significance of the existence of two different ERs is at this moment unclear. Differences in the ligand-binding properties and/or transactivation function on certain target genes may exist.

KEYWORDS: Estrogen receptor, estrogens, prostate, ovary, nuclear receptor.

## INTRODUCTION

Estrogens influence the growth, differentiation and functioning of many target tissues. These include tissues of the male and female reproductive systems such as mammary gland, uterus, ovary, testis and prostate (Clark, 1992). Estrogens also play an important role in bone maintenance, and in the cardiovascular system were estrogens have certain cardioprotective effects (Farhat, 1996; Turner, 1994). Estrogens are mainly produced in the ovaries and testis. They diffuse in and out of cells, but are retained with high affinity and specificity in target cells by an intranuclear binding protein, termed the estrogen receptor (ER). Once bound by estrogens, the ER undergoes a conformational change allowing the receptor to bind with high affinity to chromatin and to modulate transcription of target genes (Murdoch, 1991). Steroid hormone receptors consist of a hypervariable N-terminal domain that contributes to the transactivation function; a highly conserved central domain responsible for specific DNA-binding, dimerization and nuclear localization and a C-terminal domain involved in ligand binding and ligand-dependent transactivation function (Beato, 1995).

In addition to these receptors for known ligands numerous so-called orphan receptors, which are putative receptors interacting with unknown ligands, have been found. Today, the nuclear receptor superfamily contains more than 100 different members, spanning a large diversity of animal species from worm to insect to human (Mangelsdorf, 1995). In an effort to clone and characterize novel nuclear receptors or unknown isoforms of existing (steroid) receptors, we designed degenerate primers based upon conserved regions within the DNA- and ligand-binding domains of nuclear receptors (Laudet, 1992). These were used to PCR amplify cDNA from several rat tissues. Prostate was selected as an organ of interest given the high incidence of prostate cancer and benign prostatic hyperplasia. There is firm evidence that many biological processes of prostate epithelial- and stroma cells are controlled by androgens and estrogens (Gleave, 1995; Prins, 1995). Reports on the presence of ERs in the prostate have been conflicting. Although ERs could readily be detected in ligand-binding studies in rat and human prostate tissue (Jung-Testas *et al*., 1981; van Beurden-Lamers *et al*., 1974; Ekman *et al*., 1983) it was difficult if not impossible to detect ERs with the existing ER antibodies (Prins, 1995; Brolin, 1992), pointing to the possible existence in the prostate of an unknown isoform of the ER. Various naturally occuring ER isoforms have been described in breast tumors, meningiomas and rat brain (Castles, 1996; Skipper *et al*., 1993; Koehorst *et al*., 1992).

## CLONING OF ER$\beta$-cDNA

The RT-PCR method, used before to identify new members of the nuclear receptor family (Enmark *et al*., 1994; Pettersson *et al*., 1996), was used to screen rat prostatic tissue. This resulted in the cloning of DNA sequences identical to several known members of the nuclear receptor family and one putative novel nuclear receptor partial cDNA (Kuiper *et al*., 1996). This novel clone was found to be highly homologous to the cDNA of the rat ER (65%), which was previously cloned from rat uterus (Koike *et al*., 1987). The amino acid residues predicted by this novel clone suggested that this DNA fragment encoded part of the DNA-binding domain, hinge region, and the beginning of the ligand-binding domain of a nuclear receptor protein. Two PCR-primers were made (Figure 1.1) to generate a probe consisting of the hinge region of the putative novel receptor. This probe was used to screen a rat prostate cDNA library, resulting in a strongly positive clone of about 2.5 kb, which was sequenced completely (Figure 1.1).

Two in-frame ATG codons are located at nucleotides 424 and 448, preceded by an in-frame stop codon at nucleotide 319, which suggests that they are possible start codons. The open reading frame encodes a protein of 485 amino acid residues with a calculated molecular weight of 54200. Analysis of the proteins synthesized by

```
ggaattcCGGGGGAGCTGGCCCAGGGGGAGCGGCTGGTGCTGCCACTGGCATCCCTAGGC 60
ACCCAGGTCTGCAATAAAGTCTGGCAGCCACTGCATGGCTGAGCGACAACCAGTGGCTGG 120
GAGTCCGGCTCTGTGGCTGAGGAAAGCACCTGTCTGCATTTAGAGAATGCAAAATAGAGA 180
ATGTTTACCTGCCAGTCATTACATCTGAGTCCCATGAGTCTCTGAGAACATAATGTCCAT 240
CTGTACCTCTTCTCACAAGGAGTTTTCTCAGCTGCGACCCTCTGAAGACATGGAGATCAA 300
AAACTCACCGTCGAGCCTTAGTTCCCTGCTTCCTATAACTGTAGCCAGTCCATCCTACCC 360
CTGGAGCACGGCCCCATCTACATCCCTTCCTCCTACGTAGACAACCGCCATGAGTATTCA 420
GCTATGACATTCTACAGTCCTGCTGTGATGAACTACAGTGTTCCCGGCAGCACCAGTAAC 480
 M  T  F  Y  S  P  A  V  M  N  Y  S  V  P  G  S  T  S  N
CTGGACGGTGGGCCTGTCCGACTGAGCACAAGCCCAAATGTGCTATGGCCAACTTCTGGG 540
L  D  G  G  P  V  R  L  S  T  S  P  N  V  L  W  P  T  S  G
CACCTGTCTCCTTTAGCGACCCATTGCCAATCATCGCTCCTCTATGCAGAACCTCAAAAG 600
H  L  S  P  L  A  T  H  C  Q  S  S  L  L  Y  A  E  P  Q  K
AGTCCTTGGTGTGAAGCAAGATCACTAGAGCACACCTTACCTGTAAACAGAGAGACACTG 660
S  P  W  C  E  A  R  S  L  E  H  T  L  P  V  N  R  E  T  L
AAGAGGAAGCTTAGTGGGAGCAGTTGTGCCAGCCCTGTTACTAGTCCAAACGCAAAGAGG 720
K  R  K  L  S  G  S  S  C  A  S  P  V  T  S  P  N  A  K  R
GATGCTCACTTCTGCCCCGTCTGCAGCGATTATGCATCTGGGTATCATTACGGCGTTTGG 780
D  A  H  F  C  P  V  C  S  D  Y  A  S  G  Y  H  Y  G  V  W
TCATGTGAAGGATGTAAGGCCTTTTTTAAAAGAAGCATTCAAGGACATAATGATTATATC 840
S  C  E  G  C  K  A  F  F  K  R  S  I  Q  G  H  N  D  Y  I
TGTCCAGCCACGAATCAGTGTACCATAGACAAGAACCGGCGTAAAAGCTGCCAGGCCTGC 900
C  P  A  T  N  Q  C  T  I  D  K  N  R  R  K  S  C  Q  A  C
CGACTTCGCAAGTGTTATGAAGTAGGAATGGTCAAGTGTGGATCCAGGAGAGAACGGTGT 960
R  L  R  K  C  Y  E  V  G  M  V  K  C  G  S  R  R  E  R  C
GGGTACCGTATAGTGCGGAGGCAGAGAAGTTCTAGCGAGCAGGTACACTGCCTGAGCAAA 1020
G  Y  R  I  V  R  R  Q  R  S  S  S  E  Q  V  H  C  L  S  K
GCCAAGAGAAACGGTGGGCATGCACCCCGGGTGAAGGAGCTACTGCTGAGCACCTTGAGT 1080
A  K  R  N  G  G  H  A  P  R  V  K  E  L  L  L  S  T  L  S
CCAGAGCAACTGGTGCTCACCCTCCTGGAAGCTGAACCACCCAATGTGCTGGTGAGCCGT 1140
P  E  Q  L  V  L  T  L  L  E  A  E  P  P  N  V  L  V  S  R
CCCAGCATGCCCTTCACCGAGGCCTCCATGATGATGTCCCTCACTAAGCTGGCGGACAAG 1200
P  S  M  P  F  T  E  A  S  M  M  M  S  L  T  K  L  A  D  K
GAACTGGTGCACATGATTGGCTGGGCCAAGAAAATCCCTGGCTTTGTGGAGCTCAGCCTG 1260
E  L  V  H  M  I  G  W  A  K  K  I  P  G  F  V  E  L  S  L
TTGGACCAAGTCCGGCTCTTAGAAAGCTGCTGGATGGAGGTGCTAATGGTGGGACTGATG 1320
L  D  Q  V  R  L  L  E  S  C  W  M  E  V  L  M  V  G  L  M
TGGCGCTCCATCGACCACCCCGGCAAGCTCATTTTCGCTCCCGACCTCGTTCTGGACAGG 1380
W  R  S  I  D  H  P  G  K  L  I  F  A  P  D  L  V  L  D  R
GATGAGGGGAAGTGCGTAGAAGGGATTCTGGAAATCTTTGACATGCTCCTGGCGACGACG 1440
D  E  G  K  C  V  E  G  I  L  E  I  F  D  M  L  L  A  T  T
TCAAGGTTCCGTGAGTTAAAACTCCAGCACAAGGAGTATCTCTGTGTGAAGGCCATGATC 1500
S  R  F  R  E  L  K  L  Q  H  K  E  Y  L  C  V  K  A  M  I
CTCCTCAACTCCAGTATGTACCCCTTGGCTTCTGCAAACCAGGAGGCAGAAAGTAGCCGG 1560
L  L  N  S  S  M  Y  P  L  A  S  A  N  Q  E  A  E  S  S  R
AAGCTGACACACCTACTGAACGCGGTGACAGATGCCCTGGTCTGGGTGATTGCGAAGAGT 1620
K  L  T  H  L  L  N  A  V  T  D  A  L  V  W  V  I  A  K  S
GGTATCTCCTCCCAGCAGCAGTCAGTCCGACTGGCCAACCTCCTGATGCTTCTTTCTCAC 1680
G  I  S  S  Q  Q  Q  S  V  R  L  A  N  L  L  M  L  L  S  H
GTCAGGCACATCAGTAACAAGGGCATGGAACATCTGCTCAGCATGAAGTGCAAAAATGTG 1740
V  R  H  I  S  N  K  G  M  E  H  L  L  S  M  K  C  K  N  V
GTCCCGGTGTATGACCTGCTGCTGGAGATGCTGAATGCTCACACGCTTCGAGGGTACAAG 1800
V  P  V  Y  D  L  L  L  E  M  L  N  A  H  T  L  R  G  Y  K
TCCTCAATCTCGGGGTCTGAGTGCAGCTCAACAGAGGACAGTAAGAACAAAGAGAGCTCC 1860
S  S  I  S  G  S  E  C  S  S  T  E  D  S  K  N  K  E  S  S
CAGAACCTACAGTCTCAGTGATGGCCAGGCCTGAGGCGGACAGACTACAGAGATGGTCAA 1920
Q  N  L  Q  S  Q  *
AAGTGGAACATGTACCCTAGCATCTGGGGGTTCCTCTTAGGGCTGCCTTGGTTACGCACC 1980
CCTTACCCACACTGCACTTCCCAGGAGTCAGGGTGGTTGTGTGGCGGTGTTCCTCATACC 2040
AGGATGTACCACCGAATGCCAAGTTCTAACTTGTATAGCCTTGAAGGCTCTCGGTGTACT 2100
TACTTTCTGTCTCCTTGCCCACTTGGAAACATCTGAAAGGTTCTGGAACTAAAGGTCAAA 2160
GTCTGATTTGGAAGGATTGTCCTTAGTCAGGAAAAGGAATATGGCATGTGACACAGCTAT 2220
AAGAAATGGACTGTAGGACTGTGTGGCCATAAAATCAACCTTTGGATGGCGTCTTCTAGA 2280
CCACTTGATTGTAGGATTGAAAACCACATTGACAATCAGCTCATTTCGCATTCCTGCCTC 2340
ACGGGTCTGTGAGGACTCATTAATGTCATGGGTTATTCTATCAAAGACCAGAAAGATAGT 2400
GCAAGCTTAGATGTACCTTGTTCCTCCTCCCAGACCCTTGGGTTACATCCTTAGAGCCTG 2460
CTTATTTGGTCTGTCTGAATGTGGTCATTGTCATGGGTTAAGATTTAAATCTCTTTGTAA 2520
TATTGGCTTCCTTGAAGCTATGTCATCTTTCTCTCTCTCCCGgaattc 2568
```

**Figure 1.1.** Sequence of rat ERβ cDNA and predicted amino acid sequence of ERβ protein. Two potential translation start sites are indicated in boldface type. The predicted DNA-binding domain is double underlined and the PCR primers, used for generation of the probe for screening of a rat prostate cDNA library, are single underlined.

*in vitro* translation of the cDNA shown in Figure 1.1 in rabbit reticulocyte lysate revealed a doublet protein band with an apparent molecular weight of 61000 during SDS-PAGE (data not shown), confirming the open reading frame. The doublet protein band is probably caused by the use of both ATG codons for initiation of protein synthesis. It remains to be seen if the same protein heterogeneity also exists *in vivo*. Protein sequence comparison (Figures 1.2–1.4) showed that the novel nuclear receptor clone is most related to the rat ER, cloned from rat uterus (Koike, 1987), with 95% identity in the DNA-binding domain and 55% identity in the putative ligand-binding domain. It was therefore decided to tentatively name the novel nuclear receptor cDNA clone rat ER$\beta$, and consequently the previously cloned receptor rat ER$\alpha$.

A number of functional characteristics have been identified within the DNA-binding domain of nuclear receptors (Zilliacus *et al.*, 1994). The ER$\beta$ protein P-box and D-box sequences of EGCKA and PATNQ, respectively, are identical to the corresponding boxes in the ER$\alpha$ protein (Koike, 1987), thus predicting that ER$\beta$ protein binds to estrogen response element (ERE) sequences. The putative ligand-binding domain of ER$\beta$ protein shows closest homology to the ligand-binding domain of ER$\alpha$ protein (Figures 1.2 and 1.3), whereas the homology with the estrogen receptor related receptors ERR1 and ERR2 (Giguere *et al.*, 1988; Pettersson *et al.*, 1996) is considerably less. The ERR1 and ERR2 nuclear orphan receptors do not bind estradiol. Several amino acid residues described to be close to or part of the ligand-binding pocket of the human ER$\alpha$ protein (Cys 530, Asp 426, and Gly 521) are conserved in the putative ligand-binding domain of rat ER$\beta$ protein (Cys 436, Asp 333, and Gly 427) and in the ligand-binding domain of ERs

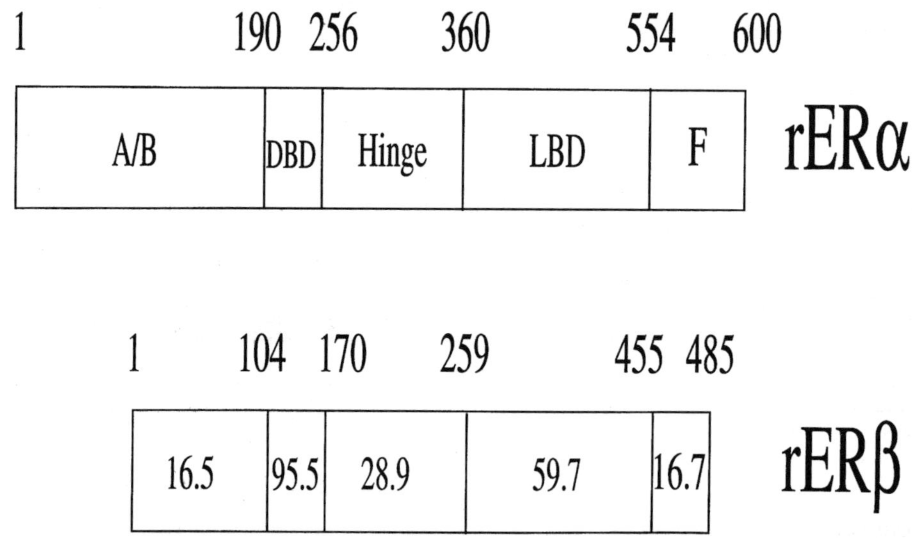


**Figure 1.2.** Comparison between rat ER$\alpha$ and ER$\beta$ protein. Percentage amino acid identity in the domains A/B (N terminus), DBD (DNA-binding), hinge, and LBD/F (ligand binding, dimerization, and ligand-dependent transactivation) are depicted.

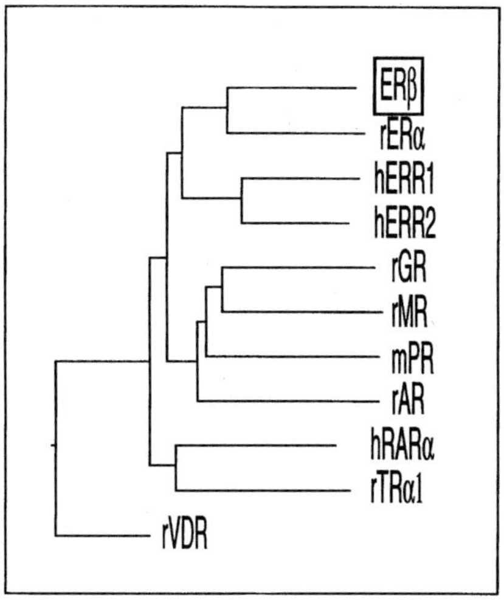


| Domain | A/B | C | D | E/F |
|---|---|---|---|---|
| rERα | 16.5 | 95.5 | 28.9 | 53.5 |
| hERR1 | 16.5 | 69.7 | 15.6 | 34.2 |
| hERR2 | 13.7 | 72.7 | 19.8 | 31.0 |
| rAR | 16.5 | 59.1 | 12.8 | 19.1 |
| rGR | 19.4 | 59.1 | 14.6 | 18.8 |
| rMR | 17.5 | 57.6 | 17.8 | 19.1 |
| mPR | 15.5 | 56.1 | 14.4 | 19.6 |
| rTRα1 | 15.4 | 50.0 | 13.3 | 17.7 |
| rRARα | 26.1 | 59.1 | 14.8 | 17.3 |
| rVDR | 21.7 | 47.0 | 15.6 | 18.8 |

**Figure 1.3.** Comparison of ERβ protein with several representative members of the nuclear receptor family. Percentage amino acid identity in the domains A/B (N terminus), C (DNA-binding), D (hinge region), and E/F (ligand binding, dimerization, and ligand-dependent transactivation function) are depicted. For the alignment and phylogenetic tree Clustal analysis of the full-length receptor sequences using the MEGALIGN/DNASTAR software was used.

```
223 LVLTLLEAEPPNVLVS-RPSMPFTEASMMMSLTKLADKELVHMIGWAKKI   rat ERß
320 M.SA..D....LIYSEYD..R..S.....GL..N...R......N...RV   rat ERα

272 PGFVELSLLDQVRLLESCWMEVLMVGLMWRSIDHPGKLIFAPDLVLDRDE   rat ERß
370 ...GD.N.H...H...CA.L.I..I..V...ME.....L...N.L...NQ   rat ERα

322 GKCVEGILEIFDMLLATTSRFRELKLQHKEYLCVKAMILLNSSMYP-LAS   rat ERß
420 ......MV.........S....MMN..GE.FV.L.SI.....GV.TF.S.   rat ERα

371 ANQEAESSRKLTHLLNAVTDALVWVIAKSGISSQQQSVRLANLLMLLSHV   rat ERß
470 TLKSL.EKDHIHRV.DKIN.T.IHLM..A.LTL...HR...Q..LI...I   rat ERα

421 RHISNKGMEHLLSMKCKNVVPVYDLLLEMLNAHTLRG 457            rat ERß
520 ..M........YN........L........D..R.HAPA 558          rat ERα
```

**Figure 1.4.** Alignment of the amino acid sequences of rat ERα protein (GenBank database Y00102) ligand-binding domain (amino acid residues 320–558), and rat ERβ protein (GenBank database U57439) ligand-binding domain (amino acid residues 223–457).

from various species (Harlow *et al.*, 1989; Fawell *et al.*, 1990). An alignment of rat ERα and rat ERβ protein ligand-binding domains (Figure 1.4) reveals several completely conserved stretches as well as a central stretch which is essentially non-conserved (boxed in Figure 1.4). Comparative studies of the ligand-binding domain of ER$\alpha$ and ER$\beta$ protein by peptide mapping using mass spectrometric techniques (Hegy *et al.*, 1996; Seielstad *et al.*, 1995) and crystal structure determination could provide more detailed information on the structural requirements for ligand binding.

The core domain of the ligand-dependent transactivation function TAF-2, identified in the ER$\alpha$ protein (Danielian *et al.*, 1992), is almost completely conserved in ER$\beta$ protein (Figure 1.4; amino acid residues 441–457). Steroid hormone receptors are phosphoproteins (Kuiper, 1994, and references therein) and ER phosphorylation has been implicated in ligand-dependent and ligand-independent transactivation functions of the ER (Ali *et al.*, 1993; Kato *et al.*, 1995). Several phosphorylation sites identified in the N-terminal domain and ligand-binding domain of the human ER$\alpha$ protein (Ali *et al.*, 1993; Arnold *et al.*, 1995) are present at similar positions in the rat ER$\beta$ protein (Ser 30, Ser 42, Ser 94 and Tyr 443). The functional significance of these putative phosphorylation sites within the ER$\beta$ protein remains to be established. The rat ER$\beta$ protein consists of 485 amino acid residues, whereas the ER$\alpha$ protein from human, mouse, and rat consists of 590–600 amino acid residues. The main difference is a much shorter N-terminal domain in ER$\beta$ protein, i.e., 103 amino acid residues as compared to 185–190 amino acid residues in the ER$\alpha$ protein. Also, the nonconserved so-called F-domain at the C-terminal end of ER$\beta$ protein is 15 amino acid residues shorter than in ER$\alpha$ proteins. Recently, the mouse and human homologues of rat ER$\beta$ were cloned in our laboratory (Enmark, Bertilsson, Grandien and Pettersson, unpublished observations) and the human ER$\beta$ cDNA also by others (Mosselman, 1996).

## LIGAND-BINDING CHARACTERISTICS OF ER$\beta$ PROTEIN

The ER protein can be isolated from the cytosol of target cell extracts as a large non-transformed (i.e. non-DNA binding) 7-8S oligomeric complex, which contains hsp90 and hsp70 (Murdoch, 1990, and references therein). It is believed that heat shock proteins function to help fold the ER protein properly and to protect the hydrophobic hormone binding domain from inappropriate interactions (Murdoch, 1990; Sabbah *et al.*, 1996). Rabbit reticulocyte lysates contain large amounts of several heat shock proteins as hsp90 and hsp70, and have been used extensively for the study of ER complex formation with hsps as well as for the study of requirements for steroid binding and interactions with DNA (Beekman *et al.*, 1993; Sabbah *et al.*, 1996). It was therefore decided to use human ER$\alpha$ and rat ER$\beta$ protein synthesized in reticulocyte lysates for the ligand binding experiments. When ER$\beta$ protein was labeled with a saturating dose of [$^3$H]-17$\beta$-estradiol and analyzed on sucrose density gradients, a single peak of specifically bound radioactivity was observed (Figure 1.5). The sedimentation coefficient of this complex was about 7S, and it shifted to 4S in the presence of 0.4 M NaCl (Figure 1.5).

In order to obtain optimal conditions for the determination of equilibrium dissociation constants (Kd) and relative binding affinities (RBA) of various ligands, the ER protein concentration in the binding assays was lowered to 10–20 pM. At these low ER protein concentrations radioligand and/or competitor depletion can be

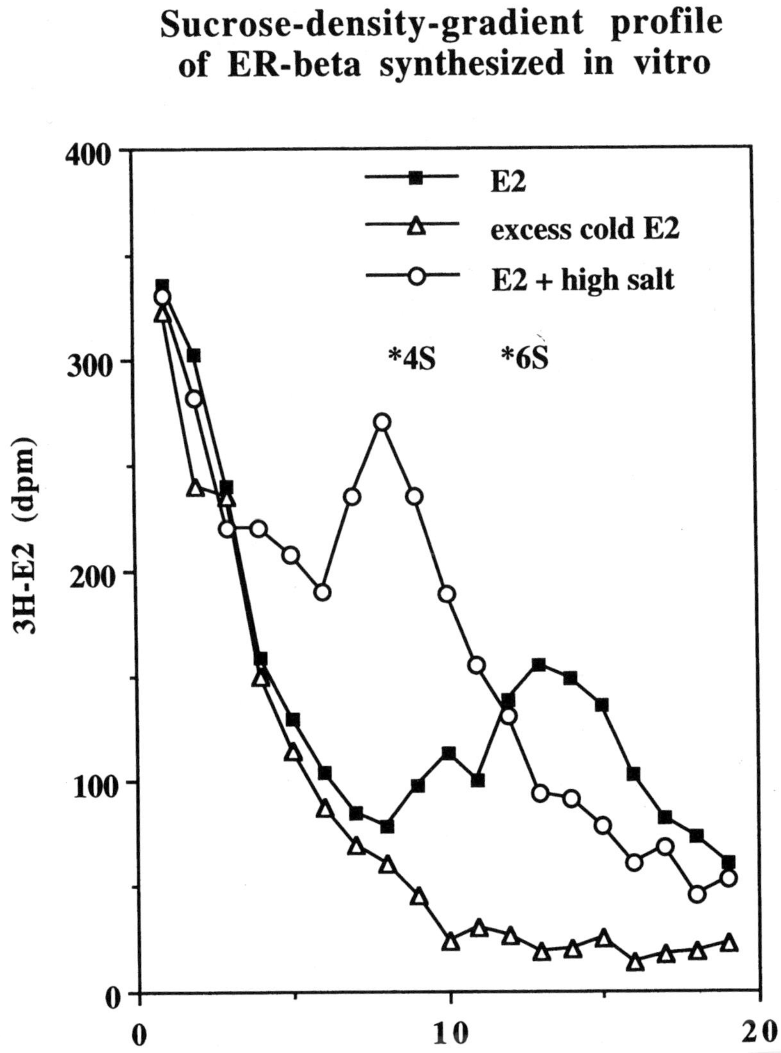


**Figure 1.5.** Sucrose density gradient analysis of rat ER$\beta$ protein synthesized *in vitro*. The ER$\beta$ protein was labelled with [$^3$H]-estradiol in the presence or absence of 200 fold unlabelled estradiol for 4 h at 8°C. Samples were run on 5–20% sucrose density gradients (with or without 0.4 M NaCl) for 20 h at 48000 rev/min in a SW-60 rotor at 8°C.

excluded, while maintaining high ER protein recovery during separation of bound and unbound ligand by the use of a gel filtration assay instead of the more traditional charcoal adsorption assay (Salomonsson, 1994). The low ER protein concentration made it necessary to use radio-iodinated estradiol as a probe since the specific radioactivity of tritiated estradiol was too low to maintain sufficient accuracy. Radio-iodinated estradiol (16α-[$^{125}$I]-iodo-estradiol) binds to the ER protein with high affinity and specificity as shown by its use in dry mount autoradiographic techniques and in various ligand binding assays (Berns, 1985; Hochberg, 1980). In Figure 1.6 the result of a saturation ligand binding assay with 16α-[$^{125}$I]-iodo-estradiol is shown. The non-specific binding was ⩽8% of total binding over the whole radioligand concentration range used. The Kd values calculated from the saturation curves (Figure 1.6) are 0.06 nM for ERα protein and 0.24 nM for ERβ protein. Linear transformation of saturation data (Scatchard plots in Figure 1.6) revealed a single population of non-interacting binding sites for 16α-iodo-estradiol with a Kd of 0.1 nM ($n=2$) for the ERα protein and 0.4 nM ($n=2$) for the ERβ protein. Although the ERβ protein has a four times lower affinity for 16α-iodo-estradiol in this system compared to the ERα protein, both Kd values are within the range (0.1–1 nM) generally reported for estradiol binding to ER in various systems (Clark, 1992).

Measurements of the equilibrium binding of the radioligand in the presence of different concentrations of unlabelled competitors provide readily interpretable information about the affinities of the latter, provided that radioligand and/or competitor depletion are avoided. Competition experiments (Table 1.1) are indicative of an ER in that only estrogens and anti-estrogens competed efficiently with

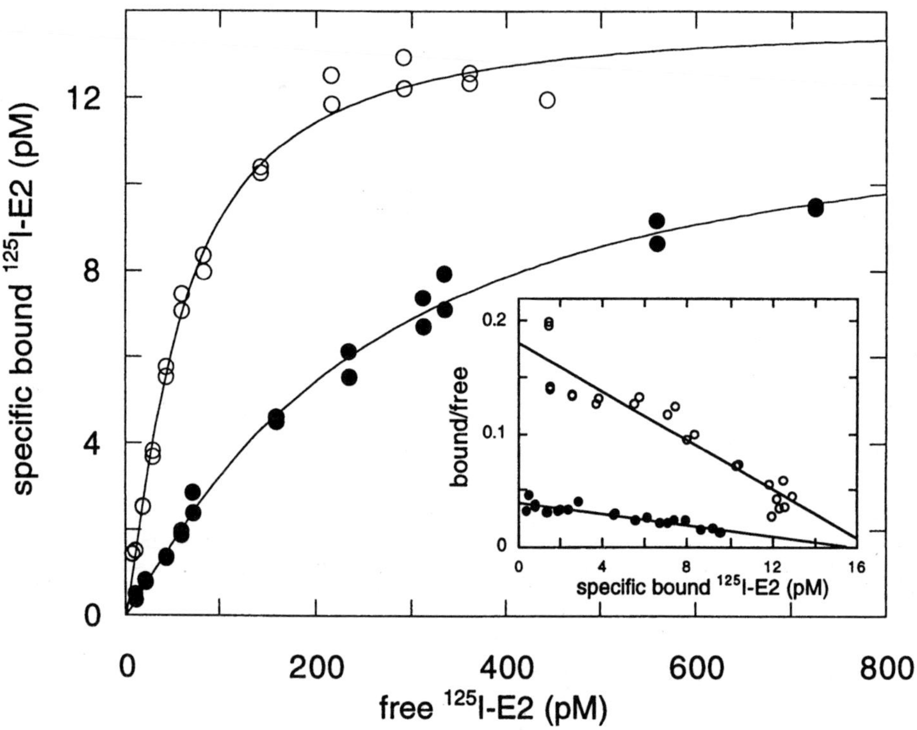


**Figure 1.6.** Binding of 16α-[$^{125}$I]-iodo-17β-estradiol to *in vitro* synthesized ERα and ERβ protein in the presence or absence of a 300-fold excess of diethylstilbestrol for 16 h at 4°C. Unbound radioactivity was removed as described (Salomonsson, 1994) and specific counts (ERα = ○; ERβ = ●) were calculated by subtracting non-specific bound counts from total bound counts. Inset shows Scatchard analysis of specific binding giving a Kd of 0.1 nM for ERα protein and a Kd of 0.4 nM for ERβ protein.

**Table 1.1** Relative binding affinity of various compounds for estrogen receptor α and β

| *Compound* | *ERα* | *ERβ* |
|---|---|---|
| 17β-estradiol | 100 | 100 |
| Diethylstilbestrol | 468 | 295 |
| Dienestrol | 223 | 404 |
| Hexestrol | 302 | 234 |
| Estrone | 60 | 37 |
| Estriol | 14 | 21 |
| Tamoxifen | 7 | 6 |
| 4-OH-tamoxifen | 178 | 339 |
| ICI-164384 | 85 | 166 |
| Progesterone | <0.1 | <0.1 |
| Testosterone | <0.1 | <0.1 |
| Corticosterone | <0.1 | <0.1 |

16α-iodo-estradiol for binding. Several differences in the relative binding affinities are present between receptor subtypes. For instance 4OH-tamoxifen and ICI-164384 have a two times higher affinity for the ERβ protein. It should be kept in mind that most if not all ER ligand binding studies performed in the past thirty years actually involved mixtures of ERα and ERβ protein. This is certainly the case for many studies in which rat uterus cytosol was used (see further). Therefore, caution is needed when comparing relative binding affinities from the individual receptors with the relative binding affinities measured involving mixtures of both ER subtypes.

A question of considerable interest is why despite the numerous ligand binding assays carried out with the ER protein, an indication for the existence of two ER subtypes was never published. Distinguishing between a mixed population of receptor subtypes and a homogenous receptor population by saturation or homologous/heterologous competition is generally difficult. This is only possible with certainty when the two subtypes differ sufficiently in affinity (10–100 fold) and the range of ligand concentrations examined is wide. Furthermore, the proportion of the two subtypes must be appropriate (Swillens, 1995). Most ER ligand binding assays have been performed with uterus extracts and breast tumor extracts or cell lines, and it could be that the right conditions are not fulfilled in these experiments (Hähnel, 1973; Notides, 1970; Clark, 1992, and references therein). It remains to be seen if the differences in relative binding affinity of some ligands for ERα and ERβ protein are also reflected in transactivation assay systems using different cellular backgrounds.

## TRANSACTIVATION FUNCTION OF ERβ PROTEIN

In order to investigate the transcriptional regulatory properties of ERβ protein cotransfection experiments in which CHO (Chinese hamster ovary) cells were transfected with an ERβ protein expression vector and an estrogen responsive reporter gene construct were performed. The pERE-ALP reporter construct contains a secreted form of the placental alkaline phosphatase gene (Berger *et al.*, 1988) and the mouse mammary tumor virus long terminal repeat in which the glucocorticoid response elements were replaced by a single consensus vitellogenin promoter estrogen response element (Martinez, 1987). In the absence of exogenously added estradiol, the ER$\beta$ protein showed considerable transcriptional activity, which could be further increased by the addition of estradiol (Figure 1.7). Simultaneous addition of a 10 fold excess of tamoxifen partially supressed the estradiol stimulated activity (Figure 1.7). The constitutive transcriptional activity of ER$\beta$ protein could be suppressed by tamoxifen or ICI-164384 (not shown). Although we have done everything possible to exclude contaminating estrogenic compounds by using phenol red free serum replacement medium for the transfection experiments, we cannot exclude the presence of very low amounts of estrogenic compounds. Ligand-independent ER$\alpha$ protein mediated transcriptional activation by dopamine and growth factors has been described (Smith, 1995; Ignar-Trowbridge *et al.*, 1996), and might also be the cause of the ER$\beta$ protein mediated constitutive activity. In dose response experiments, ER$\beta$ protein began to respond at 0.1 nM estradiol (Figure 1.8) and maximal stimulation was observed at 1 nM estradiol. The maximal stimulation factor was $2.6 \pm 0.5$ fold (mean$\pm$SD, $n=4$) as compared to incubation in the absence of estradiol. Dexamethasone, testosterone, progesterone, thyroid hormone, and all-trans-retinoic acid could not stimulate transcriptional activity of ER$\beta$ protein, even at the highest concentration (1000 nM) tested (not shown). In control

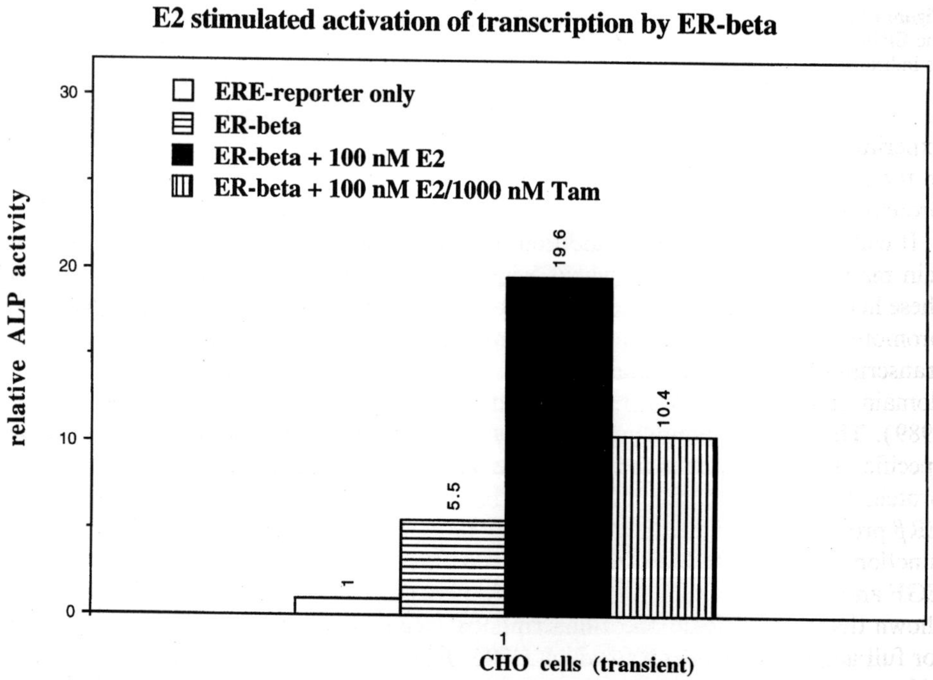


**Figure 1.7.** Activation of transcription by ER$\beta$ protein. CHO cells were transiently transfected with ERE-reporter plasmid alone (ERE-reporter) or together with a ER$\beta$ protein expression plasmid. Cells were incubated with estradiol alone or estradiol and tamoxifen as indicated.

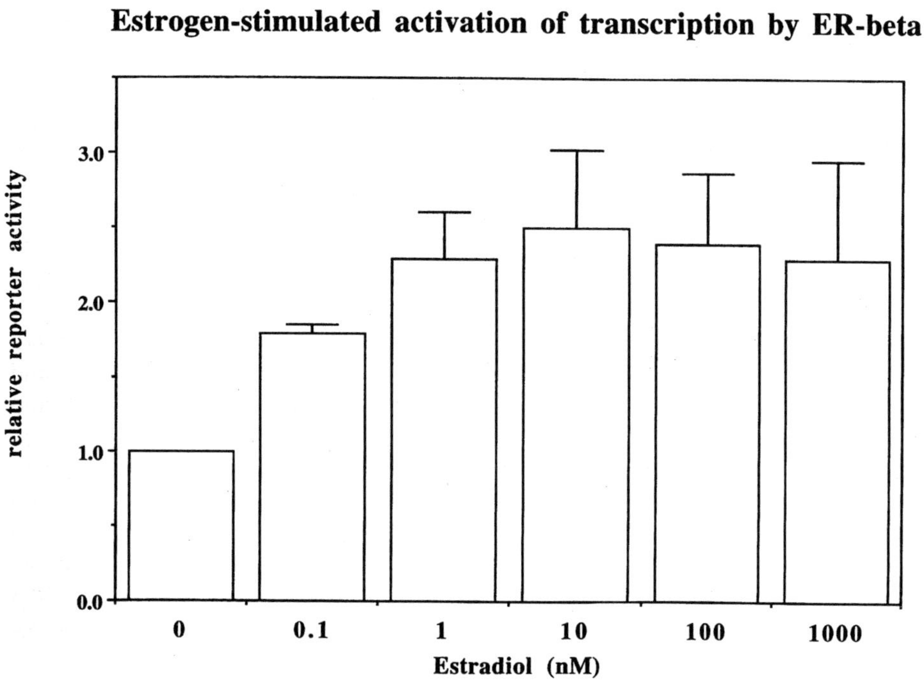


**Figure 1.8.** Estrogen-stimulated activation by ERβ protein. CHO cells were transiently transfected with the ERE-reporter plasmid and the ERβ protein expression plasmid. Cells were incubated with estradiol as indicated.

experiments, the wild type human ERα protein also showed transcriptional activity in the absence of estradiol which could be stimulated to a similar extent as for ERβ protein by estradiol.

It can be envisaged that in addition to forming homodimers, ERα and ERβ protein may form heterodimers when bound to DNA. The transcriptional activity of these heterodimers as compared to the respective homodimers on various target gene promoters remains to be investigated. The ERα protein contains two independent transcriptional activation domains; AF-1 which is located in the N-terminal A/B domain and AF-2 located in the ligand-binding domain (Lees, 1989; Tora *et al.*, 1989). The transcriptional activities of AF-1 and AF-2 are promoter and cell type specific. The N-terminal A/B domain is poorly conserved between ERα and ERβ protein (Figure 1.2) and it remains to be determined if the N-terminal domain of ERβ protein has any transcriptional activation function. The transcription activation function AF-1 is necessary for the activation of ERα protein by growth factors as EGF and IGF-I (Ignar-Trowbridge *et al.*, 1996), and in more detailed studies it was shown that phosphorylation of human ERα protein at serine residue 118 is required for full activity of AF-1 (Kato *et al.*, 1995; Bunone *et al.*, 1996). Serine residue 118 of human ERα protein is phosphorylated by MAPK *in vitro* and in cells treated with EGF and IGF-I *in vivo* (Kato *et al.*, 1995). Several potential phosphorylation sites for MAPK are present in the N-terminus of rat ERβ protein and it would be interesting to see if they are involved in the constitutive transcriptional activity of ER$\beta$ (Figure 1.7).

## EXPRESSION OF ER$\beta$ mRNA

*In situ* hybridisation with oligonucleotide probes revealed high expression of ER$\beta$ in male and female rat reproductive tissues, whereas in all other tissues the expression was much lower or below the level of detection by *in situ* hybridization with oligonucleotide probes (Kuiper *et al.*, 1996). In the prostate ER$\beta$ is expressed in the epithelial cells of the secretory alveoli, whereas in the ovary the granulosa cells in primary, secondary, and mature follicles showed expression of ER$\beta$. Examination of ER$\alpha$ expression at the cellular level, by *in situ* hybridisation, showed that ER$\alpha$ mRNA is expressed at a low level throughout the ovary with no particular cellular localization (Byers *et al.*, 1997). In rat prostate and ovary it has been difficult to demonstrate the presence of ER protein by immunostaining with the available ER protein antibodies, although specific binding of estradiol could be measured in the prostate (van Beurden-Lamers *et al.*, 1974; Jung-Testas *et al.*, 1981) and ovary (Clark, 1992; Richards, 1975). In human and rat prostate weak staining in stromal cells was detected, while the glandular epithelial cells were negative (Prins, 1995; Brolin, 1992; Ehara *et al.*, 1995), which is in contrast with our *in situ* hybridisation results (Kuiper *et al.*, 1996). In the rat ovary specific binding of estradiol was found in intact follicles and granulosa cells (Kudolo, 1984; Richards, 1975), but no ER protein could be detected with the available ER protein antibodies. These discrepancies can be explained by the fact that the most frequently used ER protein antibodies H-222 and H-226 (Greene, 1982), do not crossreact with rat ER$\beta$ protein on immunoblots (unpublished observations). The above findings and our results indicate that the ER$\beta$ protein is the predominant if not the only ER subtype in rat prostate and ovary.

In order to determine the relative distribution of ER$\alpha$ and ER$\beta$ mRNA, total RNA was isolated from rat tissues and used for RT-PCR experiments, using primers which are specific for each ER subtype (Kuiper *et al.*, 1997). Highest expression of ER$\beta$ mRNA was found in the ovary and prostate, which is in agreement with our previous *in situ* hybridisation experiments using male and female rats (Kuiper *et al.*, 1996). In addition testis, uterus, bladder and lung revealed moderate expression, while pituitary, epididymis, thymus, various brain sections (thalamus/ hypothalamus, cerebellum, olfactory lobe, brain stem) and spinal cord reveal low expression of ER$\beta$ mRNA. The ER$\alpha$ mRNA is highly expressed in epididymis, testis, pituitary, uterus, kidney and adrenal, which all showed low or no expression of ER$\beta$ mRNA. Apart from weak expression in thalamus/hypothalamus the brain sections tested are negative for ER$\alpha$ mRNA. Ovary and uterus, which are known to contain high amounts of ER protein, clearly express both ER subtypes. All organs from male rats previously described to display specific binding of estradiol to an 8S cytosolic protein i.e., lung, adrenal, pituitary, prostate, epididymis and testis showed clear expression of either or both ER subtype mRNA (van Beurden-Lamers *et al.*, 1974; Morishige, 1978).

## FUNCTION OF ERβ PROTEIN

Since the ERβ protein has been discovered very recently no definite statements of its unique function(s) can be made, only some interesting observations and speculations will be discussed. Estrogens synergize with follicle stimulating hormone in ovarian weight augmentation, which is associated with a pronounced proliferation of granulosa cells, and the growth of small- and medium-sized follicles (Richards, 1994). A mutant mouse line without a functional ERα protein was created and assessed for estrogen responsiveness (Lubahn *et al.*, 1993). Female mice were infertile and showed hypoplastic uteri and hyperemic ovaries with no detectable corpora lutea. The fact that disruption of the ERα gene did not eliminate the ability of small follicles to grow as was evident from the presence of secondary and antral follicles in the knock-out mouse ovary, pointed to the possible existence of alternative estrogen receptors mediating the intra-ovarian effects of estradiol. Indeed, in some tissues from the ERα knock-out mice residual binding of estradiol with a Kd of 0.7 nM could be measured (Lubahn *et al.*, 1993; Couse *et al.*, 1995). The authors ascribed this to a possible "splicing over" event (Couse *et al.*, 1995), resulting in the production of a smaller mutant ERα protein that could be the source of the residual estradiol binding. In retrospect it seems more likely that the remaining estradiol binding is caused by the presence of ERβ protein. The mouse and human homologues of rat ERβ have recently been cloned in our laboratory (Enmark, Bertilsson, Grandien and Pettersson, unpublished observations).

We (Byers *et al.*, 1997) have investigated the expression and regulation of the two ER subtypes in the rat ovary. Examination of ERβ mRNA expression at the cellular level, by *in situ* hybridization, showed that ERβ mRNA is expressed preferentially in granulosa cells of small, growing, and preovulatory follicles. In contrast, similar studies for ERα mRNA revealed expression at a low level throughout the ovary with no particular cellular localization. Thus, the ovarian responsiveness to estrogen in the ERα knock-out mice probably results from the expression of ERβ protein. In the rat pituitary high levels of ERα mRNA and very low levels of ERβ mRNA are expressed (Kuiper *et al.*, 1997). Therefore, the fact that in the ERα knock-out mouse the ovarian follicles do not develop beyond the follicle stimulating hormone (FSH) independent stage could be explained by a dysregulation in the release of FSH by the pituitary, and not by the absence of ERα protein in the ovary. The ERβ knock-out mice will probably show no follicular growth at all, although this, of course, remains to be proven.

The presence of estrogen receptors (ERα) in preimplantation mouse embryos (Hou, 1993; Hou *et al.*, 1996) and the absence of reported human ER mutations have been interpreted as indications for an essential role of estrogens during embryonic development. This view seemed to be challenged by the survival of the ERα knock-out mice (Lubahn *et al.*, 1993) and the existence of a male with (partial) estrogen resistance caused by a mutation in the ERα gene (Smith *et al.*, 1994). With the identification of a second ER protein, the unexpected viability of the ERα knock-out mouse and human could also be explained by complementation through ERβ protein during embryo development. On the other hand, the existence of a karyotypically female patient with pseudohermaphroditism caused by a null mutation in the aromatase cytochrome P-450 gene (Ito *et al.*, 1993), raises further questions about the validity of the lethality hypothesis.

Although we have no data on ER$\beta$-mRNA expression or protein concentration in tissues of the ER$\alpha$ knock-out mouse, the possible presence of ER$\beta$ protein should be kept in mind when interpreting experiments using the ER$\alpha$ knock-out mouse. In the brain of the ER$\alpha$ knock-out mouse specific binding of estradiol and modulation of progesterone receptor gene expression by estradiol was observed (Shughrue, 1996). Again this is most likely caused by the presence of ER$\beta$ protein, since the ER$\beta$ mRNA is broadly expressed in the rat brain (Kuiper *et al.*, 1997) and probably also in mouse brain.

Estrogen receptors are present in human and rat prostate, as evidenced by ligand-binding studies (Ekman *et al.*, 1983; van Beurden-Lamers *et al.*, 1974; Jung-Testas *et al.*, 1981). Estrogens are, besides androgens, implicated in the growth of the prostate (Griffiths *et al.*, 1991), and estrogens have been implicated in the pathogenesis of benign prostatic hyperplasia (Habenicht *et al.*, 1993). Estrogen treatment of neonatal rats was shown to down-regulate androgen receptor expression in the epithelial cells of all the three prostate lobes, thus leading to an overall prostate growth retardation (Prins, 1995). Various means of androgen deprivation are in use for the treatment of prostate cancer (Furr, 1996). Diethylstilbestrol (DES) treatment leads to long term suppression of serum testosterone due to inhibition of LH release by the pituitary, although direct effects of DES on the rat and human prostate are not excluded (Yamashita *et al.*, 1996; Cox, 1995). In co-transfection experiments ER$\alpha$ was capable of inhibiting transcriptional activation of MMTV-LTR-CAT reporter constructs by the androgen receptor in a estrogen-dependent manner (Kumar, 1994). The co-expression of androgen receptor and ER$\beta$ protein in the secretory epithelial cells of the rat prostate is of interest in this regard, and the putative inhibitory effect of ER$\beta$ should be investigated in more detail.

## ACKNOWLEDGEMENTS

GGJMK was supported, in part, by grants from the Netherlands Organization for Scientific Research (NWO) and from the Karolinska Institute. J-ÅG was supported by a grant from the Swedish Cancer Society.

## REFERENCES

Ali, S., Metzger, D., Bornert, J.-M. and Chambon, P. (1993) Modulation of transcriptional activation by ligand dependent phosphorylation of the human oestrogen receptor A/B region. *EMBO J.*, **12**, 1153–1160.

Arnold, S.F., Obourn, J.D., Jaffe, H. and Notides, A.C. (1995) Phosphorylation of the human estrogen receptor on tyrosine 537 *in vivo* and by src family tyrosine kinases *in vitro*. *Mol. Endocrinol.*, **9**, 24–33.

Bunone, G., Briand, P.-A., Miksicek, J. and Picard, D. (1996) Activation of the unliganded estrogen receptor by EGF involves the MAP kinase pathway and direct phosphorylation. *EMBO J.*, **15**, 2174–2183.

Beato, M., Herrlich, P. and Schutz, G. (1995) Steroid hormone receptors: many actors in search of a plot. *Cell*, **83**, 851–857.

Brolin, J., Skoog, L. and Ekman, P. (1992) Immunohistochemistry and biochemistry in detection of androgen, progesterone, and estrogen receptors in benign and malignant human prostatic tissue. *The Prostate*, **20**, 281–295.

Berns, E.M.J.J., Rommerts, F.G.G. and Mulder, E.M. (1985) Rapid and sensitive detection of oestrogen receptors in cells and tissue sections by autoradiography with [$^{125}$I]-oestradiol. *Histochemical J.*, **17**, 1185–1196.

Beekman, J.M., Allan, G.F., Tsai, S.Y., Tsai, M.-J. and O'Malley, B.W. (1993) Transcriptional activation by the estrogen receptor requires a conformational change in the ligand binding domain. *Mol. Endocrinol.*, **7**, 1266–1272.

Beurden-Lamers van, W.M.O., Brinkmann, A.O., Mulder, E. and van der Molen, H.J. (1974) High-affinity binding of oestradiol by cytosols from testis interstitial tissue, pituitary, adrenal, liver and accessory sex glands of the male rat. *Biochem. J.*, **140**, 495–502.

Berger, J., Hauber, J., Hauber, R., Geiger, R. and Cullen, B.R. (1988) Secreted placental alkaline phosphatase: a powerful new quantitative indicator of gene expression in eukaryotic cells. *Gene*, **66**, 1–10.

Byers, M., Kuiper, G.G.J.M., Gustafsson, J.-Å. and Park-Sarge, O.-K. (1997) Estrogen receptor-β mRNA expression in the rat ovary: down regulation by gonadotropins. *Mol. Endocrinol.*, **11**, 172–182.

Clark, J.H., Schrader, W.T. and O'Malley, B.W. (1992) Mechanisms of action of steroid hormones. In *Textbook of Endocrinology*, edited by J.D. Wilson and D.W. Foster, pp. 35–90. W.B. Saunders Company.

Castles, C.G. and Fuqua, S.A.W. (1996) Alterations within the estrogen receptor in breast cancer. In *Hormone-dependent cancer*, edited by J.R. Pasqualini and B.S. Katzenellenbogen, pp. 81–105. New York: Marcel Dekker.

Cox, R.L. and Crawford, E.D. (1995) Estrogens in the treatment of prostate cancer. *J. Urology*, **154**, 1991–1998.

Couse, J.F., Curtis, S.W., Washburn, T.F., Lindzey, J., Golding, T.S., Lubahn, D.B. *et al.* (1995) Analysis of transcription and estrogen insensitivity in the female mouse after targeted disruption of the estrogen receptor gene. *Mol. Endocrinol.*, **9**, 1441–1454.

Danielian, P.S., White, R., Lees, J.A. and Parker, M.G. (1992) Identification of a conserved region required for hormone-dependent transcriptional activation by steroid hormone receptors. *EMBO J.*, **11**, 1025–1031.

Ehara, H., Koji, T., Deguchi, T., Yoshii, A., Nakano, M., Nakane, P.K. and Kawada, Y. (1995) Expression of estrogen receptor in diseased human prostate assessed by non-radioactive *in situ* hybridization and immunohistochemistry. *The Prostate*, **27**, 304–313.

Ekman, P., Barrack, E.R., Greene, G., Jensen, E.V. and Walsh, P.C. (1983) Estrogen receptors in human prostate: Evidence for multiple binding sites. *J. Clin. Endocrinol. Metab*, **57**, 166–176.

Enmark, E., Kainu, T., Pelto-Huikko, M. and Gustafsson, J.-Å. (1994) Identification of a novel member of the nuclear receptor superfamily which is closely related to rev-erbA. *BBRC*, **204**, 49–56.

Farhat, M.Y., Lavigne, M.C. and Ramwell, P.W. (1996) The vascular protective effects of estrogen. *FASEB J.*, **10**, 615–624.

Fawell, S.E., Lees, J.A., White, R. and Parker, M.G. (1990) Characterization and colocalization of steroid binding and dimerization activities in the mouse estrogen receptor. *Cell*, **60**, 953–962.

Furr, B.J.A., Blackledge, G.R.P. and Cockshott, I.D. (1996) Casodex: preclinical and clinical studies. In *Hormone-dependent Cancer*, edited by J.R. Pasqualini and B.S. Katzenellenbogen, pp. 397–424. New York: Marcel Dekker.

Giguere, V., Yang, N., Segui, P. and Evans, R.M. (1988) Identification of a new class of steroid hormone receptors. *Nature*, **331**, 91–94.

Gleave, M.E. and Chung, L.W.K. (1995) Stromal-epithelial interaction affecting prostate tumour growth and hormonal responsiveness. *Endocrine-Related Cancer*, **2**, 243–265.

Greene, G.L. and Jensen, E.V. (1982) Monoclonal antibodies as probes for estrogen receptor detection and characterization. *J. Steroid Biochem.*, **16**, 353–359.

Griffiths, K., Davies, P., Eaton, C.L., Harper, M.E., Turkes, A. and Peeling, W.B. (1991) Endocrine factors in the initiation, diagnosis, and treatment of prostatic cancer. In *Endocrine-dependent tumors*, edited by K.D. Voigt and C. Knabbe, pp. 83–125.

Habenicht, U.-F., Tunn, U.W., Senge, Th., Schröder, F.H., Schweikert, H.U., Bartsch, G. and El Etreby, M.F. (1993) Management of benign prostatic hyperplasia with particular emphasis on aromatase inhibitors. *J. Steroid Biochem. Molec. Biol.*, **44**, 557–563.

Harlow, K.W., Smith, D.N., Katzenellenbogen, J.A., Greene, G.L. and Katzenellenbogen, B.S. (1989) Identification of cysteine 530 as the covalent attachment site of an affinity-labeling estrogen and antiestrogen in the human estrogen receptor. *J. Biol. Chem.*, **264**, 17476–17485.

Hähnel, R., Twaddle, E. and Ratajczak, T. (1973) The specificity of the estrogen receptor of human uterus. *J. Steroid Biochem.*, **4**, 21–31.

Hegy, G.B., Shackleton, C.H.L., Carlquist, M., Bonn, T., Engström, O., Sjöholm, P. and Witkowska, H.E. (1996) Carboxymethylation of the human estrogen receptor ligand-binding domain-estradiol complex: HPLC/ESMS peptide mapping shows that cysteine 447 does not react with iodoacetic acid. *Steroids*, **61**, 367–373.

Hochberg, R.B. and Rosner, W. (1980) Interaction of 16α-[$^{125}$I]iodo-estradiol with estrogen receptor and other steroid-binding proteins. *Proc. Natl. Acad. Sci. USA*, **77**, 328–332.

Hou, Q. and Gorski, J. (1993) Estrogen receptor and progesterone receptor genes are expressed differentially in mouse embryos during preimplantation development. *Proc. Natl. Acad. Sci. USA*, **90**, 9460–9464.

Hou, Q., Paria, B.C., Mui, C., Dey, S.K. and Gorski, J. (1996) Immunolocalization of estrogen receptor protein in the mouse blastocyst during normal and delayed implantation. *Proc. Natl. Acad. Sci. USA*, **93**, 2376–2381.

Ignar-Trowbridge, D.M., Pimentel, M., Parker, M.G., McLachlan, J.A. and Korach, K.S. (1996) Peptide growth factor cross-talk with the estrogen receptor requires the A/B domain and occurs independently of protein kinase C or estradiol. *Endocrinology*, **137**, 1735–1744.

Ito, Y., Fisher, C., Conte, F.A., Grumbach, M.M. and Simpson, E.R. (1993) Molecular basis of aromatase deficiency in an adult female with sexual infantilism and polycystic ovaries. *Proc. Natl. Acad. Sci. USA*, **90**, 11673–11677.

Jung-Testas, I., Groyer, M.-T., Bruner-Lorand, J., Hechter, O., Baulieu, E.-E. and Robel, P. (1981) Androgen and estrogen receptors in rat ventral prostate epithelium and stroma. *Endocrinology*, **109**, 1287–1289.

Kato, S., Endoh, H., Masuhiro, Y., Kitamoto, T., Uchiyama, S., Sasaki, H. *et al.* (1995) Activation of the estrogen receptor through phosphorylation by mitogen-activated protein kinase. *Science*, **270**, 1491–1494.

Koehorst, S.G.A., Jacobs, H.M., Thijssen, J.H.H. and Blankenstein, M.A. (1993) Wild type and alternatively spliced estrogen receptor messenger RNA in human meningioma tissue and MCF-7 breast cancer cells. *J. Steroid Biochem. Molec. Biol.*, **45**, 227–233.

Koike, S., Sakai, M. and Muramatsu, M. (1987) Molecular cloning and characterization of rat estrogen receptor cDNA. *Nucleic Acids Res.*, **15**, 2499–2513.

Kudolo, G.B., Elder, M.G. and Myatt, L. (1984) A novel oestrogen-binding species in rat granulosa cells. *J. Endocr.*, **102**, 83–91.

Kumar, M.V., Leo, M.E. and Tindall, D.J. (1994) Modulation of androgen receptor transcriptional activity by the estrogen receptor. *J. Andrology*, **15**, 534–542.

Kuiper, G.G.J.M. and Brinkmann, A.O. (1994) Steroid hormone receptor phosphorylation: is there a physiological role? *Mol. Cell. Endocrinol.*, **100**, 103–107.

Kuiper, G.G.J.M., Enmark, E., Pelto-Huikko, M., Nilsson, S. and Gustafsson, J.-Å. (1996) Cloning of a novel estrogen receptor expressed in rat prostate and ovary. *Proc. Natl. Acad. Sci. USA*, **93**, 5925–5930.

Kuiper, G.G.J.M., Carlsson, B., Grandien, K., Enmark, E., Häggblad, J., Nilsson, S. and Gustafsson, J.-Å. (1997) Comparison of the ligand binding specificity and transcript tissue distribution of estrogen receptors α and β. *Endocrinology*, **138**, 863–870.

Laudet, V., Hänni, C., Coll, J., Catzeflis, F. and Stehelin, D. (1992) Evolution of the nuclear receptor gene superfamily. *EMBO J.*, **11**, 1003–1013.

Lees, J.A., Fawell, S.E. and Parker, M.G. (1989) Identification of two transactivation domains in the mouse oestrogen receptor. *Nucl. Acids Res.*, **17**, 5477–5487.

Lubahn, D.B., Moyer, J.S., Golding, T.S., Couse, J.F., Korach, K.S. and Smithies, O. (1993) Alteration of reproductive function but not prenatal sexual development after insertional disruption of the mouse estrogen receptor gene. *Proc. Natl. Acad. Sci. USA*, **90**, 11162–11166.

Martinez, E., Givel, F. and Wahli, W. (1987) The estrogen responsive element as an inducible enhancer: DNA sequence requirements and conversion to a glucocorticoid-reponsive element. *EMBO J.*, **6**, 3719–3727.

Mangelsdorf, D.J., Thummel, C., Beato, M., Herrlich, P., Schutz, G., Umesono, K. *et al.* (1995) The nuclear receptor superfamily: the second decade. *Cell*, **83**, 835–839.

Mosselman, S., Polman, J. and Dijkema, R. (1996) ERβ: identification and characterization of a novel human estrogen receptor. *FEBS Lett.*, **392**, 49–53.

Morishige, W.K. and Uetake, C.-A. (1978) Receptors for androgen and estrogen in the rat lung. *Endocrinology*, **102**, 1827–1836.

Murdoch, F.E. and Gorski, J. (1991) The role of ligand in estrogen receptor regulation of gene expression. *Mol. Cell. Endocrinol.*, **78**, C103–C108.

Notides, A.C. (1970) Binding affinity and specificity of the estrogen receptor of the rat uterus and anterior pituitary. *Endocrinology*, **87**, 987–992.

Pettersson, K., Svensson, K., Mattsson, R., Carlsson, B., Ohlsson, R. and Berkenstam, A. (1996) Expression of a novel member of estrogen response element-binding nuclear receptors is restricted to the early stages of chorion formation during mouse embryogenesis. *Mech. Development*, **54**, 211–223.

Prins, G.S. and Birch, L. (1995) The developmental pattern of androgen receptor expression in rat prostate lobes is altered after neonatal exposure to estrogen. *Endocrinology*, **136**, 1303–1314.

Richards, J.S. (1975) Estrogen receptor content in rat granulosa cells during follicular development: modification by estradiol and gonadotropins. *Endocrinology*, **97**, 1174–1184.

Richards, J.A. (1994) Hormonal control of gene expression in the ovary. *Endocrine Rev.*, **15**, 725–751.

Sabbah, M., Radanyi, C., Redeuilh, G. and Baulieu, E.-E. (1996) The 90 kDa heat-shock protein (hsp90) modulates the binding of the oestrogen receptor to its cognate DNA. *Biochem. J.*, **314**, 205–213.

Salomonsson, M., Carlsson, B. and Häggblad, J. (1994) Equilibrium hormone binding to human estrogen receptors in highly diluted cell extracts is non-cooperative and has a Kd of approximately 10 pM. *J. Steroid Biochem. Molec. Biol.*, **50**, 313–318.

Seielstad, D.A., Carlson, K.E., Kushner, P.J., Greene, G.L. and Katzenellenbogen, J.A. (1995) Analysis of the structural core of the human estrogen receptor ligand binding domain by selective proteolysis/mass spectrometric analysis. *Biochemistry*, **34**, 12605–12615.

Shughrue, P.J., Komm, B. and Merchenthaler, I. (1996) The distribution of estrogen receptor-β mRNA in the rat hypothalamus. *Steroids*, **61**, 678–681.

Skipper, J., Young, L.J., Bergeron, J.M., Tetzlaff, M.T., Osborn, C.T. and Crews, D. (1993) Identification of an isoform of the estrogen receptor messenger RNA lacking exon four and present in the brain. *Proc. Natl. Acad. Sci. USA*, **90**, 7172–7175.

Smith, E.P., Boyd, J., Frank, G.R., Takahashi, H., Cohen, R.M., Specker, B. *et al.* (1994) Estrogen resistance caused by a mutation in the estrogen-receptor gene in a man. *New England J. Medicine*, **331**, 1056–1061.

Smith, C.L., Conneely, O.M. and O'Malley, B.W. (1995) Oestrogen receptor activation in the absence of ligand. *Biochem. Soc. Trans.*, **23**, 935–939.

Swillens, S., Waelbroeck, M. and Champeil, P. (1995) Does a radiolabelled ligand bind to a homogeneous population of non-interacting receptor sites? *Trends Pharmaceutical Sci.*, **16**, 151–155.

Tora, L., White, J., Brou, C., Tasset, D., Webster, N., Scheen, E. and Chambon, P. (1989) The human estrogen receptor has two independent non-acidic transcriptional activation functions. *Cell*, **59**, 477–487.

Turner, R.T., Riggs, B.L. and Spelsberg, T.C. (1994) Skeletal effects of estrogen. *Endocrine Rev.*, **15**, 275–296.

Yamashita, A., Hayashi, N., Sugimura, Y., Cunha, G.R. and Kawamura, J. (1996) Influence of diethylstilbestrol, leuprolin, finasteride and castration on the lobar subdivisions of the rat prostate. *The Prostate*, **29**, 1–14.

Zilliacus, J., Carlstedt-Duke, J., Gustafsson, J.-Å. and Wright, A.P.H. (1994) Evolution of distinct DNA-binding specificities within the nuclear receptor family of transcription factors. *Proc. Natl. Acad. Sci. USA*, **91**, 4175–4179.

# 2 Tissue Selective Estrogen Receptor Modulators: A Concept Reduced to Practice

**Donald P. McDonnell** * **and John D. Norris**

*Department of Pharmacology, Duke University Medical School,*
*Durham, NC 27710, USA.*
* *Corresponding author, Tel.: 001 919 684 6035, Fax: 001 919 681 7139,*
*E-mail: mcdon016@acpub.duke.edu.*

Although the general mechanism by which estrogen exerts its activity in the female reproductive system is well established, its role and mode of action in non-classical targets, such as bone and the cardiovascular system, remains elusive. The dramatic changes which occur in these latter systems following cessation of ovarian steroid production however, underscores the importance of this hormone. Not surprisingly therefore, replacement of estrogens in postmenopausal women has evolved as a major intervention strategy for osteoporosis and cardiovascular disease. Although the benefits associated with estrogen replacement therapy greatly outweigh any potential risks, the perception that this treatment option increases breast cancer risk definitely has curtailed its use. This realization, and the sheer magnitude of this public health problem, has created the need to develop tissue specific estrogens which display beneficial effects in the bone and cardiovascular system while being neutral in the breast and endometrium. Whereas initially it was not intuitive that such dissociations could be achieved, the entry of the first generation of tissue selective estrogens into the clinic has signaled a new era in steroid hormone receptor pharmacology. This paper reviews what is known about the mechanism of action of the currently available tissue selective ER modulators, and highlights how advances in the study of the molecular pharmacology of these compounds will quickly lead to the development of additional compounds for clinical evaluation.

KEYWORDS: Estrogen receptor, tamoxifen, raloxifene, transcription, osteoporosis, cancer.

## THE MECHANISM OF ACTION OF THE ESTROGEN RECEPTOR

The estrogen receptor is a member of the nuclear receptor superfamily of ligand dependent transcription factors. Until recently it was believed that all the actions of estrogen were mediated by a single high affinity estrogen receptor (ER$\alpha$) located within target cell nuclei. However, a second estrogen receptor-(ER$\beta$)-has been identified (Kuiper *et al.*, 1996). Although ER$\alpha$ and ER$\beta$ may have unique activities in estrogen responsive tissues, the high degree of amino acid sequence similarity between these two proteins suggests that their general mechanism of action is conserved (Figure 2.1). It has been determined that in the absence of ligand, ER is

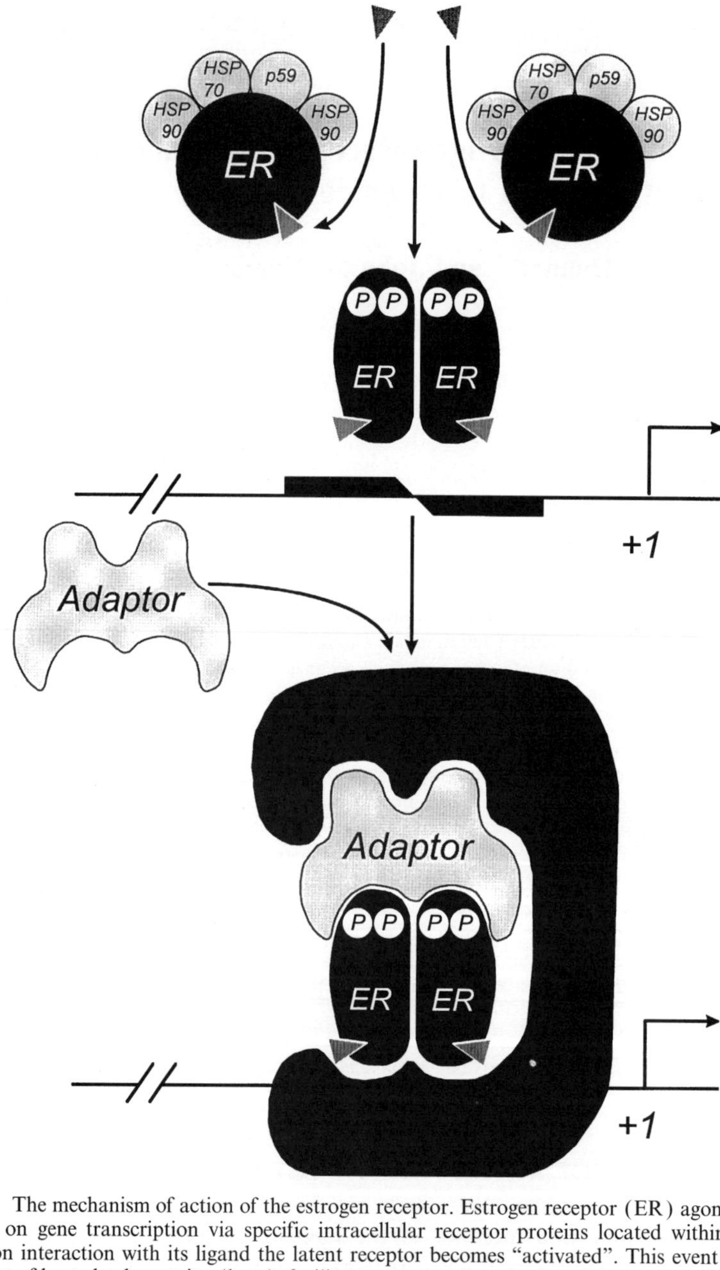


**Figure 2.1.** The mechanism of action of the estrogen receptor. Estrogen receptor (ER) agonists mediate their effect on gene transcription via specific intracellular receptor proteins located within target cell nuclei. Upon interaction with its ligand the latent receptor becomes "activated". This event permits the displacement of heat-shock proteins (hsps), facilitates receptor dimerization and promotes the interaction of the receptor with specific estrogen response elements (EREs) located within the regulatory regions of target promoters. At this location, depending on the cellular and promoter context, the ligand activated receptor can interact with the general transcription machinery (GTA) directly or indirectly through adaptor proteins. Ultimately, these interactions stabilize the transcription preinitiation complex and enhance RNA polymerase activity. Although several rounds of phosphorylation of the receptor have been shown to occur, its role in ER-signaling has yet to be determined.

located in target cell nuclei associated with a large macromolecular complex of heat shock proteins (hsps) and chaperones (Kumar and Chambon, 1988). Upon ligand binding however, the receptor undergoes a conformational change permitting spontaneous dimerization. This event facilitates the interaction of the receptor with specific DNA response elements (Estrogen Response Elements; EREs) located within the regulatory regions of target genes (Kumar and Chambon, 1988). The precise mechanism by which the promoter bound receptor influences the activity of RNA polymerase is unknown. In some cases it may contact the general transcription machinery (GTA) directly, whereas in other configurations it may contact the basal machinery indirectly by interacting with intermediary "adaptor" or "coactivator" proteins (Ing *et al.*, 1992; Halachmi *et al.*, 1994; Cavaillès *et al.*, 1995; Onate *et al.*, 1995). This plasticity may be an important part of estrogen signaling permitting ER to engage in different regulatory pathways within target cells.

Concurrent with the development of our understanding of the molecular mechanism of action of ER was the emergence of a wealth of data on the *in vivo* pharmacology of a variety of novel ER-modulators (Gottardis *et al.*, 1988; Love *et al.*, 1991; Jordan, 1992; Love *et al.*, 1992). These studies indicated that there were at least four classes of ER-ligands: Pure agonists, such as estrogen, two classes of mixed agonists of which tamoxifen and raloxifene are representative members, and pure antagonists like ICI 182,780 (McDonnell *et al.*, 1995a,b). The mixed agonist class of compounds are particularly intriguing from a mechanistic point of view. Tamoxifen, for instance, functions as an antagonist in ER-positive breast cancer cells, whereas it is a partial agonist in the uterus and appears to function as a pure agonist in bone (Love *et al.*, 1991, 1992). Raloxifene, although functioning similarly to tamoxifen in bone, breast and in the cardiovascular system, appears to be mechanistically distinct in that it exhibits minimal agonist activity in the uterus (Draper *et al.*, 1993; Yang *et al.*, 1996). It appears therefore that the established model of ER-action, where ligand is responsible for switching the receptor from an inactive to an active state, is probably an oversimplification. Rather, it is likely that the transcriptionally active receptor is part of a large complex which includes the ligand, target DNA and receptor associated proteins (Katzenellenbogen *et al.*, 1996). This complexity predicts that the biological activity of a given receptor ligand is influenced by the nature of the bound ligand, the sequence of the response element and the cell and promoter context of the DNA bound receptor (Figure 2.2). Thus, *different* compounds acting through the same receptor may display *different* activities in *different* cells; a hypothesis which stands in contradiction to classical receptor theory which holds that receptor affinity is the primary determinant of efficacy (Clark and Peck, 1979; Meyer *et al.*, 1990; Tzukerman *et al.*, 1994; McDonnell *et al.*, 1995a; Webb *et al.*, 1995).

## Definition of the Mechanisms Which Enable Cells to Distinguish Between Different Classes of ER-Ligands

We and others have applied molecular and biochemical approaches to determine at what stage in the ER-signal transduction pathway the various ER ligands manifest their unique activities. Thus far, it has been determined that all ER ligands

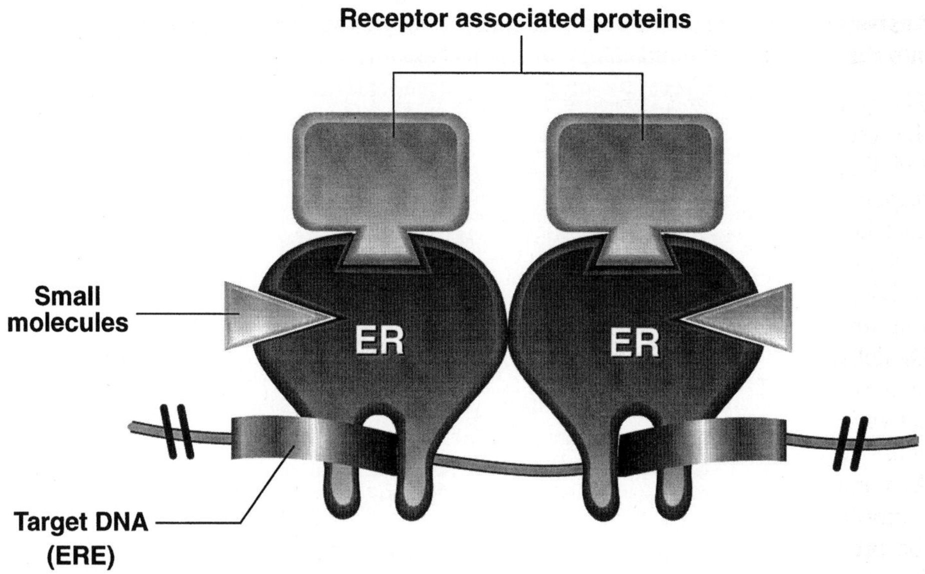


**Figure 2.2.** The transcriptional activity of the estrogen receptor is affected by each of three classes of ligands. The established models of ER action predict that the biological activity of a ligand is a direct reflection of its affinity for the receptor (Clark and Peck, 1979). Whereas this remains an important aspect of ER-action it is now only part of a larger picture in which high affinity interactions of DNA and other proteins associated with the receptor must also be considered (Katzenellenbogen *et al.*, 1996). We propose that since target DNA and associated proteins demonstrate specific, high affinity interactions with ER, that they are by definition receptor ligands. Consequently, it is the product of these three interactions which determines the biological activity of an ER-ligand.

(agonists, mixed agonists and pure antagonists) interact with the hormone binding domain within the receptor, efficiently displacing hsp's and promoting association of ER with target DNA. This was an unexpected and fundamental observation as it suggested that both agonists and antagonists were capable of "activating" the ER signal transduction pathway. This result alone indicated that ER-antagonists were not merely "competitive antagonists," blocking agonist access, but were integral parts of a complex which interfered with agonist action at several steps in the signal transduction pathway. They were in fact "active antagonists" committing the receptor down a distinct, albeit non-productive, pathway. This was sufficient to explain the activity of the pure antiestrogens such as ICI 182,780. Tamoxifen on the other hand, which can function as an antagonist, a mixed agonist or an agonist depending on the cell and promoter context, introduces another level of complexity (Meyer *et al.*, 1990; Tzukerman *et al.*, 1994; McDonnell *et al.*, 1995a). With the exception of estrogen therefore, the definition of whether a compound is an agonist or an antagonist is appropriate only in the context of an individual ER-responsive process.

### Analysis of the Structure–Function Relationships Within ER Provides Insights into the Molecular Pharmacology of Agonists and Antagonists

The cloning of ER and subsequent efforts to define the structural domains within the receptor have been well documented and reviewed extensively elsewhere (McDonnell *et al.*, 1994; 1995b; Katzenellenbogen *et al.*, 1996). In brief, domains responsible for ligand binding, DNA recognition, dimerization and interaction with heatshock proteins have been defined. Importantly, it was observed that the receptor had at least two regions which were required for maximal ER-transcriptional activity: Activating function-1 (AF-1) located towards the amino terminus and AF-2 contained within the ligand binding domain of the receptor (Meyer *et al.*, 1990; Danielian *et al.*, 1992; Tzukerman *et al.*, 1994). These AFs were shown to function in a cell specific manner and it was determined that the requirement for these activation functions varied depending on the cell and promoter context (Meyer *et al.*, 1990; Tzukerman *et al.*, 1994). Thus, depending on the context, either or both AFs may be required. We hypothesized that different compounds, by virtue of promoting different conformational changes within ER, could differentially affect the presentation of these AFs to the transcription apparatus. This prediction was found to be correct. It was determined that in contexts where AF-2 was required, either alone or with AF-1, tamoxifen functioned as an ER-antagonist (McDonnell *et al.*, 1994; 1995b; Tzukerman *et al.*, 1994). Thus, we and others have defined tamoxifen as an AF-2 antagonist (Meyer *et al.*, 1990; McDonnell *et al.*, 1995a). However, in contexts where AF-1 alone is required, we have determined that tamoxifen can function as a partial agonist manifesting 30–40% of the agonist activity of estradiol. Thus, tamoxifen functions as an AF-1 agonist (Tzukerman *et al.*, 1994; McDonnell *et al.*, 1995b). Although it is not possible, as yet, to prove that these *in vitro* mechanistic distinctions explain what occurs *in vivo*, it is striking that droloxifene, toremifene, clomiphene and nafoxidine, compounds which display agonist activity in the uterus and antagonist activity in the breast, all function as cell specific AF-1 agonists *in vitro* (Tzukerman *et al.*, 1994; McDonnell *et al.*, 1995a; and our unpublished results). Based on these results, and those of additional studies, we have concluded that these triphenylethelene-derived mixed agonists are mechanistically identical and, apart from differences in their pharmacokinetic properties, should have the same effect on ER *in vivo*.

The fact that AF-1 or AF-2 activity is sufficient for ER action in some contexts whereas both are required in others suggests that the role of ER in different cells and on different promoters is not the same. In this regard it was proposed by Webb *et al.* that in addition to the classical ER/ERE mediated signal transduction pathway ER may regulate transcription in an indirect manner by interacting with AP1 complexes associated with target gene promoters (Webb *et al.*, 1995). Thus it was proposed that ER was a component of two distinct signaling pathways within the cell. Interestingly, these investigators noted that antiestrogens were on the whole more effective at promoting ER/AP1 interactions and consequently were more efficient regulators of transcription though the ER/AP1 mechanism (Figure 2.3). It is intriguing that this non-classical pathway is very active in cultured endometrial cells where tamoxifen displays partial ER-agonist activity, whereas it is minimally active in ER positive

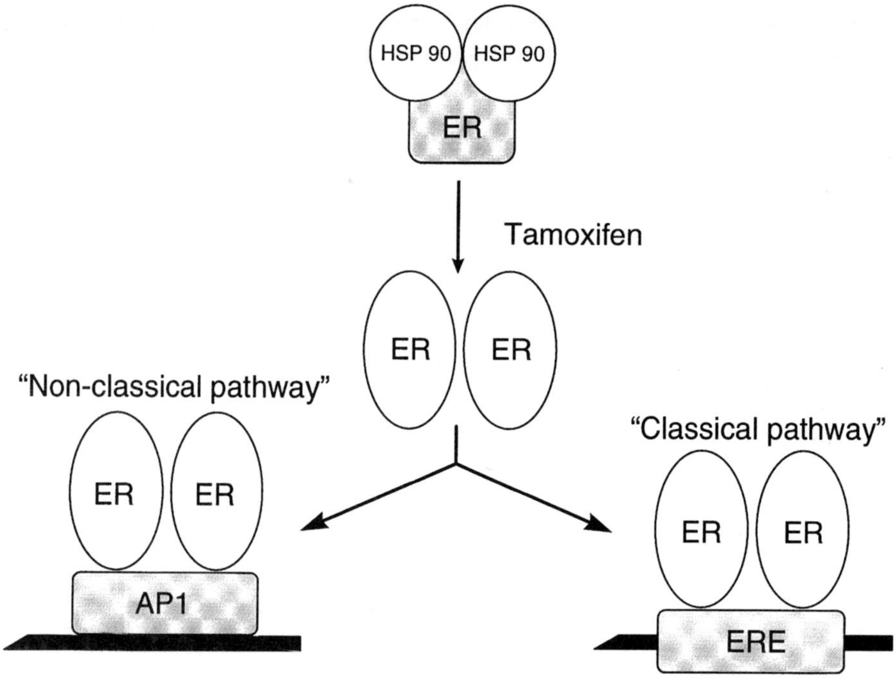


**Figure 2.3.** Potential mechanisms for the cell and promoter specific partial agonist activity of tamoxifen. It has been proposed that tamoxifen activated ER can operate through either of two distinct mechanisms within the cell. In most cells it operates as an antagonist through the "classical" signaling pathway where it interacts with the receptor, promoting its association with DNA, but in a manner which opposes the actions of the estradiol activated ER. In some cells tamoxifen activated ER can manifest partial agonist activity through a "non-classical" pathway where the ligand activated receptor interacts in a indirect manner with target gene promoters through the AP1 protein complex.

breast cancer cell lines where tamoxifen functions as a pure antagonist. Unfortunately, *in vivo* evidence in support of this non-classical pathway has not been forthcoming and as yet tamoxifen regulation of an endogenous AP1-responsive gene has not yet been reported. Nevertheless, the discovery of this novel pathway tested the dogma that all the actions of ER were manifest through the classical ER–ERE mediated pathway. These data suggested that the ER-agonist and antagonist activities of tamoxifen may not occur through the same pathway. Consequently, we attempted to develop *in vitro* systems which model the cell specific partial agonist activities of tamoxifen in order to determine whether tamoxifen could exhibit ER-partial agonist activity through a classical ERE mediated-manner. Specifically, this work led to the identification of the complement 3 (C3) promoter, an ER-responsive promoter which was positively regulated by estradiol in all cell contexts examined. Interestingly, tamoxifen partial agonist activity was only manifest in a subset of these cells reflecting the cell selective actions of this compound observed *in vivo* (Tzukerman *et al.*, 1994; McDonnell *et al.*, 1995a; Norris *et al.*, 1996). Thus, the pharmacology of ER-ligands acting on the same promoter, and through the same receptor, differed according to the cell-type. Using biochemical and genetic approaches we have been able to define the molecular mechanism of action of tamoxifen in this cell and promoter context (Tzukerman *et al.*, 1994; McDonnell *et al.*, 1995; Norris *et al.*, 1996). Specifically, dissection of the C3 promoter revealed that both estrogen and tamoxifen agonist activity occurred as a consequence of a direct interaction of ER with each of three estrogen response elements within this promoter (Norris *et al.*, 1996). We noticed also that the activity of ER on this promoter in the presence of either agonist or antagonist did not require, and was not influenced by AP-1 (Norris *et al.*, 1996). This result indicated that tamoxifen could manifest partial agonist activity through a classical ER–ERE mediated pathway. Whether or not additional pathways function *in vivo* remains an open question. However, these data clearly indicate that tamoxifen delivers ER to classical EREs within target genes and has the potential to exhibit partial agonist activity in those environments where ER-AF-1 alone is required for transcriptional activity.

From a pharmaceutical perspective this result was important as it indicated that cell restricted ER agonists and antagonists could be developed. Its importance was underscored by the accumulating evidence that ER-AF-1 agonists, such as tamoxifen, functioned as ER agonists in bone (Love *et al.*, 1992). One of the initial hypotheses we proposed to explain the agonist activity of tamoxifen in bone was that the AF-1 activity alone was required for the regulation of the ER-responsive genes involved in the maintenance of bone mass. It was predicted therefore, that any compound which displayed AF-1 activity would function as an ER-agonist in the bone. This hypothesis was supported by the observations that droloxifene and clomiphene, compounds which also demonstrate AF-1 activity, function as ER-agonists in bone (Chen *et al.*, 1995). However, although it was possible to conclude that tamoxifen was an AF-1 agonist and that this activity tracked with bone protection it was not possible to say that AF-1 function was actually required for estrogenic actions in this target. To address this issue we extended our studies to define the molecular mechanism of action of raloxifene, a compound which appeared from clinical and preclinical studies to be mechanistically distinct from tamoxifen (Draper *et al.*, 1993; McDonnell *et al.*, 1995; Yang *et al.*, 1996). Specifically, this compound functions similarly to tamoxifen in all tissues except the uterus where it demonstrates significantly less agonist activity (Yang *et al.*, 1996). We determined that raloxifene binds with high affinity to ER and induces a conformational change within the receptor, which appears to be indistinguishable from that induced by tamoxifen (McDonnell *et al.*, 1995a). However, there were clear differences in the activities of these compounds when assayed on ER-responsive promoters *in vitro* (McDonnell *et al.*, 1995a). Unlike tamoxifen, which displayed partial agonist activity in some promoter contexts, we determined that raloxifene was not an AF-1 agonist but was by all previous definitions a pure antagonist (McDonnell *et al.*, 1995a). In fact, in our *in vitro* systems its mechanism of action was similar to that of pure antagonist ICI 182,780. Thus, it appears as if compounds exhibiting AF-1 activity are in fact bone protective, however this activity is not absolutely required. In support of this hypothesis we have shown that the novel ER-antagonist GW5638, which does not demonstrate ER-AF-1 activity effectively blocks ovariectomy induced bone loss in rats (Willson *et al.*, 1994; and our unpublished results). Thus, the role of ER in bone and the properties of an ER-modulator required for bone protection are unclear. Although several classes of ER-ligand can protect against bone loss in ovariectomized rats it is unclear what property these compounds have in common which permit this activity (Chow *et al*., 1992; Love *et al*., 1992; Draper *et al*., 1993; Willson *et al*., 1994; Yang *et al*., 1996).

## MODELS FOR THE ACTIVITY OF ER AND ITS LIGANDS IN BONE

Based on our current understanding of ER-action, the interaction of the latent receptor with any ER-ligand induces a structural alteration in the receptor which promotes dissociation of the inhibitory heat-shock proteins (Beekman *et al*., 1993; McDonnell *et al*., 1995a). However, since the pure ER-antagonists effectively activate ER, but promote bone loss, it indicates that this event alone is not sufficient for ER-mediated bone protective activities (Dukes *et al*., 1994; D. Morris and T. Willson, unpublished data). Following activation by estrogen, ER undergoes spontaneous dimerization permitting it to interact specifically and with high affinity to DNA response elements within target promoters. It may be at this stage that those ER-modulators which protect against bone loss, and those that do not, diverge. Specifically, it appears as if the pure antiestrogens, like ICI 182,780, interfere with the interaction of ER with DNA. The mechanism of this inhibitory activity is complex involving a combination of changes in nuclear-cytoplasmic shuttling, impaired dimerization and acceleration of the rate of ER turnover (Fawell *et al*., 1990; Dauvois *et al*., 1992). In contrast, all the other classes of ER-ligands, including GW5638 (I. Asplin and D.P. McDonnell, unpublished results) efficiently deliver ER to DNA and do not induce turnover of the receptor (McDonnell *et al*., 1995a). Upon reviewing the activity of the known ER ligands at each stage of the estrogen signaling pathway, it is apparent that the ability of a compound to activate ER and promote its association with DNA tracks well with bone protection (Table 2.1).

There are at least three models which are compatible with available data and which could explain the mechanism of action of bone-protective ER-ligands. The first model suggests that even though raloxifene and GW5638 do not display ER-agonist activity in our reconstituted *in vitro* transcription systems, under conditions where tamoxifen and estrogen functions as agonists, it is possible that in relevant bone targets all four compounds can promote ER-mediated activation of target gene transcription. Alternatively, a second model proposes that the mechanism by which ER acts in bone may be different from that in classical ER-targets;

**Table 2.1** Correlation between the mechanism of action of ER-modulators and biological activity

| | *ICI 182,780* | *GW5638* | *Raloxifene* | *Tamoxifen* | *Estradiol* |
|---|---|---|---|---|---|
| Interaction with ER-LBD | + | + | + | + | + |
| Displacement of HSPs | + | + | + | + | + |
| Receptor dimerization | + | ± | + | + | + |
| Interaction of ER with DNA | + | ± | + | + | + |
| Manifest AF-1 agonist activity | – | – | – | + | + |
| Manifest AF-2 agonist activity | – | – | – | – | + |

therefore the definition of "agonist" must be considered only in reference to the target cell (or process). Thus, if the role of estradiol within a target cell in bone is to deliver ER to DNA and in some way alter the transcription rate of a target gene(s) by disrupting the activity of a negative regulator; then any agent which delivers ER to DNA would be an agonist. It is worth noting that the only class of ER-modulators examined to date which do not appear to protect against bone loss are the pure antiestrogens, compounds which do not efficiently promote ER–DNA interactions. A third model, for which there is a considerable amount of supporting preclinical data, is that activated ER can inhibit the actions of a positive regulator of a gene, an event requiring activated receptor, but not DNA binding. One example of this activity is the regulation of the expression of the cytokine interleukin-6 (IL-6) (Jilka *et al.*, 1992). It has been shown recently in mice that ovariectomy leads to an increase in IL-6 production and a concomitant increase in the relative number of osteoclasts with respect to osteoblasts. This event subsequently leads to a net loss of bone. The link between this activity and bone was established by the demonstration that administration of neutralizing antibodies to IL-6 was sufficient to block ovariectomy induced bone loss. Although other events in addition to IL-6 up regulation, are likely to be involved in osteoporosis, the observation that the IL-6 promoter is down regulated upon administration of estrogen makes it a useful system in which to study ER-action. Recently, the molecular basis of ER-regulation of the IL-6 promoter has been determined (Figure 2.4). Specifically, it has been

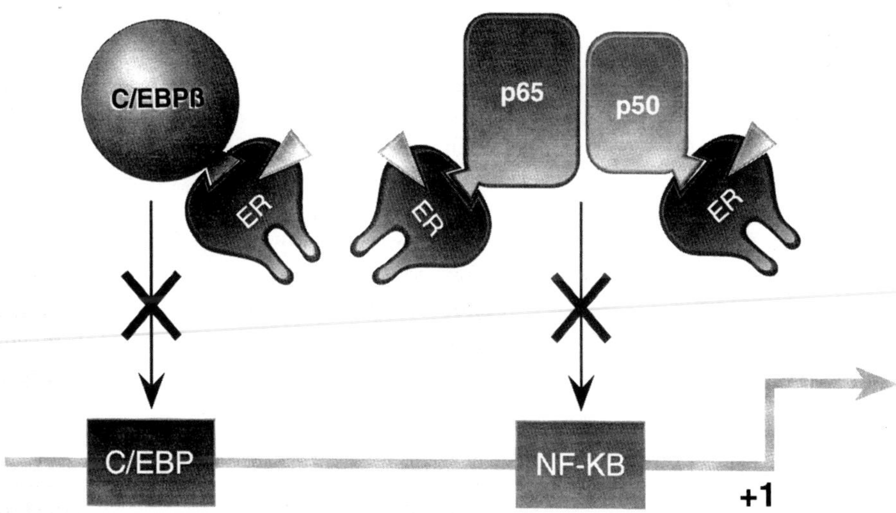


**Figure 2.4.** Estrogen mediated repression of the IL-6 promoter; a model for ERE-independent regulation of gene transcription. The IL-6 promoter is negatively regulated by estrogens (Jilka *et al.*, 1992). It is proposed that upon activation ER can physically interact with either (or both) C/EBP$\beta$ or NF$\kappa$B (p65/p50 complex) and inhibit their positive activity on this promoter (Ray *et al.*, 1994; Stein and Yang, 1995). Interestingly, this activity does not require a direct interaction of ER with DNA. It is predicted therefore, that any compound which can promote receptor activation and which does not have an effect on receptor turnover will protect against bone loss.

determined that both NF-κB and C/EBPβ are required for IL-6 promoter activity (Ray *et al.*, 1994; Stein and Yang, 1995). However ER in the presence of estradiol physically interacts with both the *rel* homology domain of NF-κB and the b-Zip region of C/EBPβ, blocking their ability to interact with DNA and subsequently lowering the transcriptional rate of the IL-6 promoter. Moreover, mutations within ER which block DNA binding do not affect the ability of the receptor to block IL-6 production (Ray *et al.*, 1994). Although the activity of tamoxifen, raloxifene or other ER ligands has not been tested in this system, it is likely that activation of the receptor by an agent which does not induce receptor turnover would manifest similar activities. Thus, we hypothesize that DNA binding activity may not actually be required (Ray *et al.*, 1994). Therefore, given what we know about the mechanism of the bone protective ER-modulators, and considering the data on IL-6 regulation, it is possible that classical ER-agonist activity is not required for bone protection.

Regardless of which model(s) is correct, the underlying observation is that the role of ER is not identical in all targets. With this understanding, the goal of developing tissue selective ER-agonists has been realized, signaling a new era in the pharmacology of estrogen receptors.

## ACKNOWLEDGMENTS

The authors would like to thank the members of our laboratory for helpful suggestions and comments. Some of the work described was supported by a research grant to DPM (NIH DK 48807).

## REFERENCES

Beekman, J.M., Allan, G.F., Tsai, S.Y., Tsai, M.-J. and O'Malley, B.W. (1993). "Transcriptional activation by the estrogen receptor requires a conformational change in the ligand binding domain." *Mol. Endo.* **7**: 1266–1274.

Cavaillès, V., Dauvois, S., L'Horset, F., Lopez, G., Hoare, S., Kushner, P.J. and Parker, M.G. (1995). "Nuclear factor RIP140 modulates transcriptional activation by the estrogen receptor." *EMBO J.* **14**(15): 3741–3751.

Chen, H.K., Ke, H.Z., Ma, Y.F., Crawford, D.T., Pirie, C.M., Simmons, H.A., Jee, W.S. and Thompson, D.D. (1995). "Droloxifene inhibits cortical bone turnover associated with estrogen deficiency in rats." *Bone* **17**: 175–179.

Chow, J., Thobias, J.H., Colston, K.W. and Chambers, T.J. (1992). "Estrogen maintains trabecular bone volume in rats not only by suppression of bone resorbtion but also by stimulation of bone formation." *J. Clin. Invest.* **89**: 74–78.

Clark, J.H. and Peck, E.J. (1979). *Female sex steroids: receptors and function.* New York, Springer-Verlag.

Danielian P.S., White, R., Lees, J.A. and Parker, M.G. (1992). "Identification of a conserved region required for hormone dependent transcriptional activation by steroid hormone receptors." *EMBO J.* **11**(3): 1025–1033.

Dauvois, S., Danielian, P.S., White, R. and Parker, M.G. (1992). "Antiestrogen ICI 164,384 reduces cellular estrogen receptor content by increasing its turnover." *Proc. Natl. Acad. Sci. USA* **89**: 4037–4041.

Draper, M.W., Flowers, D.E., Huster, W.J. and Neild, J.A. (1993). *Effects of raloxifene (LY 139481 HCL) on biochemical markers of bone and lipid metabolism in healthy post-menopausal women.* Aalborg, Handelstrykkeriet.

Dukes, M., Chester, R., Yarwood, L. and Wakeling, A.E. (1994). "Effects of a non-steroidal pure antiestrogen, ZM 189,154, on oestrogen target organs of the rat including bones." *J. Endocrinology* **141**: 335–341.

Fawell, S.E., White, R., Hoare, S., Sydenham, M., Page, M. and Parker, M.G. (1990). "Inhibition of estrogen receptor-DNA binding by the "pure" antiestrogen ICI164,384 appears to be mediated by impaired receptor dimerization." *Proc. Natl. Acad. Sci. USA* **87**: 6883–6887.

Gottardis, M.M., Robinson, S.P. and Jordan, V.C. (1988). "Estradiol-stimulated growth of MCF-7 tumors implanted in athymic nude mice: A model to study the tumoristatic action of tamoxifen." *J. Steroid Biochem.* **30**: 311–314.

Halachmi, S., Marden, E., Martin, G., MacKay, H., Abbondanza, C. and Brown, M. (1994). "Estrogen receptor-associated proteins: possible mediators of hormone-induced transcription." *Science* **264**: 1455–1458.

Ing, N., Beekman, J., Tsai, S., Tsai, M.-J. and O'Malley, B. (1992). "Members of the steroid hormone receptor superfamily interact with TFIIB (S300-II)." *J. Biol. Chem.* **267**: 17617–17623.

Jilka, R.L., Hangoc, G., Girasole, G., Passeri, G., Williams, D.C., Abrams, J.S., Boyce, B., Broxmeyer, H. and Manolagus, S.C. (1992). "Increased osteoclast development after estrogen loss: Mediation by interleukin-6." *Science* **257**: 88–91.

Jordan, V.C. (1992). "The strategic use of antiestrogens to control the development and growth of breast cancer." *Cancer* **70**: 977–982.

Katzenellenbogen, J.A., O'Malley, B.W. and Katzenellenbogen, B.S. (1996). "Tripartite steroid hormone receptor pharmacology: Interaction with multiple effector sites as a basis for the cell- and promoter-specific action of these hormones." *Mol. Endo.* **10**: 119–131.

Kuiper, G.G.J.M., Enmark, E., Pelto-Huikko, M., Nilsson, S. and Gustafsson, J.-A. (1996). "Cloning of a novel estrogen receptor expressed in rat prostate and ovary." *Proc. Natl. Acad. Sci. USA* **93**: 5925–5930.

Kumar, V. and Chambon, P. (1988). "The estrogen receptor binds tightly to its responsive element as a ligand-induced homodimer." *Cell* **55**: 145–156.

Love, R.R., Wiebe, D.A., Newcombe, P.A., Cameron, L., Leventhal, H., Jordan, V.C., Feyzi, J. and DeMets, D.L. (1991). "Effects of tamoxifen on cardiovascular risk factors in postmenopausal women." Ann. Intern. Med. **115**: 860–864.

Love, R.R., Mazess, R.B., Barden, H.S., Epstein, S., Newcomb, P.A., Jordan, V.C., Carbone, P.P. and DeMets, D.L. (1992). "Effects of tamoxifen on bone mineral density in postmenopausal women with breast cancer." *New Engl. J. Med.* **326**: 852–856.

McDonnell, D.P., Clemm, D.L. and Imhof, M.O. (1994). "Definition of the cellular mechanisms which distinguish between hormone and antihormone activated steroid receptors." *Seminars in Cancer Biology* **5**: 503–513.

McDonnell, D.P., Clemm, D.L., Herman, T., Goldman, M.E. and Pike, J.W. (1995). "Analysis of estrogen receptor function *in vitro* reveals three distinct classes of antiestrogens." *Mol. Endo.* **9**: 659–669.

McDonnell, D.P., Lieberman, B.A. and Norris, J. (1995b). Development of Tissue-Selective Estrogen Receptor Modulators. *Organ-Selective Actions of Steroid Hormones.* D.T. Baird, G. Schutz and R. Krattenmacher. Berlin, Springer-Verlag. **16**: 1–28.

Meyer, M.E., Pornon, A., Ji, J., Bocquel, M.T., Chambon, P. and Gronemeyer, H. (1990). "Agonist and antagonist properties of RU486 on the functions of the human progesterone receptor." *EMBO J.* **9**: 3923–3932.

Norris, J., Fan, D. and McDonnell, D.P. (1996). "Identification of the sequences within the human complement 3 promoter required for estrogen responsiveness provides insight into the mechanism of tamoxifen mixed agonist activity." *Mol. Endo.* **10**: 1605–1616.

Onate, S.A., Tsai, S., Tsai, M.-J. and O'Malley, B.W. (1995). "Sequence and characterization of a coactivator for the steroid hormone receptor superfamily." *Science* **270**: 1354–1357.

Ray, A, Prefontaine, K.E. and Ray, P. (1994). "Down-regulation of Interleukin-6 gene expression by 17-$\beta$-estradiol in the absence of high affinity DNA binding by the estrogen receptor." *J. Biol. Chem.* **269**: 12940–12946.

Stein, B. and Yang, M. (1995). "Repression of the interleukin-6 promoter by estrogen receptor is mediated by NF-kB and C/EBP$\beta$." *Molecular and Cellular Biology* **15**: 4971–4979.

Tzukerman, M.T., Esty, A., Santiso-Mere, D., Danielian, P., Parker, M.G., Stein, R.B., Pike, J.W. and McDonnell, D.P. (1994). "Human estrogen receptor transcriptional capacity is determined by both cellular and promoter context and mediated by two functionally distinct intramolecular regions." *Mol. Endo.* **8**: 21–30.

Webb, P., Lopez, G.N., Uht, R.M. and Kushner, P.J. (1995). "Tamoxifen activation of the estrogen receptor/AP-1 pathway: potential origin for the cell-specific estrogen-like effects of antiestrogens." *Mol. Endo.* **9**: 443–456.

Willson, T.M., Henke, B.R., Momtahen, T.M., Charifson, P.S., Batchelor, K.W., Lubahn, D.B., Moore, L.B., Oliver, B.B., Sauls, H.R. and Triantafillou, J.A. (1994). "3-[4-(1,2-Diphenylbut-1-enyl) phenyl]acrylic acid: a non-steroidal estrogen with functional selectivity for bone over uterus in rats." *J. Med. Chem.* **37**: 1550–1552.

Yang, N.N., Bryant, H.U., Hardikar, S., Sato, M., Galvin, R.J.S., Glasebrook, A.L. and Termine, J.D. (1996). "Estrogen and raloxifene stimulate transforming growth factor-$\beta 3$ gene expression in rat bone: Potential mechanism for estrogen-or raloxifene mediated bone maintenance." *Endocrinology* **137**: 2075–2084.

# 3 Cross-talk Between Growth Factors and Estrogen Receptor

**Sylvia W. Curtis [1,*], Todd Washburn [1], Charles Sewall [3,†], Richard DiAugustine [2], John F. Couse [1] and Kenneth S. Korach [1]**

[1] *Receptor Biology Section, Laboratory of Reproductive and Developmental Toxicology,*
[2] *Laboratory of Biochemical Risk Analysis,* [3] *Environmental Toxicology Program, National Institute of Environmental Health Sciences, National Institutes of Health, PO Box 12233, Research Triangle Park, NC 27709, USA.*
**Corresponding author, Tel.: 001 919 541 3429, Fax: 001 919 541 7935, E-mail: curtis.s@niehs.nih.gov.*
[†] *Present Address: Procter and Gamble, 11050 York Rd, Hunt Valley, MD 21030, USA.*

Both estrogen and peptide growth factors play important roles in mammalian sexual differentiation and development. Our previous studies had indicated an interaction or cross-talk between plasma membrane bound epidermal growth factor (EGF) receptor signaling and nuclear localized estrogen receptor signaling. Studies *in vitro* showed that EGF treatment lead to estrogen independent induction of an estrogen responsive reporter gene, but only if estrogen receptor was present. To dissect the mechanism of cross-talk *in vivo* we characterized EGF signaling both through conventional pathways as well as ER activation by EGF using the estrogen receptor knockout (ERKO) mouse. EGF receptors were present in the uterus and liver of ERKO mice and could be autophosphorylated after EGF treatment. In addition, EGF treatment resulted in increase in the mRNA for the EGF responsive c-fos in both wild type and ERKO animals. EGF caused an increase in the estrogen regulated progesterone receptor (PR) mRNA message in wild type but not ERKO mice. The PR induction could be inhibited by antiestrongen. In addition, EGF caused an increase in 3H thymidine incorporation only in wild type uteri. These results show that in the mouse uterus, the estrogen receptor is essential for EGF initiated increases in progesterone receptor mRNA and DNA synthesis, indicating activation of the membrane bound EGF receptor leads to estrogen independent activation of the nuclear localized estrogen receptor.

KEYWORDS: EGF, estrogen receptor, ER knockout, cross-talk, ligand-independent, mouse uterus.

Prenatal sexual development of mammals is the culmination of interactions and feedback amidst numerous hormones and peptides. We have been interested in the role of estrogens in the development and function of the rodent reproductive tract and other estrogen responsive tissues. Estrogen transmits its signal via a specific receptor protein, the estrogen receptor (ER). The ER has specific high affinity binding sites for estrogens as well as specific DNA enhancer sequences (EREs) that regulate estrogen responsive genes (Parker *et al.*, 1993; Gronemeyer, 1991). The ER also binds to transcriptional regulators, and has two domains that are important for regulation for transcriptional activity (AF1 and AF2) (Green and Chambon, 1991; Tora *et al.*, 1989; Lees *et al.*, 1989a,b).

The evidence thus far indicates that androgen has a directive role in reproductive development, and that its absence results in a "default" development of female phenotype (Jost, 1953). The role of estrogen in reproductive tract development is not apparent, and may be masked by androgen action. Estrogen is important, however in the growth and maintenance of the uterus in the sexually mature female. To assess the role of estrogen in the rodent reproductive tract, we evaluated the effects of estrogen signaling disruption on rodent reproductive tissue development and function using the estrogen receptor knockout (ERKO), mouse.

The ERKO females have grossly normal reproductive tracts at birth as well, again indicating a lack of a role for estrogen in prenatal female reproductive tract development. The adult females are also infertile (Lubahn *et al.*, 1993). Ovariectomized ERKO females have uteri that are one quarter the size of wildtype littermates, and show no growth in response to estrogen. The non-responsiveness of the uterus to estrogen is shown by the lack of induction of lactoferrin and progesterone mRNAs (Couse *et al.*, 1995). The ERKOs do not ovulate and the ovaries become cystic and enlarged after puberty (Lubahn *et al.*, 1993; Korach *et al.*, 1996) (unpublished data). Thus, removal of estrogen action by genetic disruption of the ER gene has allowed us to show that estrogen does not have a major role in prenatal development of reproductive structures in the mouse. In contrast, estrogen plays an essential role at the time of puberty in growth, maintenance and development of reproductive capability both in males and females.

We also have an interest in the role of epidermal growth factor (EGF) in the growth of the mouse uterus. The effects of estrogen in the rodent reproductive tract include the initiation of hypertrophy and mitosis (Quarmby *et al.*, 1984; Quarmby and Korach, 1984). As these tissues have growth factor receptors, a role for cytokines in the process have also been hypothesized. Estrogen has been shown to increase the level of both epidermal growth factor (EGF) and its receptor (EGF-R) (Das *et al.*, 1994; Stancel *et al.*, 1987; Mukku and Stancel, 1985; Lingham *et al.*, 1988; Gardner *et al.*, 1989; Stancel *et al.*, 1990; DiAugustine *et al.*, 1988; Huet Hudson *et al.*, 1990). Treatment with exogenous EGF has also been shown to mimic estrogen induced effects in the uterus, resulting in mitosis, induction of lactoferrin, and modified cell morphology (Nelson *et al.*, 1991). Treatment with an EGF-specific antibody resulted in a 60–70% reduction in the estrogen-induced epithelial proliferation in the uterus and vagina (Nelson *et al.*, 1991), illustrating the probability of the role of EGF in estrogen action as well as a cross-talk between nuclear ER and membrane-bound EGF-R. Further evidence of EGF-estrogen cross-talk was provided by showing that treatment with the pure estrogen antagonist ICI 164,384, which reduces the level of uterine ER, greatly diminished uterine response following EGF treatment (Ignar Trowbridge *et al.*, 1992). This indicates the necessity for the presence of ER for the EGF to mimic estrogen. Further studies in Ishikawa cells, a human endometrial carcinoma cell line devoid of ER, indicated an estrogen-responsive CAT reporter gene could be activated by EGF only after co-transfection with an ER-expression plasmid. Furthermore, an ER deletion mutant that lacked a hormone binding domain could still be activated by EGF, while a deletion mutant lacking the N-terminal transcription activation domain could not (Ignar Trowbridge *et al.*, 1993; 1996). These studies support a model in which EGF can play a role in ER-mediated events even in the absence of estrogen.

To prove a direct role of ER itself, EGF effects were studied in the uteri of ERKO mice. Since estrogen up-regulates the EGF-R in the mouse uterus and removal of the ER during development might alter the level of EGF-R, it was necessary to determine the levels of EGF-R in the ERKO mouse before proceeding with the studies of EGF action. As shown in Table 3.1, EGF-R was present in membranes from the uteri and liver of ERKO animals, although at a level 35% and 15% lower, respectively, than the wild-type levels. The EGF-R also undergoes autophosphorylation in both tissues in response to EGF (Figure 3.1), indicating that the EGF-R can be activated by EGF in the ERKO mouse.

**Table 3.1** $^{125}$I EGF binding to membrane fractions*

| | *Wild type* | *ERKO* |
|---|---|---|
| Uterus | $16.3 \times 10^3$ | $10.6 \times 10^3$ |
| Liver | $23.6 \times 10^4$ | $20.1 \times 10^4$ |

*The [$^{125}$I] EGF receptor binding assay was done as previously described (Sewall *et al.*, 1995). Values are the average of triplicate determinations and are expressed as counts per minute (CPM) bound per mg protein.

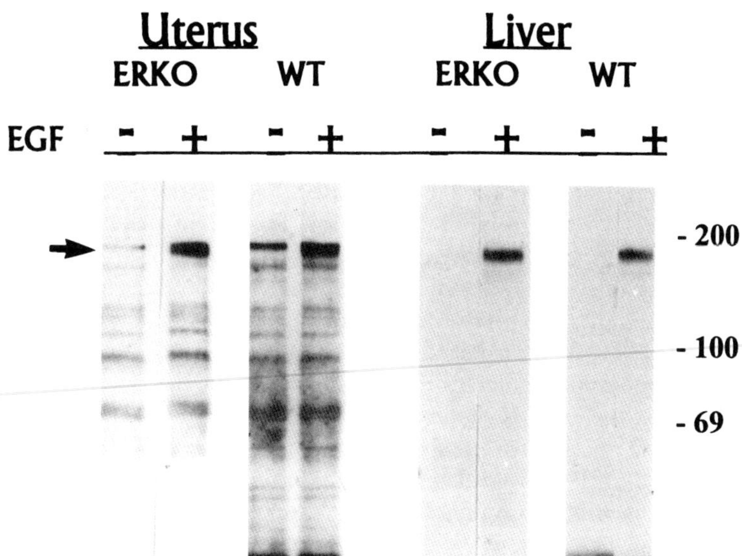


**Figure 3.1.** EGF stimulated autophosphorylation of the EGF-R in ERKO and wild type mice. Solubilized membrane from uterus and liver was subjected to an *in vitro* autophosphorylation assay by incubation in the presence (+) or absence (−) of EGF (Rubin and Earp, 1983). After stopping the reaction, equivalent amounts of total protein (25 μg/lane) were separated by SDS-PAGE. The arrow indicates the position of the EGF-R band. Migration of [$^{14}$C] molecular weight standards (Amersham, Arlington Heights, IL), in kilodaltons is indicated to the right (reproduced from Curtis *et al.*, 1996).

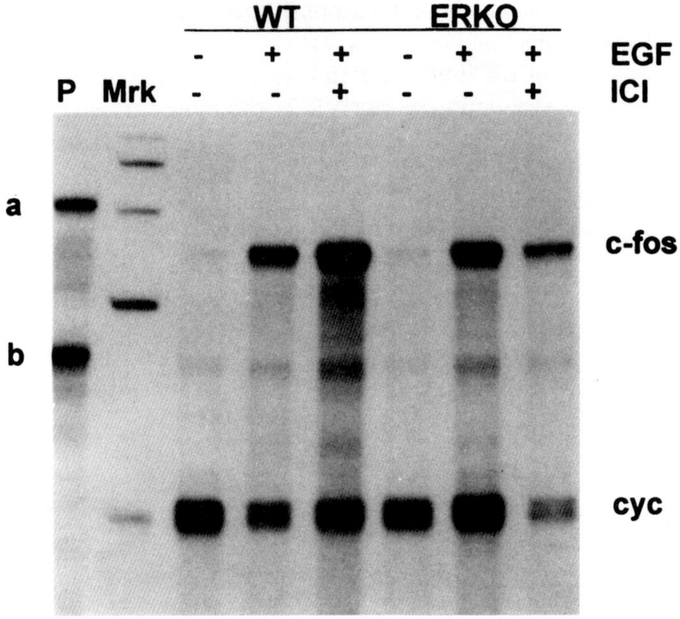


**Figure 3.2.** Uterine c-fos mRNA is induced by EGF in uteri of wild type and ERKO mice. RNase protection assay was used to measure the relative amounts of c-fos mRNA from mice treated with PBS vehicle, EGF for 30 minutes, or pretreated with ICI prior to EGF treatment. Positions of c-fos (250 nt) and cyclophilin (cyc, 103 nt) are indicated. Lane P shows the positions of the riboprobes for c-fos (a) and cyclophilin (b) before digestion. Molecular weight markers (Ambion) (Mrk) are at 500, 400, 300, 200 and 100 nt (reproduced from Curtis *et al.*, 1996).

c-fos is a nuclear transcription factor whose transcription is induced by EGF (Herschman, 1991). To determine if the level of EGF-R in the ERKO was sufficient to elicit a gene response, RNA was prepared from uteri of animals 30 minutes after subcutaneous injection of EGF, and c-fos levels were determined by RPA (Figure 3.2). Levels of c-fos mRNA were induced 18-fold in the wild type uterus and 13-fold in the ERKO, after normalizing to the level of cyclophilin mRNA. This response in both genotypes indicates that although the EGF-R level is lower in the ERKO (Table 3.1), the EGF-R signaling pathway is intact and able to mediate gene activation in ERKO mice. Since c-fos is also known to be an estrogen inducible gene (Loose Mitchell *et al.*, 1988), we wanted to determine whether the EGF induction might have an ER-mediated component via the proposed cross-talk mechanism. The antiestrogen ICI 182,780, however, did not diminish the induction of c-fos by EGF in wild type animals, indicating that ER is not involved in EGF induction of c-fos at this early time point of 30 minutes. The increased EGF induction after ICI pre-treatment is most likely due to the DMSO vehicle, since DMSO pre-treatment caused a similar enhancement of EGF activity (data not presented), possibly due to increased EGF-R autophosphorylation following DMSO treatment (Rubin and Earp, 1983).

The ERKO uterus lacks a growth response following estradiol treatment as measured by a uterotropic bioassay and $^3$H-thymidine incorporation. Estradiol treatment results in almost a 10-fold increase in $^3$H-thymidine incorporation in wild type uteri, whereas there is a minimal increase in the ERKO (Couse *et al.*, 1995; Lubahn *et al.*, 1993). To determine if EGF would initiate DNA synthesis in the ERKO uterus, ovariectomized ERKO and wild type animals were implanted with EGF releasing pellets, and DNA synthesis measured using relative $^3$H-thymidine incorporation. After 16–18 hours of EGF treatment, wild type uteri showed a 4.3-fold increase in $^3$H-thymidine incorporation over placebo treated animals, whereas the ERKO mice showed no increase in DNA synthesis compared to the placebo control (Figure 3.3).

The PR gene is also estrogen regulated in the rodent uterus (Aronica and Katzenellenbogen, 1991; Cho *et al.*, 1994), however it is not induced by estrogen in the ERKO (Couse *et al.*, 1995). To determine whether PR mRNA could be induced by EGF, RNA was prepared from uterine tissue after *in vivo* treatments with EGF pellets with or without pre-treatment with the pure antiestrogen ICI 182,780. Northern blot analysis (Figure 3.4) revealed that PR message increases after EGF treatment in the wild type animals, but not in the ERKO mice. In addition, the antiestrogen ICI 182,780 is able to block the EGF induction of the PR in the wild type, and has no effect on the ERKO.

Our studies and those of others (Ignar Trowbridge *et al.*, 1992; 1993; Nelson *et al.*, 1991; Sumida and Pasqualini, 1989) have led us to propose a model of "cross-talk" or interaction between the EGF and estrogen signaling pathways such that EGF-R activation can result in activation of ER target genes in the absence of

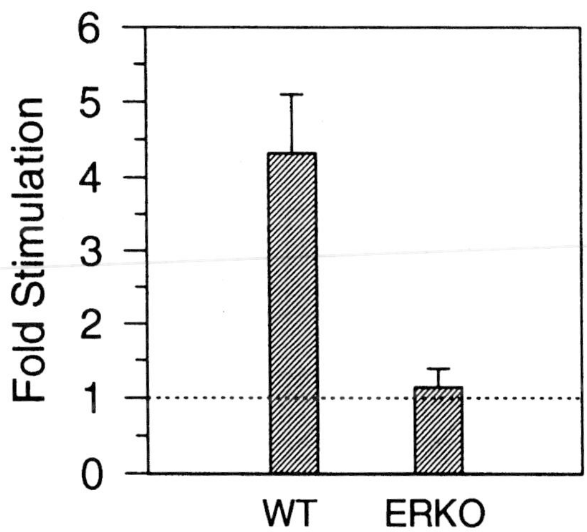


**Figure 3.3.** EGF does not increase DNA synthesis in the ERKO uterus. Relative [$^3$H]thymidine incorporation into uteri of wild type or ERKO mice was measured following insertion of a placebo pellet or an EGF-containing pellet. [$^3$H]Thymidine incorporation was calculated as DPM/μgDNA and is plotted as fold increase after EGF treatment compared to placebo control. Values are the means of two independent experiment and the bars show the ranges of values (reproduced from Curtis *et al.*, 1996).

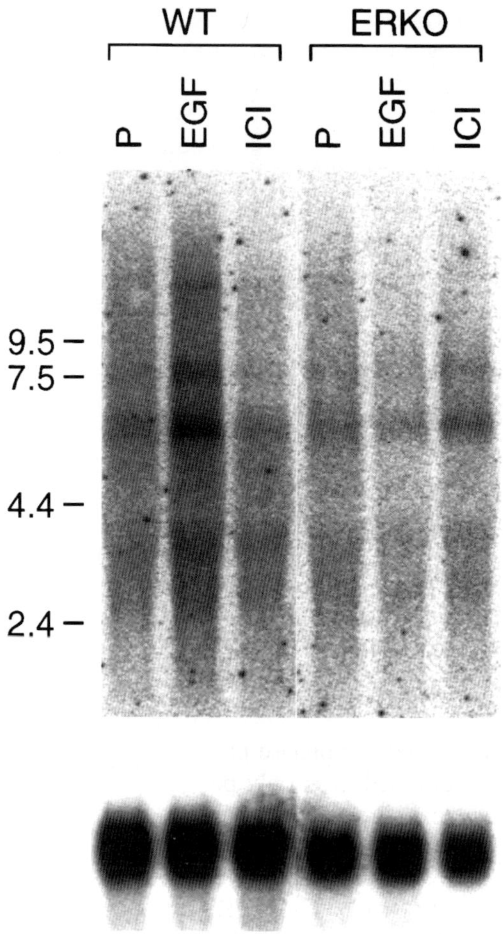


**Figure 3.4.** Progesterone receptor mRNA is increased by EGF treatment of wild type but not ERKO mice. 20 μg of total RNA isolated from uteri was analyzed by Northern blot for PR mRNA (top panel). RNA samples were from wild type (WT) or ERKO animals containing either placebo pellets (P), EGF pellets, or were injected with 1 mg/kg ICI 182,780 (ICI) 30 minutes prior to insertion of EGF pellets. The bottom panel shows the same blot re-probed with PL-7, a ribosomal protein used to confirm equal loading of samples (reproduced from Curtis *et al.*, 1996).

estrogen (Figure 3.5). In this model, the EGF-R and ER pathways can initiate distinct and separable signals that modulate responses of their respective target genes. However, it is also possible to couple the two pathways when, as depicted by the dashed arrow in Figure 3.5, the EGF-R signal activates ER-mediated response of estrogen targets. The ERKO mouse is the definitive model to demonstrate the role of ER in this mechanism since removal of the ER should disable the cross-talk. Indeed, both DNA synthesis and PR mRNA, which are estrogen responsive, could be induced by EGF in wild type mice but not in ERKO mice. Induction of c-fos, which required only the EGF-R pathway, occurred in both genotypes. Since EGF-R

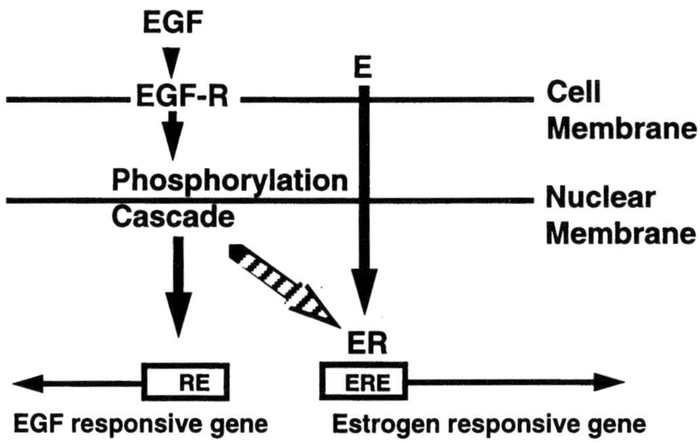


**Figure 3.5.** Model of cross-talk between EGF-R/ER signaling pathways. The model depicts the separate EGF and estrogen inducible pathways. EGF binds to its membrane-bound receptor, which initiates a cascade of phosphorylation modulating the activities of cellular kinases and nuclear transcription factors (depicted as the phosphorylation cascade in the model). The activation of transcription factors results in regulation of responsive genes. In contrast, estrogen (E) diffuses to the nucleus where it binds and activates the nuclear ER thereby modulating transcription of estrogen responsive genes. Cross-talk between the pathways (depicted by the dashed arrow) occurs when EGF initiates the phosphorylation cascade, which activates the ER, presumably via ER phosphorylation, regardless of the presence of estrogen (reproduced from Curtis *et al.*, 1996).

activates several pathways and the precise pathway involved in EGF induction via the ER has not been determined, it is only possible to definitely conclude that the c-fos induction mechanism is intact. However, it appears ER mediated responses, including EGF-R initiated cross-talk with ER, but not processes dependent only on the EGF-R signaling pathway, are disrupted in the ERKO.

The mechanism of this EGF-R mediated ER activation is unknown, but *in vitro* studies using a truncated ER illustrated the importance of ER amino terminal sequences for activation by EGF (Ignar Trowbridge *et al.*, 1993). The ER mediates transcriptional activities by utilizing two transcription activation functions or AFs (Green and Chambon, 1991; Lees *et al.*, 1989a,b; Tora *et al.*, 1989). AF-1 is a ligand independent activation function in the N-terminus and AF-2 is a ligand dependent activation function in the ligand binding domain. An ER deletion mutant lacking the hormone binding domain was activated by EGF, whereas a deletion mutant lacking the AF-1 was not (Ignar Trowbridge *et al.*, 1993). Phosphorylation in the AF-1 region of the ER is a likely mechanism of EGF-mediated activity, because EGF-R activation results in phosphorylation and consequent modulation of the activities of several cellular kinases (Heldin, 1995; Hill and Treisman, 1995; Marshall, 1995), and subsequent phosphorylation and activation of nuclear transcription factors (Heldin, 1995; Hill and Treisman, 1995; Marshall, 1995). Serines 122, 156, 158 and 298 of the mouse ER are known to be phosphorylated in COS-1 cells (Lahooti *et al.*, 1994). Serines 104, 106 and 167 of the human ER have been identified as phosphorylation sites as well (Ali *et al.*, 1993; Arnold *et al.*, 1994; Le Goff *et al.*, 1994). Mutation of serine 118 to alanine, which is located in AF-1 of the human ER, eliminated phosphorylation of this serine and resulted in reduction of transcriptional activity by 40–75% in HeLa and COS-1 cells (Ali *et al.*, 1993), indicating that phosphorylation of serine 118 may mediate transcriptional activity of ER and therefore could be a target for EGF-R activated kinase. Furthermore, recent *in vitro* data show that serine 118 of the human ER can be phosphorylated by the MAP kinase pathway (Kato *et al.*, 1995), resulting in increased estrogen dependent activity in transfected COS-1 cells. Estrogen independent activation of the human ER by EGF via the MAP kinase pathway and phosphorylation of serine 118 has been shown in HeLa and SK-Br-3 cells (Bunone *et al.*, 1996). It will be interesting to determine if serine 122 of the mouse ER, which is analogous to serine 118 in the human ER, is required for EGF activity, and whether MAP kinase mediated phosphorylation results in mouse ER transcriptional activity. Alternatively, other serines in the AF-1 region or other kinases may be required. Perhaps phosphorylation of substrates other than ER, such as steroid receptor coactivators or nuclear transcription factors, are required for ligand independent activity.

The importance of estrogen independent ER action to normal physiology is unclear. The levels of EGF and EGF-R are both increased by estradiol (Das *et al.*, 1994), which indicates EGF-R may play a role at a time when estrogen is already present. Perhaps during proestrous, when estrogen is low, EGF is able to mediate and maintain a constitutive level of estrogen responsive gene expression. In addition, the EGF may play a greater role during fetal development or postnatally before estrogen levels increase.

Alternatively, EGF-R mediated ER activation may synergize with the estrogen hormone signal by increasing the activity of the ER both in terms of the level of response of genes and broadening the spectrum of activated genes. Studies *in vitro* have suggested that AF-1 and AF-2 activities are promoter specific (Metzger *et al.*, 1992), and an increase in AF-1 activity as a result of EGF-R activation may allow for an enhanced response by genes requiring AF-1 activity.

The ERKO mouse has provided the first definitive link between growth factor pathways and ER *in vivo*. However, other membrane receptor pathways have also been shown to activate nuclear steroid receptors. Insulin and IGF-1 both increase growth and induce an estrogen responsive CAT reporter in human neuroblastoma cells that contain ER (Ma *et al.*, 1994). Membrane permeable agents that increase cellular phosphorylation, such as 8-bromo-cAMP and okadaic acid, also enhance progesterone dependent gene transcription (Denner *et al.*, 1990). Dopamine, a neurotransmitter has been shown to induce ligand independent activation of several nuclear receptors including the chicken progesterone receptor (Power *et al.*, 1991a,b) and the human ER (Smith *et al.*, 1993). Together with the activation of mouse ER by EGF, these results confirm the existence of conserved cross-talk mechanisms between membrane receptors and nuclear receptors.

## ACKNOWLEDGMENTS

We would like to thank Dr. Alan Wakeling for his gift of ICI 182,780. We also want to acknowledge the expert animal husbandry and maintenance of the ERKO colony by Lisa Padillo-Burgos, Tanya Moore, Chris Lee, and Kenneth Jeffries. In addition, we want to thank James Clark and Page Myers for their surgical expertise.

## REFERENCES

Ali, S., Metzger, D., Bornert, J.M. and Chambon, P. (1993). Modulation of transcriptional activation by ligand-dependent phosphorylation of the human oestrogen receptor A/B region. *EMBO J.* **12**, 1153–1160.

Arnold, S.F., Obourn, J.D., Jaffe, H. and Notides, A.C. (1994). Serine 167 is the major estradiol-induced phosphorylation site on the human estrogen receptor. *Molecular Endocrinology* **8**, 1208–1214.

Aronica, S.M. and Katzenellenbogen, B.S. (1991). Progesterone receptor regulation in uterine cells: stimulation by estrogen, cyclic adenosine 3',5'-monophosphate, and insulin-like growth factor I and suppression by antiestrogens and protein kinase inhibitors. *Endocrinology* **128**, 2045–2052.

Bunone, G., Briand, P.-A., Miksicek, R.J. and Picard, D. (1996). Activation of the unliganded estrogen receptor by EGF involves the MAP kinase pathway and direct phosphorylation. *The EMBO J.* **15**, 2174–2183.

Cho, H., Aronica, S.M. and Katzenellenbogen, B.S. (1994). Regulation of progesterone receptor gene expression in MCF-7 breast cancer cells: a comparison of the effects of cyclic adenosine 3',5'-monophosphate, estradiol, insulin-like growth factor-I, and serum factors. *Endocrinology* **134**, 658–664.

Couse, J.F., Curtis, S.W., Washburn, T.F., Lindzey, J., Golding, T.S., Lubahn, D.B., Smithies, O. and Korach, K.S. (1995). Analysis of transcription and estrogen insensitivity in the female mouse after targeted disruption of the estrogen receptor gene. *Molecular Endocrinology* **9**, 1441–1454.

Curtis, S.W., Washburn, T., Sewall, C., DiAugustine, R., Lindzey, J., Couse, J.F. and Korach, K.S. (1996). Physiological coupling of growth factor and steroid receptor signaling pathways: Estrogen receptor knockout mice lack estrogen-like response to epidermal growth factor. *Proceedings of the National Academy of Sciences, USA* **93**, 12626–12630.

Das, S.K., Tsukamura, H., Paria, B.C., Andrews, G.K. and Dey, S.K. (1994). Differential expression of epidermal growth factor receptor (EGF-R) gene and regulation of EGF-R bioactivity by progesterone and estrogen in the adult mouse uterus. *Endocrinology* **134**, 971–981.

Denner, L.A., Weigel, N.L., Maxwell, B.L., Schrader, W.T. and O'Malley, B.W.W. (1990). Regulation of progesterone receptor-mediated transcription by phosphorylation. *Science* **250**, 1740–1743.

DiAugustine, R.P., Petrusz, P., Bell, G.I., Brown, C.F., Korach, K.S., McLachlan, J.A. and Teng, C.T. (1988). Influence of estrogens on mouse uterine epidermal growth factor precursor protein and messenger ribonucleic acid. *Endocrinology* **122**, 2355–2363.

Gardner, R.M., Verner, G., Kirkland, J.L. and Stancel, G.M. (1989). Regulation of uterine epidermal growth factor (EGF) receptors by estrogen in the mature rat and during the estrous cycle. *J. Steroid Biochem.* **32**, 339–343.

Green, S. and Chambon, P. (1991). The oestrogen receptor: From perception to mechanism. In *Nuclear Hormone Receptors Molecular Mechanisms, Cellular Functions, Clinical Abnormalities*, M.G. Parker, ed. (San Diego, CA: Academic Press Inc.), pp. 15–38.

Gronemeyer, H. (1991). Transcription activation by estrogen and progesterone receptors. *Annu. Rev. Genet.* **25**, 89–123.

Heldin, C.-H. (1995). Dimerization of cell surface receptors in signal transduction. *Cell* **80**, 213–233.

Herschman, H.R. (1991). Primary response genes induced by growth factors and tumor promoters. *Annu. Rev. Biochem.* **60**, 281–319.

Hill, C.H. and Treisman, R. (1995). Transcriptional regulation by extracellular signals: mechanisms and specificity. *Cell* **80**, 199–211.

Huet Hudson, Y.M., Chakraborty, C., De, S.K., Suzuki, Y., Andrews, G.K. and Dey, S.K. (1990). Estrogen regulates the synthesis of epidermal growth factor in mouse uterine epithelial cells. *Mol. Endocrinol.* **4**, 510–523.

Ignar Trowbridge, D.M., Nelson, K.G., Bidwell, M.C., Curtis, S.W., Washburn, T.F., McLachlan, J.A. and Korach, K.S. (1992). Coupling of dual signaling pathways: epidermal growth factor action involves the estrogen receptor. *Proc. Natl. Acad. Sci. USA* **89**, 4658–4662.

Ignar Trowbridge, D.M., Teng, C.T., Ross, K.A., Parker, M.G., Korach, K.S. and McLachlan, J.A. (1993). Peptide growth factors elicit estrogen receptor-dependent transcriptional activation of an estrogen-responsive element. *Mol. Endocrinol.* **7**, 992–998.

Ignar-Trowbridge, D.M., Pimentel, M., Parker, M.G., McLachlan, J.A. and Korach, K.S. (1996). Peptide growth factor-cross-talk with the estrogen receptor requires the A/B domain and occurs independently of protein kinase C or estradiol. *Endocrinology*, in press.

Jost, A. (1953). *Recent Progress in Hormone Research* **8**, 379–418.

Kato, S., Endoh, H., Masuhiro, Y., Kitamoto, T., Uchiyama, S., Sasaki, H., Masushige, S., Gotoh, Y., Nishida, E., Kawashima, H., Metzger, D. and Chambon, P. (1995). Activation of the estrogen receptor through phosphorylation by mitogen-activated protein kinase. *Science* **270**, 1491–1494.

Korach, K.S., Couse, J.F., Curtis, S.W., Washburn, T.F., Lindzey, J., Kimbro, K.S., Eddy, E.M., Migiaccio, S., Snedeker, S.M., Lubahn, D.B., Schomberg, D.W. and Smith, E.P. (1996). Estrogen recpeptor gene disruption: Molecular characterization and experimental and clinical phenotypes. In *Recent Progress in Hormone Research*, P. Conn, M., ed. (Bethesda MD: The Endocrine Society), pp. 159–188.

Lahooti, H., White, R., Danielian, P.S. and Parker, M.G. (1994). Characterization of ligand-dependent phosphorylation of the estrogen receptor. *Mol. Endocrinol.* **8**, 182–188.

Le Goff, P., Montano, M.M., Schodin, D.J. and Katzenellenbogen, B.S. (1994). Phosphorylation of the human estrogen receptor. Identification of hormone-regulated sites and examination of their influence on transcriptional activity. *J. Biol. Chem.* **269**, 4458–4466.

Lees, J.A., Fawell, S.E. and Parker, M.G. (1989a). Identification of constitutive and steroid-dependent transactivation domains in the mouse oestrogen receptor. *J. Steroid Biochem.* **34**, 33–39.

Lees, J.A., Fawell, S.E. and Parker, M.G. (1989b). Identification of two transactivation domains in the mouse oestrogen receptor. *Nucleic Acids Res.* **17**, 5477–5488.

Lingham, R.B., Stancel, G.M. and Loose Mitchell, D.S. (1988). Estrogen regulation of epidermal growth factor receptor messenger ribonucleic acid. *Mol. Endocrinol.* **2**, 230–235.

Loose Mitchell, D.S., Chiappetta, C. and Stancel, G.M. (1988). Estrogen regulation of c-fos messenger ribonucleic acid. *Mol. Endocrinol.* **2**, 946–951.

Lubahn, D.B., Moyer, J.S., Golding, T.S., Couse, J.F., Korach, K.S. and Smithies, O. (1993). Alteration of reproductive function but not prenatal sexual development after insertional disruption of the mouse estrogen receptor gene. *Proc. Natl. Acad. Sci. USA* **90**, 11162–11166.

Ma, Z.Q., Santagati, S., Patrone, C., Pollio, G., Vegeto, E. and Maggi, A. (1994). Insulin-like growth factors activate estrogen receptor to control the growth and differentiation of the human neuroblastoma cell line SK-ER3. *Mol. Endocrinol.* **8**, 910–918.

Marshall, C.J. (1995). Specificity of receptor tyrosine kinase signalling: transient versus sustained extracellular signal-regulated kinase activation. *Cell* **80**, 179–185.

Metzger, D., Losson, R., Bornert, J.M., Lemoine, Y. and Chambon, P. (1992). Promoter specificity of the two transcriptional activation functions of the human oestrogen receptor in yeast. *Nucleic Acids Res.* **20**, 2813–2817.

Mukku, V.R. and Stancel, G.M. (1985). Regulation of epidermal growth factor receptor by estrogen. *J. Biol. Chem.* **260**, 9820–9824.

Nelson, K.G., Takahashi, T., Bossert, N.L., Walmer, D.K. and McLachlan, J.A. (1991). Epidermal growth factor replaces estrogen in the stimulation of female genital-tract growth and differentiation. *Proc. Natl. Acad. Sci. USA* **88**, 21–5.

Parker, M.G., Arbuckle, N., Dauvois, S., Danielian, P. and White, R. (1993). Structure and function of the estrogen receptor. *Ann. NY Acad. Sci.* **684**, 119–126.

Power, R.F., Lydon, J.P., Conneely, O.M. and O'Malley, B.W.W. (1991). Dopamine activation of an orphan of the steroid receptor superfamily. *Science* **252**, 1546–1548.

Power, R.F., Mani, S.K., Codina, J., Conneely, O.M. and O'Malley, B.W. (1991). Dopaminergic and ligand-independent activation of steroid hormone receptors. *Science* **254**, 1636–1639.

Quarmby, V.E., Fox Davies, C. and Korach, K.S. (1984). Estrogen action in the mouse uterus: the influence of the neuroendocrine-adrenal axis. *Endocrinology* **114**, 108–115.

Quarmby, V.E. and Korach, K.S. (1984). The influence of 17 beta-estradiol on patterns of cell division in the uterus. *Endocrinology* **114**, 694–702.

Rubin, R.A. and Earp, H.S. (1983). Solubilization of EGF receptor with Triton X-100 alters stimulation of tyrosine residue phosphorylation by EGF and dimethyl sulfoxide. *J. Biol. Chem.* **258**, 5177–5182.

Sewall, C.H., Clark, G.C. and Lucier, G.W. (1995). TCDD reduces rat hepatic epidermal growth factor receptor: comparison of binding, immunodetection, and autophosphorylation. *Toxicol. Appl. Pharmacol.* **132**, 263–272.

Smith, C.L., Conneely, O.M. and O'Malley, B.W. (1993). Modulation of the ligand-independent activation of the human estrogen receptor by hormone and antihormone. *Proc. Natl. Acad. Sci. USA* **90**, 6120–6124.

Stancel, G.M., Chiapetta, C., Gardner, R.M., Kirkland, J.L., Lin, T.H., Lingham, R.B., Loose Mitchell, D.S., Mukku, V.R. and Orengo, C.A. (1990). Regulation of the uterine epidermal growth factor receptor by estrogen. *Prog. Clin. Biol. Res.* **322**, 213–226.

Stancel, G.M., Gardner, R.M., Kirkland, J.L., Lin, T.H., Lingham, R.B., Loose Mitchell, D.S., Mukku, V.R., Orengo, C.A. and Verner, G. (1987). Interactions between estrogen and EGF in uterine growth and function. *Adv. Exp. Med. Biol.* **230**, 99–118.

Sumida, C. and Pasqualini, J.R. (1989). Antiestrogens antagonize the stimulatory effect of epidermal growth factor on the induction of progesterone receptor in fetal uterine cells in culture. *Endocrinology* **124**, 591–597.

Tora, L., White, J., Brou, C., Tasset, D., Webster, N., Scheer, E. and Chambon, P. (1989). The human estrogen receptor has two independent nonacidic transcriptional activation functions. *Cell* **59**, 477–487.

# 4 The Estrogen Receptor in the Cardiovascular System

**Ioakim Spyridopoulos and Douglas W. Losordo** *

*St. Elizabeths Medical Ctr., 736 Cambridge St., Brighton, MA 02135, USA.*
** Corresponding author, Tel.: 001 617 789 3346.*

The mechanisms of the atheroprotective effects of estrogen ($E_2$) have not been entirely clarified. *In vitro* and *in vivo* experiments from our laboratory found, that

(1) $E_2$ was able to inhibit smooth muscle cell proliferation in physiological concentration, mediated by its receptor.
(2) $E_2$ treatment resulted in a dose-dependent, receptor-mediated inhibition of TNF-$\alpha$ induced endothelial cell apoptosis.
(3) $E_2$ accelerates functional endothelial recovery after barotraumatic deendothelializing injury.

These findings, along with the recent demonstration of estrogen receptor expression by endothelial cells, suggest that the antiatherogenic action of estrogen may be mediated in part through direct effects on endothelial cells.

KEYWORDS: Estrogen receptor, vascular smooth muscle cells, endothelial cells, apoptosis, tumour necrosis factor alpha, reendothelialization.

Epidemiological data indicate that pre-menopausal women are at lower risk of coronary heart disease than are men of similar age (Clawson, 1941). The increased incidence of atherosclerosis in women who undergo premature menopause has also been well described (Oliver and Boyd, 1959). Finally, there is increasing evidence that treatment with replacement estrogen after menopause will reduce cardiovascular mortality (Gruchow *et al.*, 1988; Matthews *et al.*, 1989). Endogenous and exogenous estrogens have been observed to alter the level of lipids and lipid metabolism in humans. Oxidative modification of low-density lipoproteins (LDL) may be atherogenic and is inhibited by 17-$\beta$ estradiol in postmenopausal women, suggesting a beneficial effect of an estrogen-replacement therapy on cardiovascular disease (Sack *et al.*, 1994). The changes in serum lipids noted in human patients, however, fail to fully account for the discrepancy in the incidence of coronary disease between men and pre-menopausal women (Bush *et al.*, 1987). It is also clear that the plasma lipid-independent antiatherogenic effect of estradiol in rabbits is not mediated through a change in aortic permeability to LDL (Haarbo *et al.*, 1994). The demonstration of a functional estrogen receptor in human vascular smooth muscle cells (Karas *et al.*, 1994; Losordo *et al.*, 1994) and in human endothelial cells (Kim-Schulze *et al.*, 1996; Venkov *et al.*, 1996) suggests a direct effect of estrogen on coronary and peripheral arteries is possible. Following these results estrogen was found to inhibit smooth muscle cell proliferation and migration, regardless of gender (Dai-Do *et al.*, 1996; Kolodgie *et al.*, 1996). *In vivo* data from mouse carotid arteries confirmed a direct effect of estrogen on VSMC proliferation in injured vessels (Sullivan *et al.*, 1995). Most recent data in ovariectomized female and castrated male rabbits disclosed the beneficial effect of estrogens on cellular proliferation and intimal thickening in experimental atherosclerosis in female rabbits only (Hanke *et al.*, 1996). Much of lipid-independent effects in the vessel wall seem to be attributed to a direct effect of estrogen on smooth muscle cells. With the data obtained in our own laboratory we hope to explore the beneficial effects of estrogen on coronary artery disease in another structure of the vessel: the endothelium. If estrogen directly acts upon endothelial cells, another facet of its complex mechanism in protecting from atherosclerosis could be revealed.

## BIOCHEMISTRY OF THE ESTROGEN RECEPTOR

The steroid/thyroid hormone receptors are members of a very large family of nuclear ligand-activated transcription factors that includes the receptors for estrogen (ER), glucocorticoids, thyroid hormone, retinoids and vitamin D (Weigel, 1996). These receptors play key roles both as transcriptional activators and as repressors. Steroid receptor monomers become quickly bound to the abundant 90 kDa heat shock protein (HSP 90) dimer. Binding of hormone to the receptor releases the HSP 90 molecule in its dimered form. The receptor, now in its activated state, is capable of both homodimerization and DNA binding. Ligand binding leads to dissociation of heat-shock proteins, receptor dimerization and binding to a specific steroid responsive elements, whose consensus sequence is an inverted repeat or so called palindrome separated by three nucleotides. Previous studies have identified the 13 basepair (bp) sequence (5′-GGTCAnnnTGACC-3′) that is the minimal DNA sequence sufficient for specific binding of the ER (Klein-Hitpass *et al.*, 1988). This DNA sequence is also termed ‘estrogen responsive element’ (ERE) and can be found as part of genes, which are transcriptionally activated by estrogen. Although steroid hormone receptors are ligand-activated receptors, they are phosphoproteins at the same time implying that phosphorylation can regulate their function. Different study groups have found either phosphotyrosine ($Tyr^{537}$) or serine phosphorylation sites ($Ser^{118}$, $Ser^{167}$) in the ER. $Ser^{118}$ can be phosphorylated *in vitro* by MAP kinase, and MAP kinase activation is capable of enhancing ER phosphorylation *in vivo* at the same position. Phorbol esters and $p21^{ras}$ expression are also able to phosphorylate serine. In addition, src kinase has been shown to phosphorylate $Tyr^{537}$, showing that ER phosphorylation is regulated by multiple kinases and thus ER can be activated by signal transduction pathways in the absence of ligand.

Cross-talk between polypeptide growth factors and steroid hormone receptors is another way of regulating normal cell growth and differentiation by interaction between the epidermal growth factor signaling pathway and the classical estrogen receptor (Ignar-Trowbridge *et al*., 1992). Administration of diethylstilbestrol or EGF to adult ovariectomized mice by slow release pellets led to induction of uterine DNA synthesis while this effect was blocked by co-administration of the estrogen receptor antagonist ICI 164,384.

## THE DIRECT EFFECT OF ESTROGEN ON VASCULAR TISSUE

A direct effect of estrogen on the arterial wall is suggested by animal and human experimental data. Vasodilation in response to estrogen administration was first noted in the rabbit ear artery (Reynolds and Foster, 1940), and later in umbilical (Silva de Sa and Meirelles, 1977) and primate coronary arteries (Williams *et al*., 1993). Estrogen receptor has been demonstrated in canine peripheral (Horwitz and Horwitz, 1982) and coronary arteries (Harder and Coulson, 1979), in cultured rat aortic VSMC (Nakao *et al*., 1981) and in human endothelial cells (Kim-Schulze *et al*., 1996; Venkov *et al*., 1996). The actual function of the ER and the cellular mechanisms which translate $E_2$–ER interaction into physiological effect on the arterial wall remain to be defined.

### Human Vascular Smooth Muscle Cells Contain Functional Estrogen Receptor

Smooth muscle cells at low passage are known to retain ER expression of the parent tissue. Cultures of hVSMC were grown by the explant outgrowth technique as reported previously (Pickering *et al*., 1992). This technique has been used successfully in our laboratory to derive cultures of hVSMC from normal arteries as well as from atherosclerotic plaque obtained by directional atherectomy. The identity of the cells in culture is confirmed by staining with antibody to smooth muscle $\alpha$-actin.

To assay for ER in this system, hVSMC in their second passage were exposed to tritiated estrogen in concentrations from 1 to 40 nmol/L, alone or in the presence of a 200-fold concentration excess of unlabelled diethylstilbestrol (DES). Cells were incubated for 30 min at 37°C and then placed on ice. After washing the cells in PBS, $E_2$–ER binding was solubilized by the addition of ethanol and bound radioactivity was assayed by liquid scintillation counting. As shown in Figure 4.1, a characteristic saturation plot was generated from these studies. From these data the precise quantity of ER is calculated to equal $1.16 \times 10^{-13}$ mol/L per $10^6$ hVSMC. These data also represent a high-affinity ER as indicated by the calculated $K_d$ of $8 \times 10^{-9}$ and have since been confirmed by Karas *et al*. (1994).

As mentioned above, the ERE has been shown to bind the ER only when the latter is induced by its ligand. The interaction of hormone-bound ER with the ERE has been shown to alter the transcriptional activity of ERE-containing genes. Therefore binding of ligand-induced ER to the ERE is a useful measure of the functional integrity of putative receptors. We examined the ability of ER from hVSMC to bind

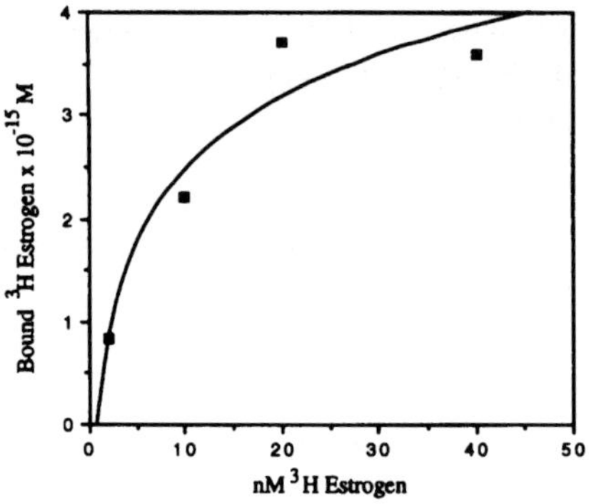


**Figure 4.1.** Binding curve generated by incubation of cultured hVSMC with $^3$H-estradiol. Specifically bound labeled estrogen (calculated by subtracting binding which occurred in the presence of unlabelled 1000-fold excess estradiol from $^3$H-estradiol binding in cells incubated with only labeled hormone) is plotted against concentration of labeled estradiol.

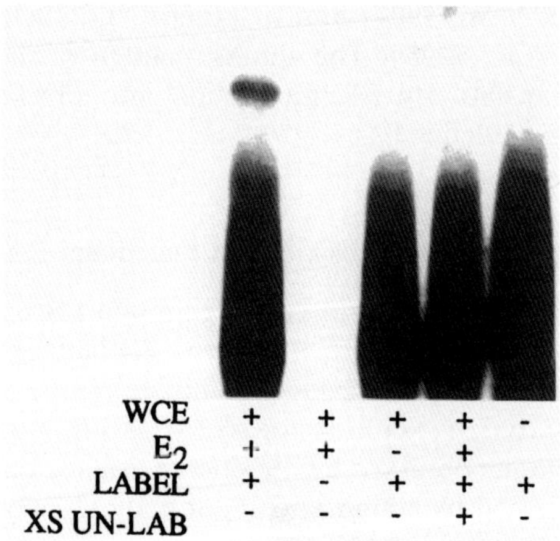


**Figure 4.2.** Gel mobility assay. Band in left lane corresponds to complex of estrogen receptor and labeled estrogen response element. Controls included exclusion of labeled oligonucleotide probe, exclusion of estradiol, addition of excess unlabelled oligonucleotide and exclusion of cell extract. All controls were appropriately negative, indicated by the absence of the band seen in lane 1. $E_2$ = estradiol, label = labeled oligonucleotide probe, WCE = whole cell extract, XS UN-LAB = excess unlabelled oligonucleotide probe.

ERE using gel mobility assays. This assay is based on the fact that protein-bound DNA migrates more slowly through a polyacrylamide gel than unbound DNA. In the presence of estrogen, the ER should form a complex with the ERE. By labeling the response element with $^{32}$P, the complexes can be identified by their altered electrophoretic mobility. Ten percent charcoal stripped fetal bovine serum in phenol red free M199 was used as a basal medium to exclude any significant steroid-contamination of the media. Figure 4.2 shows a gel mobility assay, where addition of estrogen leads to a isolated shifted band of higher molecular weight, induced by a complex of labeled ERE with the ER. Thus, confirmation of the presence and functional integrity of ER in hVSMC suggests that a direct effect of estrogen on human arteries is a possibility.

### Antiproliferative Effect of Estrogen on Human Vascular Smooth Muscle Cells is Mediated by the Estrogen Receptor

Having established that hVSMC express functional ER, our next goal was to identify a physiological effect of estrogen on these cells. A primary goal was to test the influence of estrogen on smooth muscle cell proliferation. Therefore human VSMC in their second passage were subcultured and grown in medium that was supplemented with serum containing varying concentrations of estradiol: 10% pooled serum from human male donors ($E_2$ concentration $= 40\,pg/mL \approx 1.5 \times 10^{-10}\,mol/L$), 10% pooled serum from human female donors ($E_2$ concentration $= 270\,pg/mL \approx 1.5 \times 10^{-9}\,mol/L$) and 10% pooled serum from male donors with the addition of estradiol in concentrations ranging from $10^{-10}$ to $10^{-7}\,mol/L$. Figure 4.3 demonstrates a decrease of thymidine incorporation in those cells treated with increased amounts of estrogen. This antiproliferative effect was completely abrogated by simultaneous introduction of the specific estrogen-receptor inhibitor ICI 182,780 ($10^{-9}$ M). Double labeling of cultured hVSMC for bromo-deoxyuridine (BrdU) incorporation and estrogen receptor expression showed a lack of positive staining for ER in cells with highly proliferative activity. Thus, there is a dissociation of ER expression and evidence of proliferation.

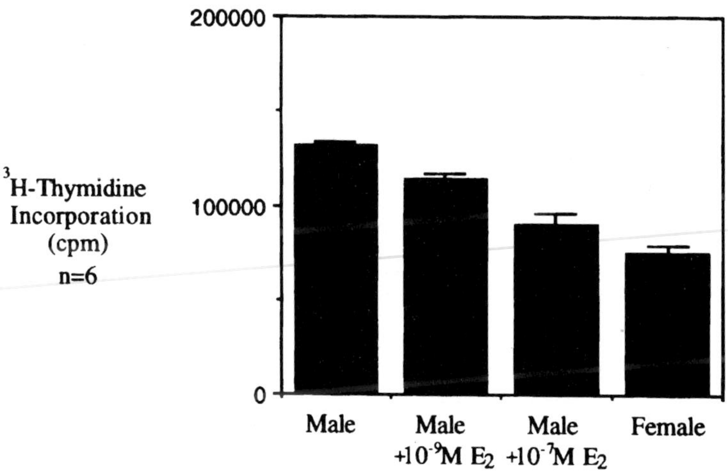


**Figure 4.3.** Thymidine incorporation of cultured hVSMC. Thymidine incorporation of cultured hVSMC incubated in serum from males (estradiol concentrations $10^{-10}\,mol/L$), before and after the addition of supplemental estradiol. Thymidine incorporation of the same hVSMC (derived from the same donor tissue) incubated with serum from premenopausal females (estradiol concentrations $10^{-9}\,mol/L$) is shown for comparison. Decremental thymidine incorporation is seen with incremental concentrations of estradiol added to the male serum ($n = 6$).

**Table 4.1** Results of immunostaining for estrogen receptor in coronary arteries

| | *ER+* | *ER−* | *p value* |
|---|---|---|---|
| *All patients* | | | |
| Normal coronary arteries | 15 | 6 | |
| Atherosclerotic coronary arteries | 6 | 13 | 0.0117 |
| *Pre-menopausal women* | | | |
| Normal coronary arteries | 10 | 2 | |
| Atherosclerotic coronary arteries | 1 | 5 | 0.0062 |
| *Post-menopausal women* | | | |
| Normal coronary arteries | 5 | 4 | |
| Atherosclerotic coronary arteries | 4 | 8 | NS |

Distribution of coronary arteries examined according to presence or absence of atherosclerosis and immunohistological evidence of estrogen receptor in all patients, in pre-menopausal females and in post-menopausal females is shown.

### Demonstration of Decreased Estrogen Receptor Expression in Atherosclerotic Coronary Arteries

To test whether premature atherosclerosis in female patients may be mediated by a failure of target cells in the vessel wall to adequately express ER, we examined coronary artery specimens from 40 female patients. The coronary arteries of these patients were classified as normal in 21 subjects and atherosclerotic in 19 subjects, based on the histological examination of the specimens. ER expression in the coronary arteries of all subjects studied is shown in Table 4.1. Of 21 normal arteries studied, 15 showed evidence of ER expression, while in 19 atherosclerotic arteries, 13 showed no evidence of ER expression when assayed immunohistochemically. Contingency table analysis revealed that the differences between these groups were statistically significant with $p = 0.0117$. Interestingly, when these results were analysed separetely, the distribution of ER in post-menopausal women was statistically not different in contrast to pre-menopausal women ($p = 0.0062$). Pre-menopausal women showed a positive staining for ER in 84% of subjects with normal coronary arteries, but only in 17% with atherosclerotic arteries. These results demonstrate a strong association between ER expression and the abscence of coronary atherosclerosis in premenopausal women.

## ESTRADIOL INHIBITS PROGRAMMED CELL DEATH IN ENDOTHELIAL CELLS

Disruption of the integrity of the endothelial barrier has been postulated as a mechanism for the initiation of atherogenesis (Kugiyama *et al.*, 1990; Ross, 1993; Steinberg *et al.*, 1989). Recent data also suggest a role for tumor necrosis factor alpha (TNF-$\alpha$) in the pathogenesis of atherosclerotic lesions and restenosis (Barath *et al.*, 1990; Kaartinen *et al.*, 1996). Since TNF-$\alpha$ can induce apoptosis in endothelial cells (Spyridopoulos *et al.*, 1996; 1997), protection of endothelial cells from cytokine induced injury and apoptosis may therefore represent one potential mechanism for prevention of atherosclerosis.

### TNF-α Induces Apoptosis in HUVECs

Electron microscopy was performed in the present study to specifically document that the light microscopic features of apoptosis seen in TNF-α-treated human endothelial cells were accompanied by the appropriate ultrastructural morphology. Under control conditions, normal cellular anatomy was present. In contrast, cells exposed to TNF-α demonstrated typical morphological features of apoptosis. Condensation of chromatin was noted at the periphery of the nucleus, and blebbing and fragmentation of the cytoplasm was seen (Figure 4.4A). Staining with acridine orange revealed similar results by usage of fluorescent microscopy (Figure 4.4B). Biochemical features of apoptosis consist of nucleosomal DNA fragmentation, which can be seen as a DNA laddering on agarose gels. Finally we established a dose–response relationship for TNF-α-induced cell death using the colorimetric MTS assay. After 24 hours of treatment at the highest dose of 40 ng/ml, 35% of cells died from apoptosis, whereas the lowest concentration of 0.1 ng/ml still induced 4% cell death over the same amount of time.

### Apoptosis and ICE

The mechanisms underlying programmed cell death were initially investigated in the developmentally deterministic nematode worm Caenorhabditis elegans (Fraser and Evan, 1996). During development of the hermaphrodite worm 1090 cells are born and 131 die. Three principal genes (ced genes) orchestrate each cell death. Ced-3

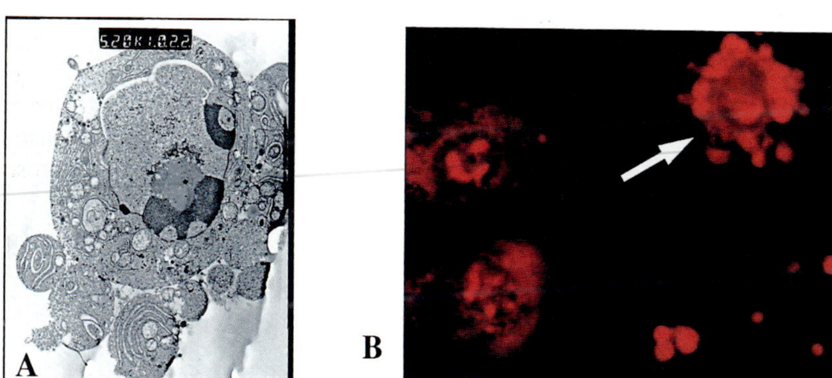


**Figure 4.4.** TNF-α induces apoptosis in human endothelial cells. (A) Transmission electron microscopy: After exposure to TNF-α, HUVEC display characteristic ultrastructural features of apoptosis, including condensation and margination of nuclear chromatin and disruption of cytoplasmic architecture with formation of cytoplasmic "blebs". (B) Acridine orange staining: At high power, a single apoptotic cell (arrow) is shown exhibiting intensified fluorescence and cytoplasmic protrusions characteristic of apoptosis. Two neighboring cells appear morphologically intact.

and ced-4 act to kill target cells, while ced-9 acts to suppress the killer genes. Ced-9 is a structural and functional homologue of the human proto-oncogene bcl-2, one of a family of genes intimately involved in the regulation of vertebrate apoptosis. Ced-3 encodes a cysteine protease that is related to an emerging family of vertebrate cysteine proteases of which the prototype is the eponymous interleukin-1$\beta$-converting enzyme (ICE). ICE cleaves the $Asp^{116}$-$Ala^{117}$ bond on pro interleukin-1$\beta$ (IL-1$\beta$) to generate the mature form of IL-1$\beta$ (Thornberry *et al.*, 1992). IL-1$\beta$ in contrast to IL-1$\alpha$ is inactive on IL-1 receptors until it is cleaved into a mature 17.5 kDa species. Interleukins-1$\alpha$ and -1$\beta$ are proinflammatory cytokines that stimulate leukocyte accumulation and prostaglandin synthesis, and are implicated in chronic inflammatory diseases such as rheumatoid arthritis. The mammalian ICE-related protease family now comprises some nine members: ICE, ICE rel II and III, CPP32, Mch2-4, ICH-1 and Nedd-2 (Fraser and Evan, 1996; Lippke *et al.*, 1996).

### TNF-$\alpha$ Induces ICE mRNA Expression in Human Endothelial Cells

Northern blot analysis demonstrates induction if ICE mRNA expression (Figure 4.5(*left*)) under treatment with TNF-$\alpha$. This is consistent with the upregulation of ICE protein after 24 hours of TNF-$\alpha$ exposure (Figure 4.5(*right*)). To provide further evidence of a functional role for ICE in human endothelial cell apoptosis, ICE protein expression was examined by immunofluorescence microscopy. Cells were simultaneously examined for biochemical features of apoptosis by use of the TUNEL method to fluorescently label DNA strand breaks. Finally, fluorescent counterstaining with Hoechst 33342 of HUVEC nuclei was also performed to permit the examination of cells for morphological features of apoptosis. As shown in Figure 4.6, expression of ICE protein localized to cells that also exhibited the morphological and biochemical features of apoptosis. The nucleus of the cell that stained positive for

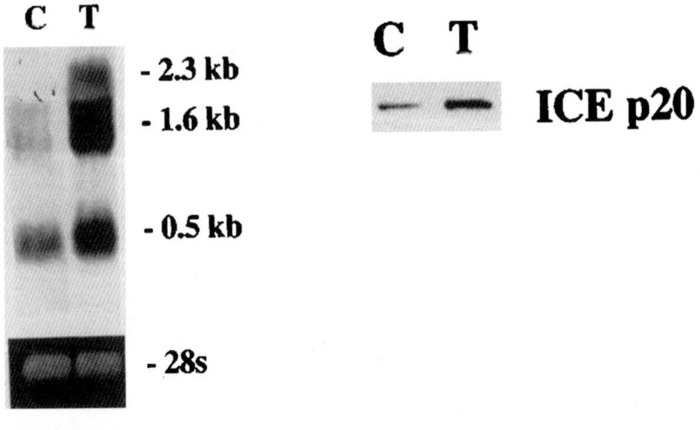


**Figure 4.5.** ICE western and northern blotting. HUVEC were exposed to TNF-$\alpha$ for 24 hours (T) or cultured under control conditions (C). Expression of ICE mRNA for all three transcripts of the *ICE* gene is markedly induced by TNF-$\alpha$ (*left*). Western blot analysis confirms the upregulation under TNF-$\alpha$ also at the protein level (*right*).

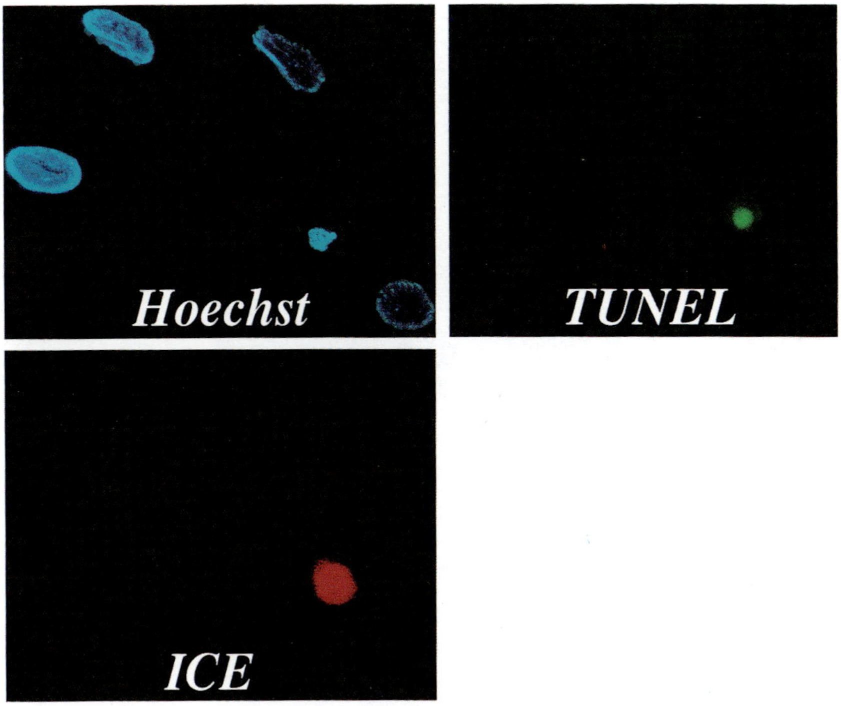


**Figure 4.6.** Immunofluorescence microscopy localizes ICE expression to apoptotic HUVEC. HUVEC were exposed to TNF-α for 24 hours and then stained for ICE protein (red fluorescence), TUNEL (green fluorescence) to detect apoptotic cells, and with the nuclear dye Hoechst 33342 (blue fluorescence). Most of the nuclei in the microscope field have normal morphology, with the exception of the cell in the lower right, which has a shrunken and deformed nucleus consistent with apoptosis. The same cell counterstains for ICE and TUNEL, confirming colocalization of apoptotic endothelial cells with ICE.

ICE was shrunken and deformed. In addition, this cell showed evidence of DNA strand breaks manifested as green fluorescence, indicative of terminal deoxy transferase labeling of 3′ DNA ends (TUNEL). The absence of ICE expression in the morphologically normal-appearing cells suggests that ICE expression was limited to cells undergoing apoptosis.

### Estrogen Inhibits TNF-α Induced Apoptosis in Human Endothelial Cells

Adherent HUVEC demonstrate typical cobblestone morphology under control conditions (Figure 4.7A). After exposure to TNF-α, the cells become rounded and partially detach while demonstrating the lobulated appearance of apoptotic cells (Figure 4.7B). At the same time the density of the adherent cells is decreased as compared with control conditions, indicating detachment of HUVEC cells exposed to TNF-α. The presence of $10^{-9}$ M estradiol together with TNF-α reversed the apoptotic morphology in most of the cells (Figure 4.7C), while cell density was increased as compared to HUVEC exposed to TNF-α.

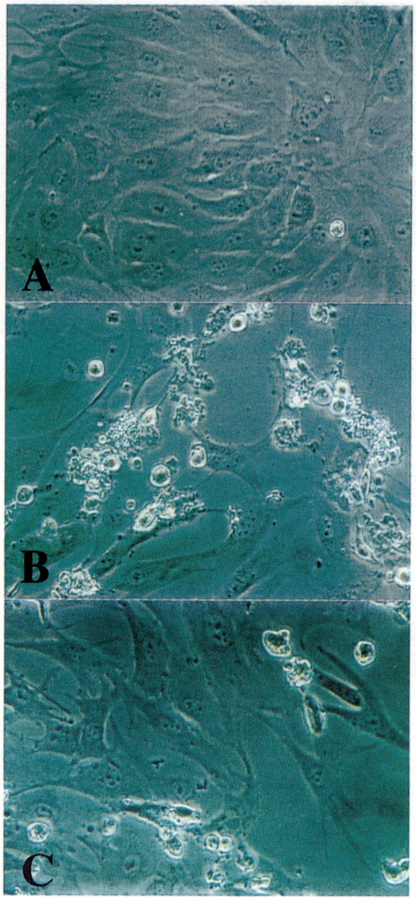


**Figure 4.7.** Estrogen inhibits TNF-$\alpha$ induced apoptosis in human endothelial cells. Phase-contrast micrographs of HUVEC under control conditions (A), after exposure to TNF-$\alpha$ for 20 hours (B), and after exposure to TNF-$\alpha$ in the presence of estrogen ($10^{-7}$ M) (C).

## Human Endothelial Cells Contain Functional Estrogen Receptor

Two independent studies have demonstrated the presence of a functional estrogen receptor in human endothelial cells, including HUVECs, coronary artery and aortic endothelial cells (Kim-Schulze *et al.*, 1996; Venkov *et al.*, 1996). Northern blots verified that human endothelial cells express abundant amount of mRNA for the ER. Ligand-binding studies estimated $2–6 \times 10^4$ receptors per endothelial cell and scatchard analysis of radioligand binding showed a $K_d$ of $\approx 5$ nmol/L. Nuclear extracts from HUVECs interacted with a consensus estrogen response element in an electrophoretic mobility shift assay and antibody against the ER supershifted the protein-DNA complex to prove specificity. Figure 4.8A shows equal ER protein expression in HUVECs under control conditions and in TNF-$\alpha$ treated cells by Western blotting.

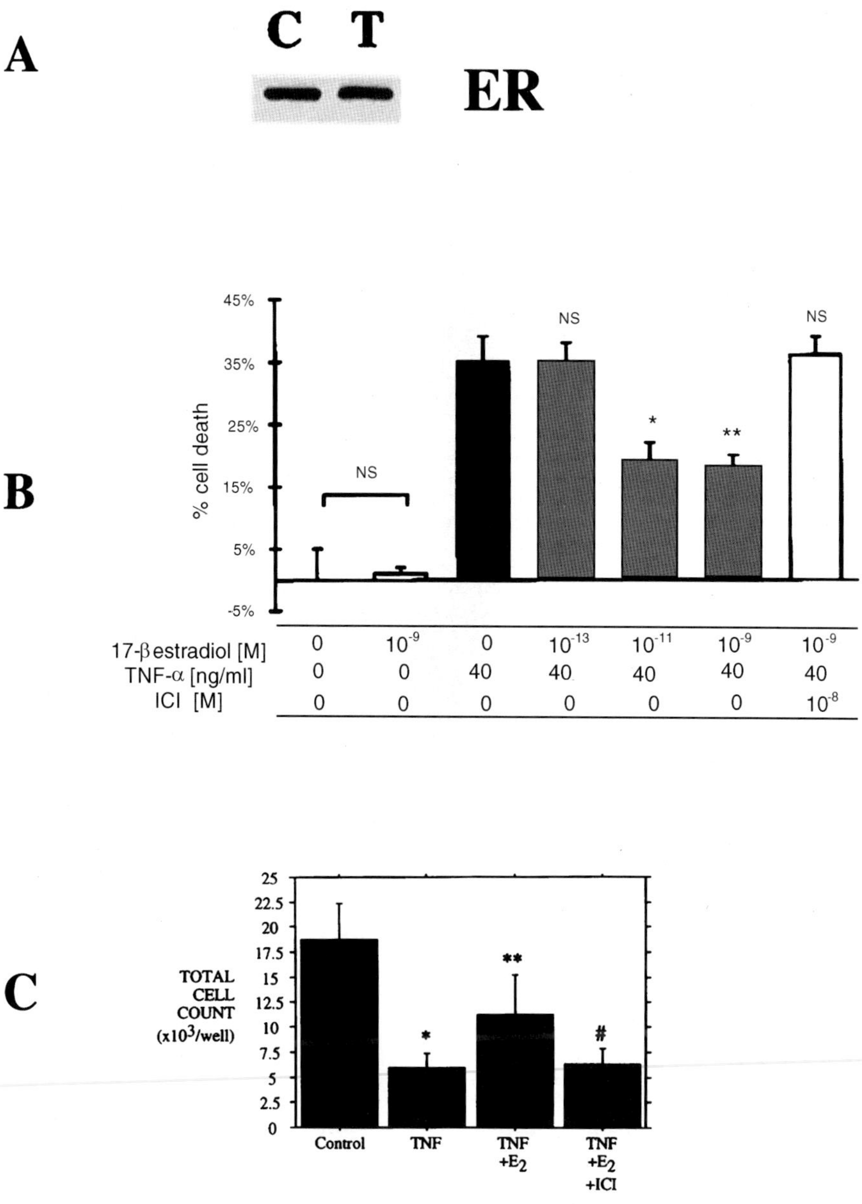


**Figure 4.8.** Estrogen-receptor mediated inhibition of human endothelial cell apoptosis. (A) Western Blot ER: Estrogen-receptor (ER) protein expression is evident in HUVEC under all conditions. (B) Viability assays: Apoptosis was induced by 24 hours of treatment with TNF-α alone or in combination with $E_2$ in the concentrations noted. TNF-α resulted in cell death in 35±4% of HUVEC after 24 hours (solid bar). Simultaneous treatment of HUVEC with $E_2$ resulted in a dose-dependent inhibition of TNF-α induced cell death (gray bars). The survival effect of $E_2$ was completely blocked by the specific estrogen-receptor antagonist ICI 182,780 (open bar). Estrogen alone did not alter viability compared to control cells. (C) Total cell count: Total cell count determined after 24 hours of exposure to control conditions, TNF-α alone, and TNF-α + $E_2$($10^{-9}$ mol/L) ± ICI 182,780($10^{-9}$ mol/L).

### Estrogen-Receptor Mediated Inhibition of Human Endothelial Cell Apoptosis

To assess cell viability we used a nonradioactive cell-proliferation assay containing the tetrazolium compound MTS and an electron coupling reagent, PMS. MTS is reduced by viable cells to formazan, whose yellow color can be measured with a spectrophotometer by the amount of 490-nm absorbance. Formazan production is time dependent and proportional to the number of viable cells (Buttke *et al.*, 1993). The advantage of this assay is its evaluation of mitochondrial function, which can detect cell death at earlier stages than other techniques (Tepper *et al.*, 1995). In the present study, TNF-$\alpha$ alone resulted in $35 \pm 4\%$ cell death after 24 hours (Figure 4.8B), as assessed by the MTS assay. Simultaneous treatment with estradiol demonstrated a dose-dependent increase in cell survival. At an estrogen concentration of $10^{-9}$ mol/L, the percentage of dying cells decreased to $18 \pm 2\%$ ($p = 0.0004$ vs. TNF-$\alpha$ alone). The specific estrogen-receptor antagonist ICI 182,780 completely abrogated the protective effect of estrogen. Treatment with estrogen alone for 24 hours did not result in a significant increase in cell number, which shows that the increase in survival due to estrogen is not based on cell proliferation over that short period of time.

We have also used flow cytometry to determine the proportion of human endothelial cells in culture that are apoptotic. This method utilizes intact cells and results in highly quantitative information regarding the degree of apoptosis in this population. Fixing the cells in 70% ethanol gives best results for apoptotic cells due to the tendency of the cytoplasm of these cells to contract. Staining with propidium iodide, an intercalative DNA binding dye, allows excitation with the 488-nm beam of an argon laser and detection at 620–700 nm. Control HUVEC cells showed 6.5% apoptotic events after being in 20% charcoal-stripped growth media. Addition of 40 ng/ml TNF-$\alpha$ led to 39.3% apoptosis whereas simultaneous treatment with estrogen decreased apoptosis by over 50% to 16.6%.

Cell counting of HUVEC exposed to study conditions in 24-well plates revealed similar findings. As shown in Figure 4.8C total cell count decreased from $18.7 \pm 3.7 \times 10^3$ cells per well under control conditions to $5.9 \pm 1.4 \times 10^3$ cells per well after 24 hours of TNF-$\alpha$ exposure. Treatment with estradiol ($10^{-9}$ mol/L) resulted an increase in cell count to $11.2 \pm 4.0 \times 10^3$ ($p < 0.001$ vs. TNF-$\alpha$ alone). This beneficial effect was abolished when the specific estrogen-receptor antagonist ICI 182,780 ($10^{-9}$ mol/L) was coincubated: the cell count decreased to $6.3 \pm 1.6 \times 10^3$ cells per well ($p < 0.001$ vs. TNF-$\alpha$ + estradiol, $p =$ NS vs. TNF-$\alpha$ alone).

The mechanism of the survival effect of estrogen on endothelial cells remains to be clarified, but the ability of estrogen to protect endothelial cells from cytokine-induced injury suggests another possible mechanism of its atheroprotective effect. The benefit of this effect *in vivo* has still to be shown.

## ESTRADIOL ACCELERATES FUNCTIONAL ENDOTHELIAL RECOVERY AFTER ARTERIAL INJURY

Disruption of the anatomic and functional integrity of the endothelium has long been postulated as a mechanism for the initiation of atherosclerosis (Ross, 1993). Recent data also suggests that acceleration of reendothelialization after arterial injury is associated with attenuation of intimal hyperplasia. Estrogen has also been shown to promote angiogenesis *in vitro* as well as *in vivo* (Morales *et al.*, 1995). Moreover, estrogen has also been shown to inhibit neointimal thickening after arterial injury in a rat model (Chen *et al.*, 1996).

To test whether estrogen can accelerate reendothelialization, we treated ovariectomized female Sprague-Dawley rats either with subcutaneously implanted estradiol-pellets or placebo in a model of carotid artery balloon injury (Krasinski *et al.*, 1997). Serum estradiol levels measured 2 weeks after balloon injury ranged at 345 ± 93 pg/ml as compared to levels below 20 pg/ml in the ovariectomized and placebo-treated group. Balloon denudation was accomplished by withdrawing a 2F Fogarty balloon inflated with saline from the omohyoid muscle to the carotid bifurcation.

## Estrogen Replacement Therapy Results in Dose-Dependent Acceleration of Reendothelialization

Evans blue staining identifies areas of non-reendothelialized artery with a blue stain. Figure 4.9 shows an example of arteries one and two weeks after injury in the control and estrogen treated animals. The total area of reendothelialization measured at one

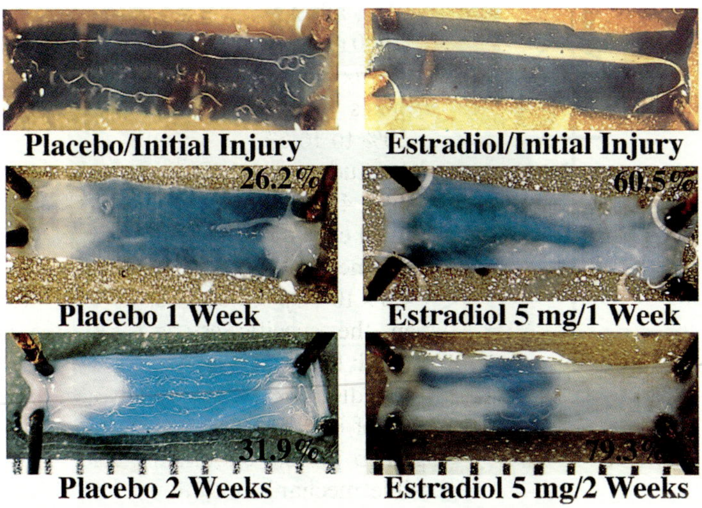


**Figure 4.9.** Evans blue staining. Examples of arteries harvested immediately (*top*) and 1 week (*middle*) or 2 weeks (*bottom*) after denudation injury. Evans blue staining identifies segments of each artery that have not been recovered by endothelium. At both time points, the estradiol-treated arteries (*right*) have a significantly larger area of recovered endothelium (white area on each specimen). The percentage of area covered by endothelium is noted.

pathogens in nonspecific immune responses. NO is derived from the guanidino nitrogen atom of L-arginine through a reaction catalyzed by NO synthase (NOS). Up to now, three distinct NOS cDNAs have been cloned: neuronal NOS (nNOS), endothelial NOS (eNOS) and inducible NOS (iNOS). Estrogen has been reported to increase endothelium-dependent relaxation, and female rabbit aorta seem to release more NO than that of male (Hishikawa *et al.*, 1995). The effect of estrogen on eNOS mRNA level is controversial in bovine and porcine endothelial cells. To determine whether functional recovery of the endothelium was accelerated by estradiol, the production of nitric oxide by excised arterial segments was measured as previously described (van der Zee *et al.*, 1997). Excised arteries were placed in an organ bath and nitric oxide was measured after 15 min according to the Griess reaction. Nitric oxide production by the estradiol-treated arteries was significantly greater than that of placebo-treated arteries 1 week after injury ($8.55 \pm 1.7$ vs. $2.8 \pm 0.18\,\mu\text{mol}\,L^{-1}\,\text{mm}^{-2}$ per 15 min, $p < 0.0001$). The acceleration of anatomical recovery, as identified by Evans blue staining, was also accompanied by estrogen-induced acceleration of functional endothelial recovery.

## VEGF Expression by Vascular Smooth Muscle Cells is Increased by Estrogen

A role for gonadal steroids in modulating endothelial cell behavior is supported by studies of blood vessel formation in the primate endometrium. It is likely that neovascularization in reproductive tissues is under control of steroid hormones, particularly 17$\beta$-estradiol in women. It is known, that estrogen enhances angiogenesis by promoting cell attachment, migration, proliferation, and differentiation (Morales *et al.*, 1995) Estrogen may enhance activity by augmenting the actions of angiogenic growth factors. VEGF production is found in tissues noted for high estrogenic content or abundant estrogen responsiveness. TNF-$\alpha$ is a necessary stimulus for expression of certain adhesion molecules by HUVECs but endothelial cell responses are greatly enhanced by estradiol. VEGF has also been shown to accelerate reendothelialization after arterial injury in the rat carotid artery-model (Asahara *et al.*, 1995). Although vascular endothelial cells contain functional receptors for VEGF, namely KDR and tie-1, VEGF itself is secreted by vascular smooth muscle cells in the vessel wall. To test, if estradiol influences VEGF expression, primary cultures of rat aortic vascular smooth muscle cells were grown according to the explant outgrowth technique as previously described (Losordo *et al.*, 1994). Cells in passage 2–4 were grown to confluence and placed in phenol red-free medium supplemented with 0.5% charcoal stripped serum for 48 hours. Cells were then exposed to one of the following for 24 hours: 0.5% stripped serum, 10% stripped serum supplemented with $10^{-7}$ mol/L estradiol, or 10% stripped serum without estradiol. Northern analysis revealed increased expression of VEGF mRNA by rat vascular smooth muscle cells exposed to estradiol (Figure 4.12).

The mechanisms by which estradiol accelerates functional endothelial recovery remain undefined. Estrogen has previously been shown to exert direct angiogenic effects on endothelial cells (Morales *et al.*, 1995), raising the possibility that this effect could be mediated through direct action of estrogen on endothelial cells. Since VEGF by itself has a strong angiogenic potential and at the same time has been

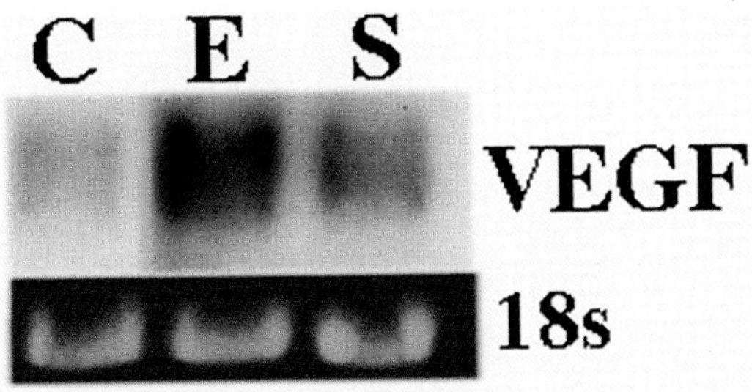


**Figure 4.12.** Estrogen increases VEGF mRNA expression (Northern). Northern analysis of VEGF mRNA expression by rat vascular smooth muscle cells. After 24 hours in the medium supplemented with estradiol $10^{-7}$ mol/L (E), VEGF mRNA expression was enhanced as compared to control (0.5% serum; C) or 10% serum (S) treated cells.

demonstrated to accelerate reendothelialization after traumatic injury, its upregulation by estrogen might contribute to the diminished neointimal lesion formation presented.

Thus, evidence from our laboratory disclosed the inhibition of smooth muscle cell proliferation, the protection of endothelial cell injury *in vitro* and finally, the acceleration of functional endothelial recovery after arterial injury, all by estrogen. These findings not only demonstrate the direct effect of estrogen on vascular tissue, but also provides several potential mechanisms to explain the relative freedom from coronary heart disease enjoyed by pre-menopausal women in contrast to their male counterparts.

## REFERENCES

Asahara, T., Bauters, C., Pastore, C., Kearney, M., Rossow, S., Bunting, S., *et al.* (1995). Local delivery of vascular endothelial growth factor accelerates reendothelialization and attenuates intimal hyperplasia in balloon-injured rat carotid artery. *Circulation*, **91**, 2793–2801.

Barath, P., Fishbein, M.C., Cao, J., Berenson, J., Helfant, R.H. and Forrester, J.S. (1990). Detection and localization of tumor necrosis factor in human atheroma. *American Journal of Cardiology*, **65**, 297–302.

Bush, T.L., Barrett-Connor, E. and Cowan, L.D. (1987). Cardiovascular mortality and non-contraceptive estrogen use in women: Results from the lipid reasearch clinics' program follow-up study. *Circulation*, **75**, 1002–1009.

Buttke, T.M., McCubrey, J.A. and Owen, T.C. (1993). Use of an aqueous soluble tetrazolium/formazan assay to measure viability and proliferation of lymphokine-dependent cell lines. *Journal of Immunological Methods*, **157**, 233–240.

Chen, S.J., Li, H., Durand, J., Oparil, S. and Chen, Y.F. (1996). Estrogen reduces myointimal proliferation after balloon injury of rat carotid artery. *Circulation*, **93**, 577–584.

Clawson, B.J. (1941). The incidence of types of heart disease among 30,265 autopsies, with special reference to age and sex. *American Heart Journal*, **22**, 607.

Dai-Do, D., Espinosa, E., Liu, G., Rabelink, T.J., Julmy, F., Yang, Z., *et al.* (1996). 17$\beta$-estradiol inhibits proliferation and migration of human vascular smooth muscle cells: similar effects in cells from postmenopausal females and in males. *Cardiovascular Research*, **32**, 980–985.

Fraser, A. and Evan, G. (1996). A license to kill. *Cell*, **85**, 781–784.

Gruchow, H.W., Anderson, A.J., Barboriak, J.J. and Sobocinski, K.A. (1988). Postmenopausal use of estrogen and occlusion of coronary arteries. *American Heart Journal*, **115**, 954–963.

Haarbo, J., Nielsen, L.B., Stender, S. and Christiansen, C. (1994). Aortic permeability to LDL during estrogen therapy. *Arteriosclerosis, Thrombosis & Vascular Biology*, **14**, 243–247.

Hanke, H., Hanke, S., Finking, G., Muhic-Lohrer, A., Mück, A.O., Schmahl, F.W., *et al.* (1996). Different effects of estrogen and progesterone on experimental atherosclerosis in female versus male rabbits. *Circulation*, **94**, 175–181.

Harder, D.R. and Coulson, P.B. (1979). Estrogen receptors and effects of estrogen on membrane electrical properties of coronary vascular smooth muscle. *Journal of Cellular Physiology*, **100**, 375–382.

Hishikawa, K., Nakabi, T., Marumo, T., Suzuki, H., Kato, R. and Saruta, T. (1995). Up-regulation of nitric oxide synthase by estradiol in human aortic endothelial cells. *FEBS Letters*, **360**, 291–293.

Horwitz, K.B. and Horwitz, L.D. (1982). Canine vascular tissues are targets for androgens, estrogens, progestins, and glucocorticoids. *Journal of Clinical Investigation*, **69**, 750–758.

Ignar-Trowbridge, D.M., Nelson, K.G., Bidwell, M.C., Curtis, S.W., Washburn, T.F., McLachlan, J.A., *et al.* (1992). Coupling of dual signaling pathways: Epidermal growth factor action involves the estrogen receptor. *Proceedings of the National Acedemy of Science USA*, **89**, 4658–4662.

Kaartinen, M., Pentitilä, A. and Kovanen, P.T. (1996). Mast cells in rupture-prone areas of human coronary atheromas produce and store TNF-α. *Circulation*, **94**, 2787–2792.

Karas, R.H., Patterson, B.L. and Mendelsohn, M.E. (1994). Human vascular smooth muscle cells contain functional estrogen receptor. *Circulation*, **89**, 1943–1950.

Kim-Schulze, S., McGowan, K.A., Hubchak, S.C., Cid, M.C., Martin, M.B., Kleinman, H.K., *et al.* (1996). Expression of an estrogen receptor by human coronary artery and umbilical vein endothelial cells. *Circulation*, **94**, 1402–1407.

Klein-Hitpass, L., Ryffel, G.U., Heitlinger, E. and Cato, A.C. (1988). A 13 bp palindrome is a functional estrogen responsive element and interacts specifically with estrogen receptor. *Nucleic Acids Research*, **16**, 647–663.

Kolodgie, F.D., Jacob, A., Wilson, P.S., Carlson, G.C., Farb, A., Verma, A., *et al.* (1996). Estradiol attenuates directed migration of vascular smooth muscle cells *in vitro*. *American Journal of Pathology*, **148**, 969–976.

Krasinski, K., Spyridopoulos, I., Asahara, T., van der Zee, R., Isner, J.M. and Losordo, D.W. (1997). Estradiol accelerates functional endothelial recovery after arterial injury. *Circulation*, **95**, 1768–1772.

Kugiyama, K., Kerns, S.A., Morrisett, J.D., Roberts, R. and Henry, P.D. (1990). Impairment of endothelium-dependent arterial relaxation by lysolecithin in modified low-density lipoproteins. *Nature*, **344**, 160–162.

Lippke, J.A., Gu, Y., Sarnecki, C., Caron, P.R. and Su, M.S. (1996). Identification and characterization of CPP32/Mch2 homolog 1, a novel cysteine protease similar to CPP32. *Journal of Biological Chemistry*, **271**, 1825–1828.

Losordo, D.W., Kearney, M., Kim, E.A., Jekanowski, J. and Isner, J.M. (1994). Variable expression of the estrogen receptor in normal and atherosclerotic coronary arteries of premenopausal women. *Circulation*, **89**, 1501–1510.

Matthews, K.A., Meilahn, E., Kuller, L.H., Kelsey, S.F., Caggiula, A.W. and Wing, R.R. (1989). Menopause and risk factors for coronary artery disease. *New England Journal of Medicine*, **321**, 641–646.

Morales, D.E., McGowan, K.A., Grant, D.S., Maheshwari, S., Bhartiya, D., Cid, M.C., *et al.* (1995). Estrogen promotes angiogenic activity in human umbilical vein endothelial cells *in vitro* and in a murine model. *Circulation*, **91**, 755–763.

Nakao, J., Chang, W.C., Murota, S.I. and Orimo, H. (1981). Estradiol-binding sites in rat aortic smooth muscle cells in culture. *Atherosclerosis*, **38**, 75–80.

Oliver, M.F. and Boyd, G.S. (1959). Effect of bilateral ovariectomy on coronary-artery disease and serum-lipid levels. *Lancet*, **2**, 690–694.

Pickering, J.G., Weir, J., Rosenfield, K., Stetz, J., Jekanowski, J. and Isner, J.M. (1992). Smooth muscle cell outgrowth from human atherosclerotic plaque: Implications for the assessment of lesion biology. *Journal of the American College of Cardiology*, **20**, 1430–1439.

Reynolds, S.R.M. and Foster, F.I. (1940). Peripheral vascular action of estrogen, observed in the ear of the rabbit. *Journal of Pharmacology and Experimental Therapy*, **68**, 173–184.

Ross, R. (1993). The pathogenesis of atherosclerosis: A perspective for the 1990s. *Nature*, **362**, 801–809.

Sack, M.N., Rader, D.J. and Cannon, R.O. (1994). Estrogen and inhibition of oxidation of low-density lipoproteins in postmenopausal women. *Lancet*, **343**, 269–270.

Silva de Sa, M.F. and Meirelles, R.S. (1977). Vasodilation effect of estrogen on the human umbilical artery. *Gynecology Investigations*, **8**, 307–313.

Spyridopoulos, I., Sullivan, A.B., Kearney, M., Isner, J.M. and Losordo, D.W. (1997). Estrogen-receptor-mediated inhibition of human endothelial cell apoptosis. Estradiol as a survival factor. *Circulation*, **95**, 1505–1514.

Spyridopoulos, I., Wang, J., Andres, V., Sullivan, A., Kearney, M. and Losordo, D.W. (1996). TNF-induced apoptosis in endothelial cells is associated with cell cycle arrest and decreased expression of Cdk inhibitor $p27^{Kip1}$. *Circulation*, **94**, I-282.

Steinberg, D., Parthasarathy, S., Carew, T.E., Khoo, J.C. and Witztum, J.L. (1989). Beyond cholesterol: modifications of low-density lipoproteins that increase its atherogenicity. *New England Journal of Medicine*, **320**, 915–924.

Sullivan, T.R., Jr., Karas, R.H., Aronowitz, M., Faller, G.T., Ziar, J.P., Smith, J.J., *et al.* (1995). Estrogen inhibits the response-to-injury in a mouse carotid artery model. *Journal of Clinical Investigation*, **96**, 2482–2488.

Tepper, C.G., Jayadev, S., Liu, B., Bielawska, A., Wolff, R., Yonehara, S.H., *et al.* (1995). Role for ceramide as an endogenous mediator of Fas-induced cytotoxicity. *Proceedings of the National Academy of Sciences of the United States of America*, **92**, 8443–8447.

Thornberry, N.A., Bull, H.G., Calacay, J.R., Chapman, K.T., Howard, A.D., Kostura, M.J., *et al.* (1992). A novel heterodimeric cysteine protease is required for interleukin-1$\beta$ processing in monocytes. *Nature*, **356**, 768–774.

van der Zee, R., Murohara, T., Luo, Z., Zollmann, F., Passeri, J., Lekutat, C., *et al.* (1997). Vascular endothelial growth factor (VEGF)/vascular permeability factor (VPF) augments nitric oxide release from quiescent rabbit and human vascular endothelium. *Circulation*, **95**, 1030–1037.

Venkov, C.D., Rankin, A.B. and Vaughan, D.E. (1996). Identification of authentic estrogen receptor in cultured endothelial cells. A potential mechanism for steroid hormone regulation of endothelial function. *Circulation*, **94**, 727–733.

Weigel, N.L. (1996). Steroid hormone receptors and their regulation by phosphorylation. *Biochem. J.*, **319**, 657–667.

Williams, J.K., Adams, M.R., Herrington, D.M. and Clarkson, T.B. (1993). Shortterm administration of estrogen and vascular responses of atherosclerotic coronary arteries. *JACC*, **20**, 452–457.

# 5 Role of Estrogen in Endothelial Cell Behavior

**M. Lourdes Ponce [1], Maria C. Cid [2], Seunghee Kim-Schulze [3], Katherine M. Malinda [1], Kelly A. McGowan [1], Derrick S. Grant [4], H. William Schnaper [3] and Hynda K. Kleinman [1,*]**

[1] *National Institute of Dental Research, NIH, 30 Convent Drive, Bethesda, MD 20892-3470, USA,*
[2] *Dept. Internal Medicine, Hospital Clinic, Provincial, Barcelona, Spain,*
[3] *Dept. Pediatrics, Northwestern School of Medicine, Chicago, Illinois, USA,*
[4] *Thomas Jefferson College of Medicine, Philadelphia, USA.*
[*] *Corresponding author, Tel.: 001 301 496 4069, Fax: 001 301 402 0897, E-mail: kleinman@yoda.nidr.nih.gov.*

It is well known that women suffer more chronic inflammatory disorders of the vasculature and yet have a decreased incidence of cardiovascular disease. Various studies have now demonstrated a direct effect of estrogen on vascular endothelial and smooth muscle cells. The interaction of leukocytes with endothelial cells is critical to the pathogenesis of chronic inflammation. Estrogen promotes the binding of leukocytes to TNF (tumor necrosis factor)-stimulated endothelial cell monolayers via an increase in surface expression of adhesion molecules including E-selectin, ICAM-1 and VCAM-1. Increased levels of mRNA for these adhesion molecules are also observed in the presence of estrogen. Estrogen was also found to increase endothelial cell adhesion to matrix substrates, migration, proliferation and protease expression, in part due, to an increase in cell surface expression of cell-substrate adhesion receptors known as integrins. Furthermore, estrogen increased endothelial cell tube formation *in vitro* and angiogenesis *in vivo* in a subcutaneous murine model. A functional estrogen receptor was defined on endothelial cells using an estrogen antagonist, binding assays, immunohistochemistry and RNase protection and mobility shift assays. Estrogen has potent biological effects on endothelial cells which may explain why women suffer more inflammatory disease. The increase in cell migration, proliferation and angiogenesis suggest that estrogen could have a role in vessel repair and provide a protective mechanism against ischemic disease.

KEYWORDS: Estrogen, endothelium, angiogenesis, adhesion, migration, inflammation.

## INTRODUCTION

Women suffer more chronic inflammatory diseases than men and less cardiovascular disease (Ahmed *et al.*, 1985; Lahita, 1985; Grossman, 1989; Beeson, 1994). The molecular mechanisms for the gender differences in these conditions is not known but likely involves sex hormones. It is well known that chronic inflammatory diseases are exacerbated by pregnancy and oral contraceptives and reduced after menopause (Sandrez-Guerrero *et al.*, 1995). Likewise the relative protection from atherosclerotic cardiovascular disease enjoyed by women is lost after menopause (Stampfer *et al.*, 1991). Estrogen exerts important regulatory roles in both normal and pathological

**Table 5.1** Endothelial cell biological activities enhanced by estradiol

| |
|---|
| Leukocyte adhesion to endothelial cell monolayers |
| Adhesion to extracellular matrix proteins |
| Migration |
| Proliferation |
| Tube formation on basement membrane matrix |
| bFGF induced angiogenesis *in vivo* |

processes. Much has been reported on the effects of estrogen on smooth muscle cells (reviewed in Farhat *et al.*, 1996). Recently we have found that estrogen has potent biological effects on the endothelial cells (Table 5.1) that line vessels and are the first site of interaction of inflammatory cells with the vessel wall. We find that estrogen treatment of endothelial cells increases leukocyte adhesion to vascular endothelial cells *in vitro* via an increase in surface expression of adhesion receptors on endothelial cells (Cid *et al.*, 1994). Endothelial cells are also important in new vessel formation and actually migrate away from the parent vessel to initiate formation of new structures (Ausprunk *et al.*, 1977). Many cell functions involved in the formation of new vessels, including basement membrane matrix degradation, cell migration, adhesion, and proliferation, are regulated by estrogen (Morales *et al.*, 1995). This chapter will summarize the effects of estrogen on the biological activities of endothelial cells and present evidence for a functional estrogen receptor.

## ROLE OF ESTROGEN IN INFLAMMATION

Autoimmune diseases, such as Lupus, Takayasu's arteritis, Sjogren's Syndrome, rheumatoid arthritis, are more common in women than in men (Ahmed *et al.*, 1985; Lahita, 1985; Shelhamer *et al.*, 1985; Grossman, 1989; Wenger *et al.*, 1993; Beeson, 1994). The mechanisms by which sex hormones activate the immune system and sustain the inflammatory processes are not understood. The interaction of leukocytes with endothelial cells is a critical and early event in the inflammatory process. Leukocytes must first bind to the surface of endothelial cells and subsequently transmigrate through the vessel wall into the tissues (Osborn, 1990; Springer, 1994). Surface receptors on both endothelial cells and leukocytes mediate this interaction (Springer, 1990).

We have investigated the effects of sex hormones *in vitro* on leukocyte adhesion to endothelial cell monolayers (Cid *et al.*, 1994). Using human umbilical vein endothelial cells in monolayer culture, we find that 17-beta estradiol, progesterone and testosterone alone did not affect leukocyte adhesion. Tumor necrosis factor (TNF) is known to stimulate leukocyte adhesion to endothelial cell monolayers. If endothelial cells were preincubated with this inflammatory cytokine, estradiol and progesterone each caused a further two- to three-fold increase in leukocyte adhesion over that observed with TNF alone whereas testosterone was without effect (Figure 5.1). The effect of estradiol was dose- (maximum at 2 to 5 ng/ml) and time- (maximum at 6 hours) dependent. Other cytokines, such as IL-1 and IL-4, did not have any effect in the presence or absence of estradiol whereas interferon gamma

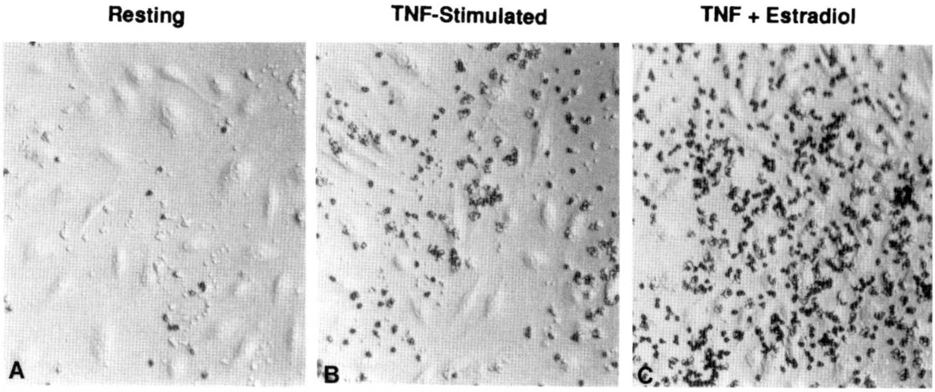


**Figure 5.1.** Effect of estradiol and TNF on leukocyte adhesion to endothelial cell monolayers. TNF stimulates leukocyte adhesion and this adhesion is further increased by estradiol. Reprinted with permission from Cid *et al.*, 1994.

**Table 5.2** Endothelial cell proteins regulated by estradiol

| *Protein* | *Effect* |
|---|---|
| E-selectin | Increased TNF-induced mRNA and surface expression |
| V-CAM-1 | Increased TNF-induced mRNA and surface expression |
| ICAM-1 | Increased TNF-induced mRNA and surface expression |
| Integrins $\alpha5, \alpha6, \beta1$ | Increased mRNA and surface expression |
| PAI-1 | Decreased mRNA in activated cells |
| uPA | Increased mRNA in activated cells |
| tPA | Increased mRNA in activated cells |
| Laminin | Increased protein and mRNA |
| Estrogen receptor | Increased mRNA |

had a slight stimulatory effect in the presence of estradiol. Endothelial cells from human fat tissue and from aorta were also activated for adhesion by estradiol with the fat cells being less responsive than the umbilical vein cells and the aortic cells being more responsive. It is well known that different endothelial cells have different activities (Auerbach, 1992) and this heterogeneity of responses was not unexpected. Thus, in the presence of an inflammatory cytokine, estradiol causes increased leukocyte adhesion. These data may explain at least in part why premenopausal women suffer more inflammatory diseases than men.

The mechanism for the enhanced leukocyte adhesion was found to be an increase in surface receptors for leukocytes on the surface of the endothelial cells (Table 5.2). TNF stimulated surface expression of E-selectin, ICAM-1 and VCAM-1 and this expression was further increased in the presence of estradiol. This was demonstrated by both FACS analysis and immunohistochemistry. The levels of mRNA for these adhesion molecules were also found to increase in a similar pattern. Furthermore, the estradiol-induced adhesion was blocked by antibodies to these adhesion molecules (see below). These data demonstrate an important regulatory role for estrogen in leukocyte–endothelial cell interactions. Such interactions are increased by estradiol and mediated by an increase in the surface adhesion receptors. This is the first demonstration of a mechanism defining the role of estrogen and explaining, at the molecular level, how it contributes to inflammatory diseases.

Various surface receptors on endothelial cells mediate the adhesion of leukocytes. Blocking antibodies to surface adhesion molecules were tested for their effects in inhibiting estradiol-induced leukocyte adhesion. Anti E-selectin and anti ICAM-1 inhibited the estradiol-induced PMN binding to TNF-stimulated endothelial cells. Anti VCAM-1 antibodies were without effect on PMNs. Anti VCAM-1 and anti ICAM-1 but not anti E-selectin reduced the estradiol induced increase in adhesion of PBMC to TNF-stimulated endothelial cells. These data demonstrate that PMN and PBMC recognize different cellular receptors and that binding is modulated by estradiol.

## ESTROGEN INCREASES INTEGRIN SURFACE EXPRESSION AND ADHESION

The interaction of endothelial cells with the underlying matrix is important in maintaining vessel integrity. *In vivo*, the basal surface of endothelial cells are in contact with an extracellular matrix, the basement membrane, which is composed of collagen IV, laminin, perlecan, and other components. The interaction of endothelial cells with the basement membrane is mediated, in part, by a family of cell surface receptors known as integrins (Hynes, 1992). These dimeric molecules mediate the binding of endothelial cells to collagen IV, laminin and possibly other components. The interaction of integrins with matrix molecules is important in cell adhesion, migration, growth and differentiation (Lampugnani *et al.*, 1991). Antibodies to integrins have been found to block various activities of endothelial cells including the formation of new vessels i.e. angiogenesis (Grant *et al.*, 1989; Gamble, 1993; Brooks *et al.*, 1994).

We have found that the surface expression of integrins on endothelial cells is regulated by estradiol (Cid *et al.*, 1994a). In this case, estradiol has a direct effect and does not require the presence of an inflammatory cytokine. We find that estradiol upregulates surface expression of integrins $\beta 1$, $\alpha 5$ and $\alpha 6$. This increase in surface expression is preceded by an increase in mRNA levels. The amount of the increase in integrin surface expression and mRNA is modulated also by the status of the cells. Estradiol-mediated upregulation is increased the most by either subconfluency, cell scraping or trypsinization. Since integrins are important in various biological activities, it is not surprising that adhesion to matrix components is increased by estradiol and some alignment of the cells is observed (Figure 5.2). Estradiol causes the greatest increase in adhesion to and spreading on laminin, collagens I and IV, and fibronectin. The response is time- and dose-dependent. It is not clear why the increase in adhesion is observed before the effects on integrins. This suggests that estradiol may also be affecting a signaling event through the integrin receptors. Since integrins have an important role in mediating endothelial cell attachment, migration, growth and differentiation (Grant *et al.*, 1989;

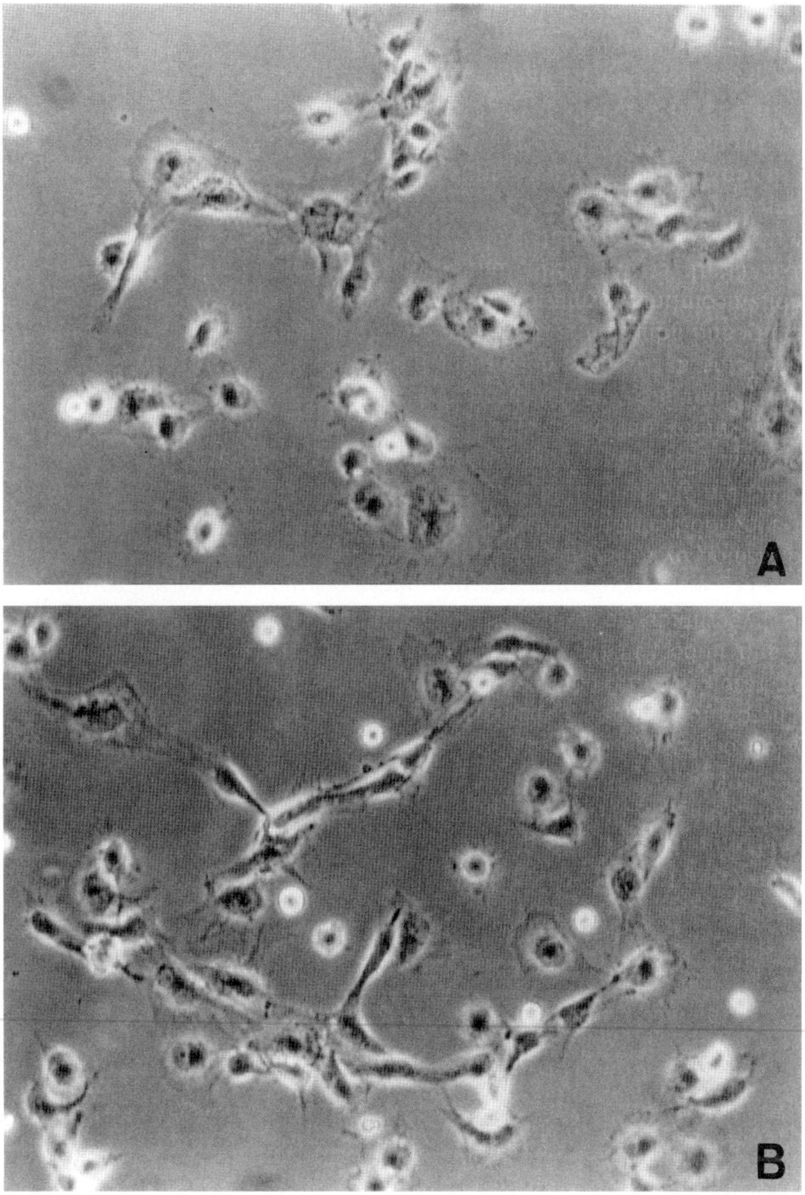

**Figure 5.2.** Effect of estradiol on endothelial cell adhesion to plastic. Increased spreading and cell alignment are observed in the presence of estradiol.

Luscinskas *et al.*, 1994), the increase in integrin expression and function may be an important mechanism by which estrogen regulates angiogenesis and repair. The protective effects of estrogen in cardiovascular disease could, in part, be explained by an accelerated repair of damaged vessels.

## ESTROGEN PROMOTES ENDOTHELIAL CELL MIGRATION, GROWTH AND ANGIOGENESIS

The formation of new blood vessels is important in both normal and pathological processes including nourishment of the implanted embryo, developing tissues, tumor growth and repair of injury (Folkman, 1995). The importance of estrogen in controlling vessel formation in pregnancy is well documented (Findlay, 1986). New blood vessel formation involves breakdown of the vessel wall, migration of the endothelial cells, proliferation and formation of new basement membrane (Folkman, 1995).

Estrogen has potent effects on endothelial cell migration and proliferation (Morales *et al.*, 1995). Using either a Boyden chamber assay or scratching a monolayer of endothelial cells, two- to three-fold increases in cell migration are observed. The rate of migration is dose-dependent and is within the physiological range of estradiol concentrations. Likewise cell proliferation is increased several-fold in the presence of estradiol. The increased migration and proliferation in response to estradiol is important in promoting new vessel formation.

The effect of estrogen on extracellular matrix proteases that are important for endothelial cell migration, differentiation and tissue remodeling are complex. These proteases are tightly regulated posttranslationally by their endogenous inhibitors. For example, the plasminogen activators tPA and uPA are inhibited by plasminogen activator inhibitor (PAI-1). When endothelial cells are confluent (quiescent), estrogen enhances PAI-1 mRNA expression but inhibits tPA and uPA mRNA expression shifting the balance between proteases and inhibitor to favor inhibition of protease activity. In contrast, when the cells are injured by scraping and therefore activated, estrogen increases tPA and uPA mRNA expression but decreases PAI-1 mRNA expression, favoring protease activity. Thus, the effect of estrogen on the net balance in this extracellular matrix pathway depends upon the status of the cells (McGowan, 1994).

Estradiol promotes new vessel formation both *in vitro* and *in vivo* (Morales *et al.*, 1995). When endothelial cells are plated on a layer of gelled basement membrane matrix (Matrigel), the cells attach, migrate and form tube-like structures with a lumen (Kubota *et al.*, 1989). This morphological differentiation mimics some of the steps involved in angiogenesis and has been a useful *in vitro* assay for screening for angiogenic and antiangiogenic components. Estradiol increases the formation of capillary-like tubes on basement membrane (Matrigel) two-fold (Figure 5.3). The effect of estradiol is dose-dependent with maximal activity at 1 to 2 ng/ml. This observation suggested that angiogensis *in vivo* might be controlled by estradiol. A murine angiogenesis assay has been developed using subcutaneously implanted basement membrane containing test compounds (Passaniti *et al.*, 1992). Basic fibroblast growth factor (bFGF), a potent angiogenic factor, has been found to promote endothelial cell migration in the basement membrane and vessel formation when mixed with the implant. This angiogenesis model was tested on ovarectomized mice, some of which received estrogen replacement with slow release pellets (Morales *et al.*, 1995). Sham operated animals were found to have the expected angiogenic response to implants containing bFGF. The ovarectomized mice receiving bFGF

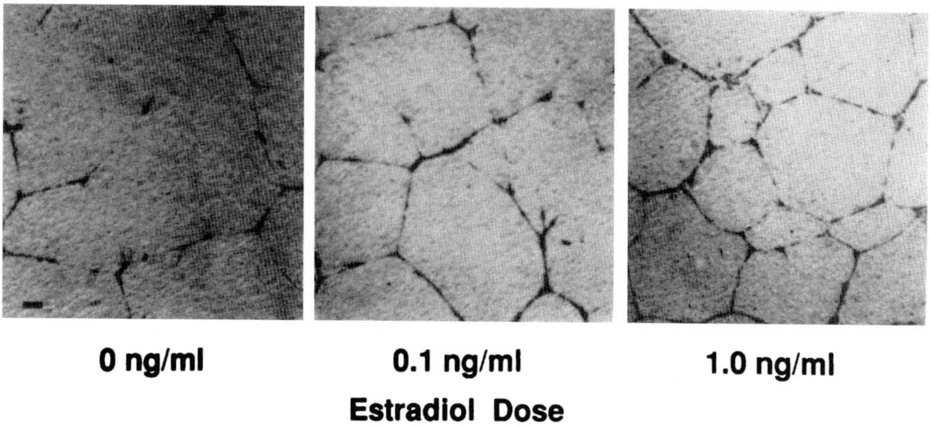


**Figure 5.3.** Effect of estradiol on the differentiation of endothelial cells into tube-like structures on Matrigel. The morphological differentiation of endothelial cells is enhanced by estradiol. Reprinted with permission from Morales *et al.*, 1995.

showed a greatly reduced response to bFGF (1.67 average cell area per field) over that observed in mice receiving the estrogen supplements (2.29 average cell area per field). The mice receiving estrogen supplements had a comparable angiogenic response to bFGF as that observed with sham operated animals. These data demonstrate that estrogen is important in the bFGF-induced angiogenesis and provide direct evidence for the role of estrogen in vessel formation and in the various steps of the process. Understanding the role of estrogen in angiogenesis may have important clinical applications in the treatment of vascular diseases.

## EFFECTS OF ESTRADIOL ON ENDOTHELIAL CELLS ARE MEDIATED BY A FUNCTIONAL ESTROGEN RECEPTOR

Estrogen receptors have been described on smooth muscle cells and in aorta and coronary artery cells *in vivo* (Karas *et al.*, 1994; Lesordo *et al.*, 1994). There is some evidence to suggest that estrogen can upregulate its own receptor activity since estrogen causes a redistribution of the receptor from the cytoplasm to the nucleus in aorta (Lin and Shain, 1985). Much of the beneficial effects of estrogen on vessel disease are demonstrated by its effect on smooth muscle cells where calcium influx and proliferation are inhibited (Farhat *et al.*, 1996). The presence of a functional estrogen receptor in modulating endothelial cell behavior has been demonstrated by a number of techniques (Table 5.3) (Kim-Shulze *et al.*, 1996; Johns *et al.*, 1996). In human umbilical vein endothelial cells and coronary artery cells *in vitro* RNase protection assay and immunohistochemistry showing nuclear staining confirm the presence of an estrogen receptor. Radioligand binding assays demonstrated $3–6 \times 10^4$ estrogen receptors per cell while scatchard analysis indicated $6–8 \times 10^4$ per cell and a Kd of 5 nmol/L. These data are consistent with what has been found with

**Table 5.3** Evidence of functional estrogen receptor on endothelial cells

- Migration, proliferation and tube formation *in vitro* are blocked by estrogen receptor antagonist ICI 182,780
- Receptor demonstrated by nuclease protection assay and immunohistochemistry
- Gel shift assay with extracts of estrogen treated endothelial cells and supershift with antibody to the estrogen receptor
- Scatchard analysis/ligand binding assays showed $3\text{–}8 \times 10^4$ receptors per cell and Kd of 5 nM

smooth muscle cells. Endothelial nuclear extract proteins bound to a radiolabeled consensus estrogen response element (ERE); this binding was blocked by excess unlabeled ERE. In addition the complex was supershifted by an antibody to the estrogen receptor (Kim-Schulze *et al.*, 1996). These data suggest that estrogen regulates endothelial function via classical receptor-mediated effects of gene transcription and regulation. A direct role of the estrogen receptor on endothelial cell activity has been demonstrated using the estrogen receptor antagonist ICI 182,780. This antagonist has been found to block estrogen-induced proliferation, migration in the scratch assay, and tube formation on the basement membrane matrix Matrigel (Kim-Schulze *et al.*, 1996; Morales *et al.*, 1995). Recently, it has been found in the estrogen receptor knockout mouse that estradiol-induced angiogenesis in response to bFGF is reduced (Johns *et al.*, 1996). These data suggest that the effect of estradiol on endothelial cells is mediated directly by a functional estrogen receptor. It is surprising that estrogen mediates so many activities such as attachment, migration, proliferation, and tube formation which are important in angiogenesis. The mechanism for the action of estrogen is not yet known but does not appear to involve secretion of bFGF. The increased expression of E-selectin, which is induced by estrogen and is important in promoting angiogenesis, may be one additional mechanism by which estrogen acts. Thus, estrogen does not appear to be inducing new activities but rather increasing existing activities.

## SUMMARY

The prevalence of chronic inflammatory vascular disease and the lower susceptibility of cardiovascular disease for premenopausal women suggests a role for sex hormones in the biology of the vasculature. Until now, the effects of estrogen on smooth muscle cells have been investigated but little was known about the effects of estrogen on endothelial cells. Estrogen has a number of effects on endothelial cells which are important in inflammation, repair and angiogenesis (Figure 5.4). Some of these effects of estrogen on leukocyte adhesion, endothelial cell adhesion to matrix, migration, proliferation and angiogenesis are mediated by a classical estrogen receptor mechanism. Although the molecular mechanisms involved are not yet known, a greater understanding of the activities induced by estrogen helps to define its role in normal and pathological processes. Part of the protective effect of estrogen on vascular disease may be related to the ability of estrogen to induce rapid repair. This may have important clinical applications.

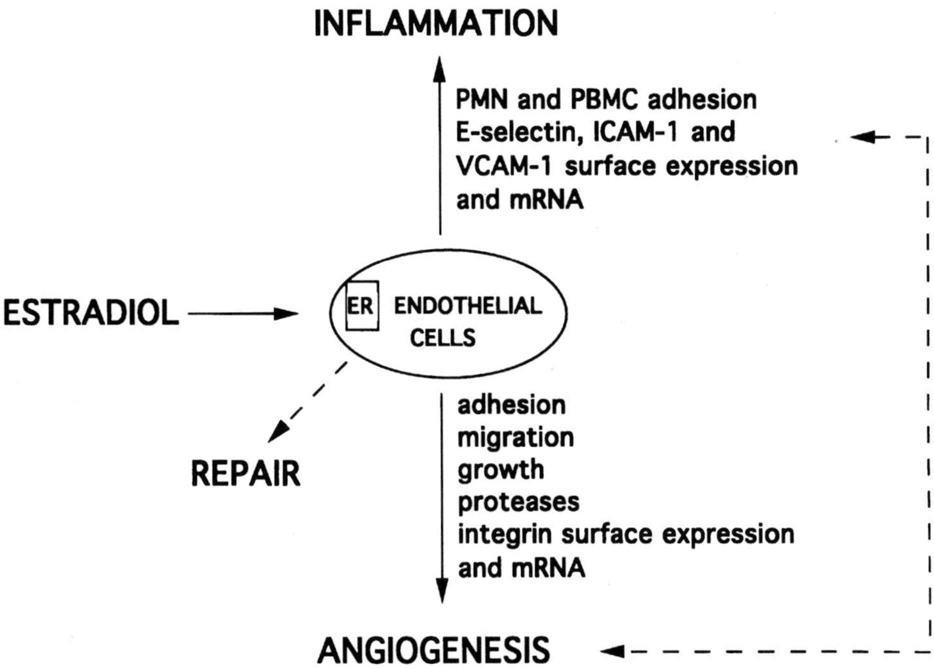


**Figure 5.4.** Schematic diagram of the effect of estradiol on endothelial cell behavior and its role in inflammation and angiogenesis.

## ACKNOWLEDGMENTS

K.A. McGowan was funded by the Howard Hughes Medical Scholars Program.

## REFERENCES

Ahmed, S.A., Penhale, W.J. and Talal, N. (1985) Sex hormones, immune responses, and autoimmune diseases. Mechanisms of sex hormone action. *Amer. J. Pathol.* **121**, 531–551.

Auerbach, R. (1992) Endothelial cell heterogeneity: Its role as a determinant of selective metastasis. In: *Endothelial Cell Dysfunction*, edited by N. Simonescu and O. Simonescu, pp. 427–437. Plenum Press.

Ausprunk, D.H. and Folkman, J. (1977) Migration and proliferation of endothelial cells in preformed and newly formed vessels during tumor angiogenesis. *Microvasc. Res.* **14**, 53–65.

Beeson, B.A. (1994) Age and sex associations of 40 autoimmune diseases. *JAMA* **96**, 457–462.

Brooks, P.C., Clark, R.A. and Cheresh, D.A. (1994) Requirement of vascular integrin alpha v beta 3 for angiogenesis. *Science* **264**, 569–571.

Cid, M.C., Kleinman, H.K., Grant, D.S., Schnaper, H.W., Fauci, A.S. and Hoffman, G.S. (1994) Estradiol enhances leukocyte binding to tumor necrosis factor (TNF)-stimulated endothelial cells via an increase in TNF-induced adhesion molecules E-selectin, intercellular adhesion molecule type 1, and vascular cell adhesion molecule type 1. *J. Clin. Invest.* **93**, 17–25.

Cid, M.C., Esparza, J., Grant, D.S., Morales, D., McGowan, K.A., Urbano-Marguez, A., Schnaper, W.H. and Kleinman, H.K. (1994a) Estradiol increases endothelial cell attachment to extracellular matrix proteins through an increase in integrin expression. *Clin. Res.* **42**, 131A.

Cockerill, G.W., Gamble, J.R. and Vadas, M.A. (1996) Angiogenesis: models and modulators. *Int. Rev. Cytol.* **159**, 113–160.

Farhat, M.Y., Lavigne, M.C. and Ramwell, P.W. (1996) The vascular protective effects of estrogen. *FASEB J.* **10**, 615–624.

Findlay, J.K. (1986) Angiogenesis in reproductive tissues. *J. Endocrinol.* **111**, 357–366.

Folkman, J. (1995) Angiogenesis in cancer, vascular, rheumatoid and other diseases. *Nature Med.* **1**, 27–31.

Gamble, J.R. (1993) Regulation of *in vitro* capillary tube formation by anti-integrin antibodies. *J. Cell Biol.* **12**, 931–943.

Grant, D.S., Tashiro, K.-I., Segui-Real, B., Yamada, Y., Martin, G.R. and Kleinman, H.K. (1989) Two different domains of laminin mediate the differentiation of human endothelial cells into capillary-like structures *in vitro*. *Cell* **58**, 933–943.

Grossman, C. (1989) Possible underlying mechanisms of sexual dimorphism in the immune response, fact and hypothesis. *J. Steroid Biochem.* **43**, 241–251.

Hynes, R.O. (1992) Integrins: versatility, modulation , and signaling in cell adhesion. *Cell* **69**, 11–25.

Johns, A., Freay, A.D., Fraser, W., Korach, K.S. and Rubanyi, G.M. (1996) Disruption of estrogen receptor gene prevents 17 beta estradiol-induced angiogenesis in transgenic mice. *J. Endocrinol.*, **10**, 4511–4513.

Karas, R.H., Patterson, B.L. and Mendelsohn, M.E. (1994) Human vascular smooth muscle cells contain functional estrogen receptor. *Circulation* **89**, 1943–1950.

Kim-Schulze, S. McGowan, K.A., Hubchak, S.C., Cid, M.C., Marin, M.B., Kleinman, H.K., Greene, G.L. and Schnaper, H.W. (1996) Expression of an estrogen receptor by human coronary artery and umbilical vein endothelial cells. *Circulation*, **94**, 1402–1407.

Kubota, Y., Kleinman, H.K., Martin, G.R. and Lawley, T.J. (1988) Role of laminin and basement membrane in the morphological differentiation of human endothelial cells in capillary-like structures. *J. Cell Biol.* **107**, 1589–1598.

Lahita, R.G. (1985) Sex steroids and the rheumatic diseases. *Arthritis and Rheum.* **28**, 121–126.

Lampugnani, M.G., Resnati, M., Dejana, E. and Marchisio, P.C. (1991) The role of integrins in the maintenance of endothelial monolayer integrity. *J. Cell Biol.* **112**, 479–490.

Lin, A.L. and Shain, S.A. (1985) Estrogen mediated cytoplasmic and nuclear distribution of rat cardiovascular estrogen receptors. *Arteriosclerosis* **5**, 668–677.

Losordo, D.W., Kearney, M., Kim, E.A., Jekanowski, J. and Isner, J.M. (1994) Variable expression of estrogen receptor in normal and atherosclerotic coronary arteries of premenopausal women. *Circulation* **89**, 1501–1510.

McGowan, K.A., Kleinman, H.K. and Schnaper, H.W. (1994) Expression of mRNA for extracellular matrix (ECM) proteases and their inhibitors by migrating endothelial cells: augmentation of response by estrogen. *J. Am. Soc. Nephrol.* **5**, 813a.

Morales, D.E., McGowan, K.A., Grant, D.S., Maheshwari, S., Bhartiya, D., Cid, M.C., Kleinman, H.K. and Schnaper, W.H. (1995) Estrogen promotes angiogenic activity in human umbilical vein endothelial cells *in vitro* and in a murine model. *Circulation* **91**, 755–763.

Osborn, L. (1990) Leukocyte adhesion to endothelium in inflammation. *Cell* **62**, 3–6.

Passaniti, A., Taylor, R.M., Pili, R., Guo, Y., Long, P.V., Haney, J.A., Pauly, R.R., Grant, D.S. and Martin, G.R. (1992) A simple quantitative method for assessing angiogenesis and antiangiogenic agents using reconstituted basement membrane, haprin and FGF. *Lab. Invest.* **67**, 519–529.

Sanchez-Geurrero, J., Liang, M.M., Karlson, E.W., Hunter, D.J. and Colditz, G.A. (1995) Postmenopausal estrogen therapy and the risk for developing systemic lupus erythematosis. *Ann. Internal Medicine* **122**, 430–433.

Shelhamer, J.H., Volkman, D.J., Parillo, J.E., Lawley, T.J., Johnston, M.R. and Fauci, A.S. (1985) Takayasu's arteritis and its therapy. *Ann. Intern. Med.* **103**, 121–126.

Springer, T.A. (1990) Adhesion receptors in the immune system. *Nature* **346**, 425–433.

Springer, T.A. (1994) Traffic signals for lymphocyte recirculation and leukocyte emigration: The multistep paradigm. *Cell* **76**, 301–314.

Stampfer, M.J., Colditz, G.A., Willett, C., Mauson, J.E., Rosner, B., Speizer, F.E. and Henneken, C.H. (1991) Postmenopausal estrogen therapy and cardiovascular disease. Ten year follow-up from the nurses' health study. *NEJM* **325**, 756–762.

Wenger, N.K. and Speroff, L., and Packard, B. (1993) Cardiovascular health and disease in women. *NEJM* **329**, 247–256.

# 6 Modulation of Cellular Adhesion Molecules and Endothelial Nitric Oxide Synthase by Estrogen

**Teresa Caulin-Glaser [1,3], William C. Sessa [2,3] and Jeffrey R. Bender [1,3,*]**

[1] *Division of Cardiovascular Medicine and the Raymond and Beverly Sackler Foundation Laboratory,* [2] *Department of Pharmacology, and* [3] *Molecular Cardiobiology Program, Boyer Center for Molecular Medicine, Yale University School of Medicine, New Haven, CT 06536-0812, USA.*
[*] *Corresponding author: Jeffrey R. Bender, Yale University School of Medicine, Boyer Center for Molecular Medicine, 295 Congress Avenue, Room 454, New Haven, CT 06536-0812, USA, Tel.: 001 203 737 2223, Fax: 001 203 737 2293, E-mail: jeffrey_bender@quickmail.yale.edu.*

Observations from many investigators have demonstrated that endothelial cell dysfunction plays an etiologic role in the initiation of vascular disease and development of clinical complications. The studies that focus on this area provide insight to the understanding of the pathophysiology of vascular disease as well as potential new ways to address therapy. The exact extent to which estrogen explains age-related differences in coronary artery disease has become the focus of many laboratories. The lack of estrogen associated with menopause is associated with altered lipid profiles, coagulation factors as well as endothelial dysfunction.

The incidence of coronary atherosclerosis in premenopausal women is half that observed in age-matched males (Gossland *et al.*, 1987). In contrast, Wegner *et al.* (1985) suggested postmenopausal women are not "protected," although estrogen replacement can reproduce those differences in cardiovascular disease seen in the premenopausal age group. This apparent protective effect of estrogen on the vascular wall is poorly understood. Estrogen affects cholesterol metabolism and deposition, generally inhibiting atherosclerotic plaque formation. Decreased estrogen production with menopause signals an increase in LDL-C and lipoprotein(a) levels in women. However, Adams *et al.* (1990) reported that the magnitude of the lipid changes cannot fully account for the observed cardiovascular protective effects. Coagulation abnormalities have also been demonstrated with menopause. Clotting factor VIIa is increased, suggesting a chronic activation of the thrombolytic system. Estrogen studies in animal models have demonstrated inhibition of platelet aggregation (Sarrel, 1988), diminished lipoprotein-induced arterial smooth muscle cell proliferation (Wolinsky, 1992), inhibition of myointimal proliferation associated with mechanical injury or stress (Williams *et al.*, 1990), decreased foam cell formation (Wagner *et al.*, 1991) and increased arterial smooth muscle prostacyclin production (Wakasugi *et al.*, 1989). The significance of these effects in human atherosclerosis is not known.

KEYWORDS: Estrogen, estrogen receptor, vascular endothelium, nitric oxide, nitric oxide synthase, atherosclerosis.

## INTRODUCTION

Histopathologic similarities exist between chronic inflammation and atherosclerosis. Early atherosclerotic lesions involve an inflammatory process consisting of intimal

T lymphocytes and lipid-laden macrophages, also known as the fatty streak in both human and animal models. The endothelium is a dynamically regulated tissue which actively contributes to the development of local inflammatory and vascular immune responses. Intercellular adhesion between vascular endothelial cells (EC) and circulatory leukocytes constitutes the initial trigger for many of these responses (Ross, 1986). These interactions are mediated by an array of leukocyte adhesion receptors and their respective EC counter-receptors, many of which are cytokine-inducible (Cotran and Pober, 1990). They demonstrated in the cholesterol-fed, Watanabe-heritable hyperlipidemic rabbit model of atherosclerosis, that an induced EC adhesion molecule, vascular cell adhesion molecule-1 (VCAM-1), is expressed in a localized fashion by aortic EC overlying very early atherosclerotic lesions. The presence of subendothelial macrophages that presumably adhere to and transmigrate the endothelium are associated with these areas of EC activation.

Palmer *et al.* (1991) and Ignarro *et al.* (1987) have demonstrated that nitric oxide (NO) is an important endothelium-derived vasodilator. Vascular smooth muscle relaxation is stimulated by NO through its release of soluble guanylate cyclase. NO is a vasodilator as well as possessing many antiatherogenic properties. Platelet aggregability and adhesion is decreased by NO (Radomski *et al.*, 1990; 1989). NO also inhibits neointima formation (von der Leyen *et al.*, 1995), prevents monocyte chemotaxis (Bath *et al.*, 1991), limits vascular smooth muscle cell proliferation (Garg *et al.*, 1989), and inhibits leukocyte adhesion to the endothelium Kubes *et al.*, 1991). Increased NO synthesis may therefore inhibit atherogenesis. NO is produced through the enzyme nitric oxide synthase (NOS). There are several isoforms of the enzyme: neuronal (nNOS, NOS 1), inducible (iNOS, NOS 2), and endothelial NOS (eNOS, NOS 3) (Sessa, 1994). The enzymes catalyze the oxidation of L-arginine to NO in the presence of cofactors and cosubstrates. The distinct NOS isoforms have different subcellular localizations and requirements for calcium. Rosenkranz-Weiss *et al.* (1993) demonstrated in endothelial cells that the regulation of NO release occurs at the level of eNOS activity which is controlled by cofactors. Michel *et al.* (1993), Hirata *et al.* (1995), Ohara *et al.* (1995), have all shown the importance of phosphorylation in eNOS activation. García-Cardeña *et al.* (1996), Sessa *et al.* (1995) have demonstrated the targeting of eNOS to specific intracellular membranes by fatty acylation.

The protective effects of estrogen ($E_2$) on cardiovascular function has been demonstrated in clinical and epidemiological data. The cellular and molecular mechanisms by which $E_2$ exerts its effect are still unclear. There are many studies ranging from clinical observations to cell culture models all hoping to provide a common basis for understanding and elucidating the effects of estrogen on the vascular system. The vascular endothelium and underlying smooth muscle play a large role in vascular control, therefore the focus of many investigations is the mechanisms by which $E_2$ may influence these cell types and the interaction between them.

The purpose of this review is to specifically address the effects of $E_2$ on cellular adhesion molecule function, endothelial nitric oxide production, and endothelial nitric oxide synthase activity in an attempt to provide further insight into the pathophysiology of vascular regulation.

## ENDOTHELIAL CELL ADHESION MOLECULES

It has been appreciated that the endothelial cell (EC) contributes actively to the development of local vascular immune and inflammatory response. The modulation of leukocyte-endothelial adhesion is thought to be mediated by constitutive or cytokine-induction of cellular adhesion molecules (CAMs) which include: E-selectin, vascular cell adhesion molecule-1 (VCAM-1) and intercellular adhesion molecules (ICAM-1 and 2). Studies have suggested the ECs can direct the traffic of leukocytes into inflamed areas via the regulated expression of CAMs (Cotran and Pober, 1990). Leukocytes also express proteins on their surface that mediate their specific localization to sites of inflammation. It has been proposed that as a consequence of activation, the circulating monocytes and lymphocytes adhere to the endothelium, migrate through subendothelial matrix to participate in the inflammatory reaction (Figure 6.1). This is a proposed mechanism for atherosclerosis (Ross, 1986). Studies of atherosclerotic lesions in human coronary arteries and abdominal aortas demonstrate an increased expression of ICAM-1 and E-selectin on

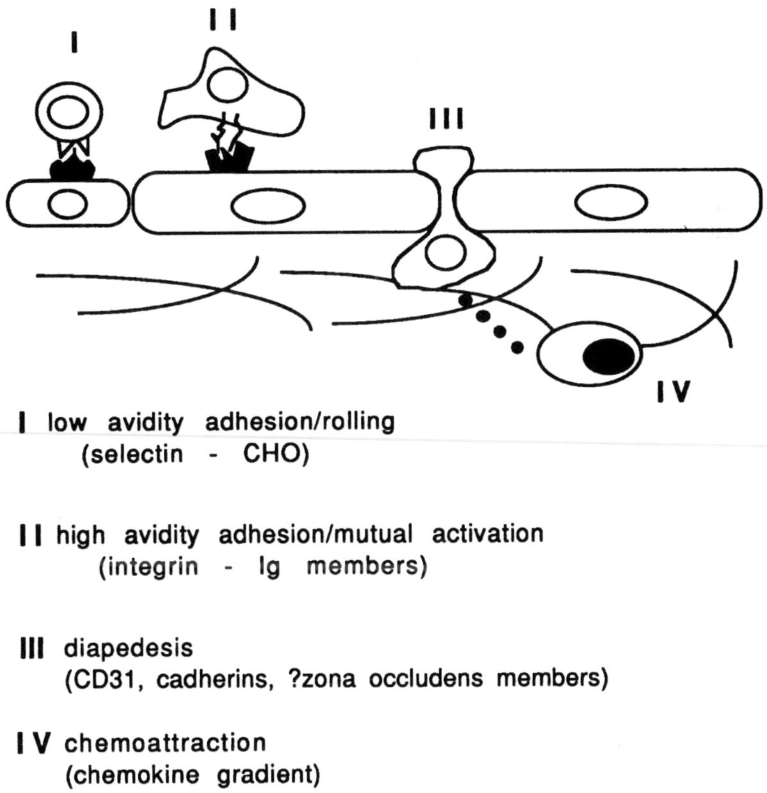


**Figure 6.1.** Leukocyte adhesion and migration.

ECs adjacent to sites of prominent inflammatory cell infiltrates (van der Wal *et al.*, 1992). The expression of these activation molecules is associated with enhanced leukocyte adhesion as well as enhanced surface thrombogenicity and intravascular coagulation, features which are thought to occur in early atherosclerosis. Studies on ECs in culture have demonstrated that many of the functional and morphologic changes seen *in vivo* in inflammation can be induced *in vitro* with purified mediators such as cytokines.

E-selectin is a cytokine induced adhesion molecule expressed on activated ECs and venules. It binds primarily neutrophils, monocytes and some subsets of T cells. Studies of the molecular mechanism regulating E-selectin gene expression demonstrated three regions in the E-selectin promoter which are cytokine-responsive. Two of these involve NF-κB family members as transacting factors, whereas one of them is inducibly bound by ATF2 and c-jun (Montgomery *et al.*, 1991). VCAM-1 is expressed on the surface of activated ECs and binds primarily monocytes and T cells. Studies of the promoter region of the VCAM-1 gene have demonstrated the presence of GATA sites as well as an NF-kB site (Iadermarco *et al.*, 1992). ICAM-1 is constitutively expressed on ECs as well as other cell types. It is hyperinduced with cytokine stimulation. Structural analysis of the promoter region of this gene has also demonstrated the presence of a NF-kB binding site (Degitz *et al.*, 1991). NF-kB is a ubiquitously expressed transcription factor which is involved in the expression of many cytokine induced genes and may play a role in the initiation of atherosclerotic lesions. Analysis of the human CAM genes has also demonstrated the presence of perfect, imperfect, and half ERE's.

## NITRIC OXIDE AND NITRIC OXIDE SYNTHASE

Abnormalities of coronary vasomotion, thrombosis and leukocyte-endothelial cell adhesion play varying roles in acute and chronic coronary artery disease. Nitric oxide (NO) is a diverse molecule with vasodilator, anti-aggregatory and anti-adhesive properties as previously mentioned. NO is produced by multiple different cell types including vascular endothelial cells, vascular smooth muscle cells, neurons, mononuclear phagocytes and hepatocytes. In all of these cells, NO is generated by the conversion of L-arginine to L-citrulline by the action of the enzyme nitric oxide synthase (NOS) (Figure 6.2).

eNOS activity and endothelium-dependent relaxation is not regulated by classical inducers of gene transcription therefore eNOS is considered a “constitutive” NOS. The human eNOS promoter contains two half palindromic estrogen response elements, GATA motifs, as well as an AP-1 site which may influence eNOS expression and NO release (Marsden *et al.*, 1993).

Rees *et al.* (1990) demonstrated that the basal release of NO regulates vascular tone in experiments showing that specific inhibitors of NOS elicited a prolonged pressor response and reduced regional blood flows *in vivo* and inhibit endothelium-dependent relaxations *in vitro*. Awolesi *et al.* (1994), demonstrated that cyclic strain increases endothelial nitric oxide synthase activity (Figure 6.3). Chronic changes in local blood flow increases endothelium-dependent relaxation, *in vitro* (Miller *et al.*, 1987).

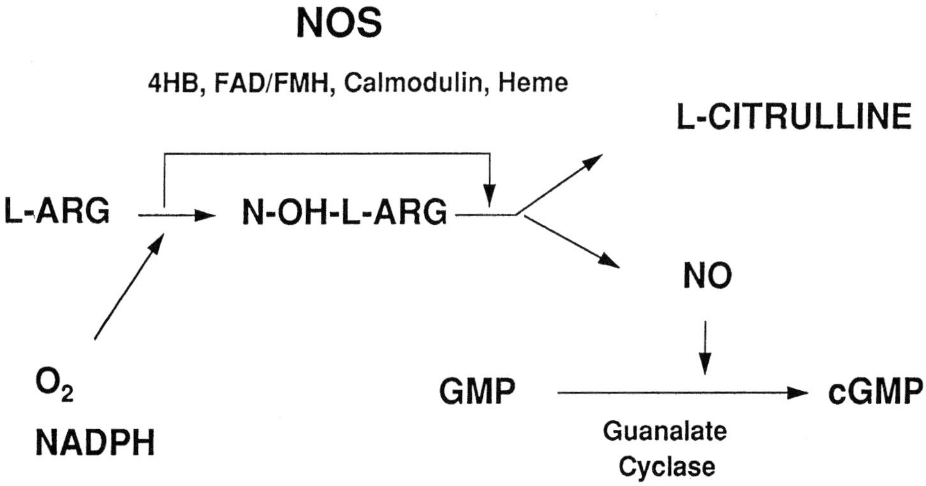


**Figure 6.2.** Generation of nitric oxide from the oxidation of L-arginine in the presence of cofactors and cosubstrates.

Mechanisms by which flow and shear-stress affect NO release and eNOS activity are unclear but may be related to the physical forces to which the endothelium is exposed.

A great variety of substances regulate eNOS activity and NO release acutely. Receptor-dependent agonists such as bradykinin, acetylcholine, histamine, ADP, and others, results in an increase in intracellular calcium, a cofactor in this calmodulin-dependent enzyme and acutely stimulates endothelial release of NO. Subsequent eNOS phosphorylation and subcellular localization are interdependent however their exact relationship remains unclear. Many stimuli acutely increase eNOS activity but few factors are known to regulate eNOS mRNA expression. Shear stress (Nishida *et al.*, 1992) and exercise training (Sessa *et al.*, 1994) have been shown to increase eNOS mRNA levels, while tumor necrosis factor-$\alpha$ decreases eNOS mRNA expression (Nishida *et al.*, 1992), suggesting that a variety of stimuli may regulate the transcription of eNOS gene expression.

## ESTROGEN AND ESTROGEN RECEPTOR

Estrogen is a steroid hormone whose physiological response is mediated by specific intracellular receptors (Katzenellenbogan, 1980). Steroid hormone receptors are members of a superfamily of ligand-induced transcription factors that include the receptors for retinoic acid, vitamin D and thyroid hormone (Evans, 1988). Estrogen may initiate a large series of physiologic responses by binding to its receptor in the target cell. The hormone passes through the cell membrane, binds to its receptor and forms an active transcription factor (Higgins *et al.*, 1973).

Expression of estrogen receptors in the vasculature has been controversial, reflecting tissue and species variability as well as limitations of the various techniques

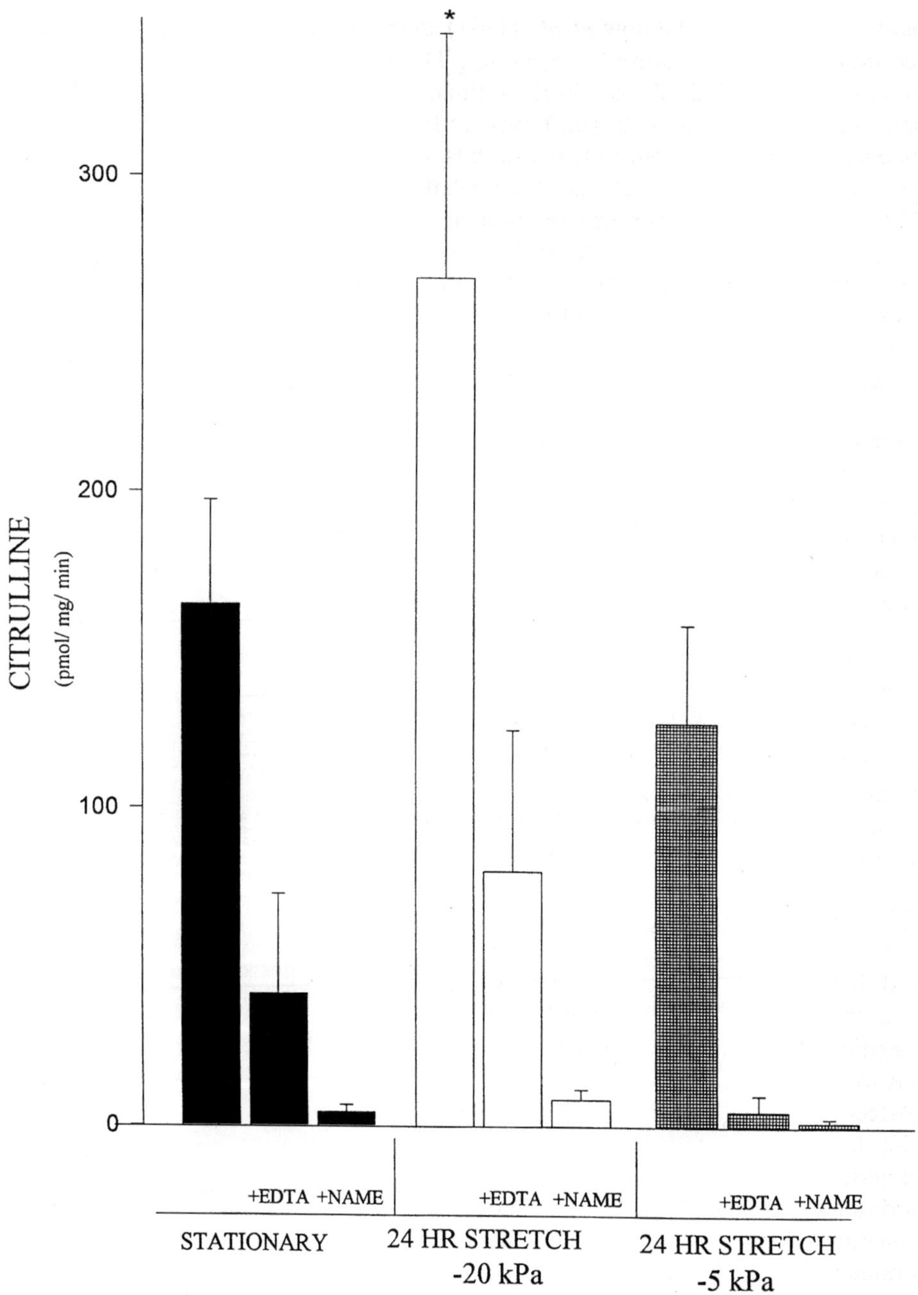


**Figure 6.3.** Assessment of eNOS activity. EC exposed to cyclic strain on membranes stretched with −20 kPa vacuum at 60 cycles/min for 24 h were analyzed for production of L-citrulline, a byproduct of eNOS activity. A significant increase occurred in eNOS activity compared with EC exposed to cyclic strain on membranes stretched with −5 kPa vacuum at 60 cycles/min or to stationary controls. Results are mean ± SE; $n = 3–7$; $^{*}p < 0.05$ compared to stationary controls. Reproduced from *Surgery* 1994, **116**: 439–445 by copyright permission.

used for detection. Malinow *et al.* (1963) demonstrated the presence of estrogen receptors in the vasculature by injecting [$^3$H]-estradiol into patients and observing uptake of the labeled estrogen in the intima, endothelium and adventitia of normal and abnormal arteries. Estrogen receptors have since been documented in various models such as aorta from rodents, rabbits, dogs, as well as human smooth muscle cells from coronary arteries and endothelial cells harvested from umbilical cords. The expression of the estrogen receptor in the vasculature has been demonstrated using several different methods such as radioactive ligand binding, immunocytochemistry, Northern analysis and Western blots. Recently, studies on both the presence and functional status of the estrogen receptor in the vasculature have been reported. Karas *et al.* (1994) have demonstrated with transfection assays the presence and functional activity of the ER in vascular smooth muscle cells. Orimo *et al.* demonstrated ER mRNA and protein in cultured rat aortic vascular smooth muscle cells using Northern blot analysis and anti-ER monoclonal antibodies. The presence of estrogen receptor mRNA in HUVEC cultured in our laboratory has been demonstrated by Northern blot analysis (Caulin-Glaser *et al.*, 1996). We have also demonstrated that the ER is functional in HUVEC as well as affected by changes in hormonal milieu. Another recent report has also suggested that receptor levels with in the vasculature are altered in disease states; ER levels were found to be diminished in the coronary arteries of premenopausal women with atherosclerosis when compared to premenopausal women without disease.

The ER protein can be divided into six regions based on sequence homology and consists of a number of functionally independent domains. The DNA binding domain is highly conserved and consists of two zinc finger motifs, each coordinates a zinc atom with four conserved cysteine residues (Evans and Hollenberg, 1988). Functional analyses have shown that these receptors act as transcriptional factors by interacting with inducible regulatory sequences (response elements, RE), which trigger a cascade of biological processes through modulation of transcription of specific gene targets (Beato, 1989). The estrogen response element (ERE) consists of a 13 base pair palindromic sequence. Since the RE are based on palindromic sequences and have 2-fold rotational symmetry, it has been speculated that the nuclear receptors bind to DNA as protein dimers. The perfect ERE has been shown to confer specific estrogen responsiveness by direct activation and requires tight binding of the ER to an ERE (Linstedt *et al.*, 1986). Half EREs and EREs which deviate from the perfect ERE by a base pair (imperfect EREs) exist and have also been shown to be able to confer $E_2$ responsiveness (Kumar and Chambon, 1988). These half and imperfect EREs occur frequently, therefore other factors are likely involved in addition to sequence homology for them to confer $E_2$ responsiveness. Indirect regulation appears to require other factors themselves bound to the RE as has been demonstrated in ER activation of the ovalbumin gene involving the AP-1 complex. Another recent report demonstrated that $E_2$ can activate kB-specific proteins in the uterus (Shyamala and Guiot, 1992). Current hypotheses on the mechanisms of $E_2$ mediated gene regulation include: (1) ER and other factors (i.e. AP-1 or kB proteins) directly interacting to effect transcription; (2) ER activating another gene whose product may modulate the function of the other transcription factors; (3) ER effecting an enhancer or inhibitor of the activity of AP-1 or other transcription factors.

$E_2$ has been shown to induce c-jun and c-fos in rat uteri and avian liver, and markedly reduce the steady state mRNA levels for the c-myc and c-jun protooncogenes in the avian oviduct (Gaub *et al.*, 1990; Maminta *et al.*, 1991). Studies in human osteoblasts have documented that $E_2$ has the ability to inhibit cytokine mediated IL-6 production (Girasole *et al.*, 1992). These reports demonstrate the ability of estrogen to both stimulate as well as inhibit gene expression. A recent study has shown that $E_2$-treated cells have decreased expression of vascular adhesion molecule-1. The mechanism of $E_2$ regulation of adhesion molecule transcription however is unknown.

There are also several reports which demonstrate a rapid vasodilator response to $E_2$ treatment which is not thought to be transcriptionally mediated. Several studies have reported that $E_2$ administration increases endothelium-dependent vasodilation in response to acetylcholine infusion into coronary arteries. Gillian *et al.* (1994) have shown that infusion of $E_2$ into postmenopausal women results in vasodilation of the brachial artery. Rosano *et al.* (1993) reported that sublingual $E_2$ administration to postmenopausal women with exercise-induced ischemia resulted in vasodilation of the coronary vessels angiographically. These findings have also been supported by *in vitro* studies. $E_2$ treatment of vasopressin-contracted coronary vessels harvested from rabbits results in vessel dilation. In a primate model, Williams *et al.* have shown that acetylcholine infusion into the coronary arteries of ovariectomized female animals resulted in vasoconstriction of the vessels. When the animals were treated with $E_2$ replacement therapy the infusion of acetylcholine into the coronary arteries caused vasodilation. The mounting data suggests therefore that $E_2$ is also capable of inducing rapid cellular events within seconds to minutes of treatment. The beneficial effect of $E_2$ on vasculature has been, in part, attributed to its direct effect on endothelial function via the production of a variety of vasoactive substances, such as nitric oxide, which regulate vascular tone and smooth muscle proliferation.

Estrogen receptors are not only transcriptional regulators of the hormone effect, but as postulated by Schule and Evans (1991), there are emerging reports that suggest they also guide the convergence of multiple signal transduction pathways. Numerous investigations have demonstrated that cell treatment with protein kinase activators, neurotransmitters, growth factors and phosphatase inhibitors can effect the activity of the ER. The modulation of the transcriptional activity of the ER by dopamine *in vitro* has been demonstrated by Powers *et al.* (1991). Estrogen and progesterone modulated the sexual behavior of the female rodent and this effect could be mimicked by dopamine therefore this was postulated as the mechanism of action. The effects of estrogen and dopamine could be blocked by antiprogestins, suggesting that the hormone and the neurotransmitter signaling pathways converge on the progesterone receptor *in vivo*. Insulin-like growth factor-1 (IGF-1), epidermal growth factor (EGF), and transforming growth factor-$\alpha$ (TGF$\alpha$) which are polypeptide growth factors have the ability of activating response reporters in cells transfected with recombinant ER as reported by Ignar-Trowbridge *et al.* (1993). EGF or IGF-1 can mimick estrogen induction of uterine proliferation. The effect can be blocked with the ER-antagonist ICI 164,384 in mice (Aronica *et al.*, 1993). Estrogen also increases cAMP levels in breast cancer cells and uterine cells *in vitro* as well as *in vivo*. Aronica *et al.* (1993) have shown that increased cAMP levels are due to enhanced adenylate cyclase activity by a mechanism which does not involve transcription or translation; the estrogen-induced increase in cAMP activates transcription from cAMP response element-containing genes. These reports have a profound impact on our understanding of the mechanism of steroid hormone action. The mechanisms of signal transduction cross-talk needs to be further studied to understand the estrogen-modulated mechanism of ER action and assess its importance *in vivo*.

## EFFECTS OF ESTROGEN ON CELLULAR ADHESION MOLECULES

Interactions between the estrogen receptor complex (ERC) and specific genomic sequences are well described and believed to be a dominant mechanism by which the ERC exerts its molecular effects. ERs preferentially bind to the ERE in the DNA. Kumar *et al.* (1988) reported that this element can be sufficient to enhance transcription, as can "imperfect" or half-EREs, which can be located a great distance from the transcription start site. In reviewing Genbank sequences, perfect, imperfect or half-EREs can be found in the VCAM-1, E-selectin and ICAM-1 genes. Traditionally thought to be enhancing, recent reports of estradiol negatively affecting transcription have emerged as demonstrated by Ray *et al.* (1994). These investigations have focused on ERE interactions as the possible mechanism for the observed effects. The inhibitory effect of $E_2$ on IL-1-mediated IL-6 gene activation has recently been reported by Ray *et al.* (1994). They demonstrated that although the negative effect involved the ER, high affinity ER binding to the IL-6 promoter could not be demonstrated, nor did recombinant ER bind to the critically involved 225 bp promoter fragment in gel mobility shift assays. Furthermore, they have shown that overexpression of the NF-$\kappa$B subunit p65 or, to a lesser extent, NF-IL6, overcomes $E_2$'s inhibitory effect. The report suggests that $E_2$-activated ER interacts with critical transacting factors, to form an inactive intermolecular complex. In addition, ER interactions with ubiquitous transcription factors exist, including SP1 (Krishnan *et al.*, 1994) and AP-1 (Weisz and Rosales, 1990). Shyamala and Guiot (1992) have reported $E_2$ induction of $\kappa$B-like proteins in rat uterus. It is likely that tissue, cell-type and gene specificity's exist which determine whether such induced interactions exert positive or negative regulatory effects.

Few studies have investigated the hormonal regulation of CAM expression. Cronstein *et al.* (1992) demonstrated that dexamethasone interacts with the glucocorticoid receptor, and results in reduction of IL-1-induced E-selectin mRNA levels. Nakai *et al.* (1994) reported reduced VCAM-1 mRNA and membrane protein levels in $E_2$-pretreated, IL-1-activated cells. We have previously demonstrated that $E_2$ markedly inhibits IL-1-mediated endothelial CAM transcription (Figure 6.4), also reflected in reduced membrane protein expression (Figure 6.5) and IL-1-induced leukocyte adhesion (Figure 6.6). The specificity of $E_2$'s effect is evident in that transcription is not generally inhibited, 17$\alpha$-estradiol at equimolar concentrations was ineffective, and the estrogen antagonist ICI 164,384 efficiently abrogates the inhibition (Figures 6.7A and 6.7B). The antagonist efficacy indicates that $E_2$'s inhibitory effect(s) is mediated, at least in part, via the endothelial ER

(Caulin-Glaser *et al.*, 1996). In contrast however, enhancement in TNF-mediated endothelial CAM induction by estradiol was reported by Cid *et al.* (1994). The assessment of hormone responses *in vitro* is complicated by incomplete definition of critical culture conditions. The discrepancies in study results may be related to culture conditions, cell type and passage number.

As reported by Kaszubska *et al.* (1993), DelLuca *et al.* (1994), Neish *et al.* (1992), Hou *et al.* (1994), and Read *et al.* (1994), E-selectin, ICAM-1, and VCAM-1, 5'-regulatory regions contain distinct elements, variably involved in induced gene activation. However, all 3 promoters contain at least 1 $\kappa$B site which are required for cytokine-mediated gene activation. Ray *et al.* (1994) studies involving the IL-6 promoter are consistent with the notion that ER complexes interact with

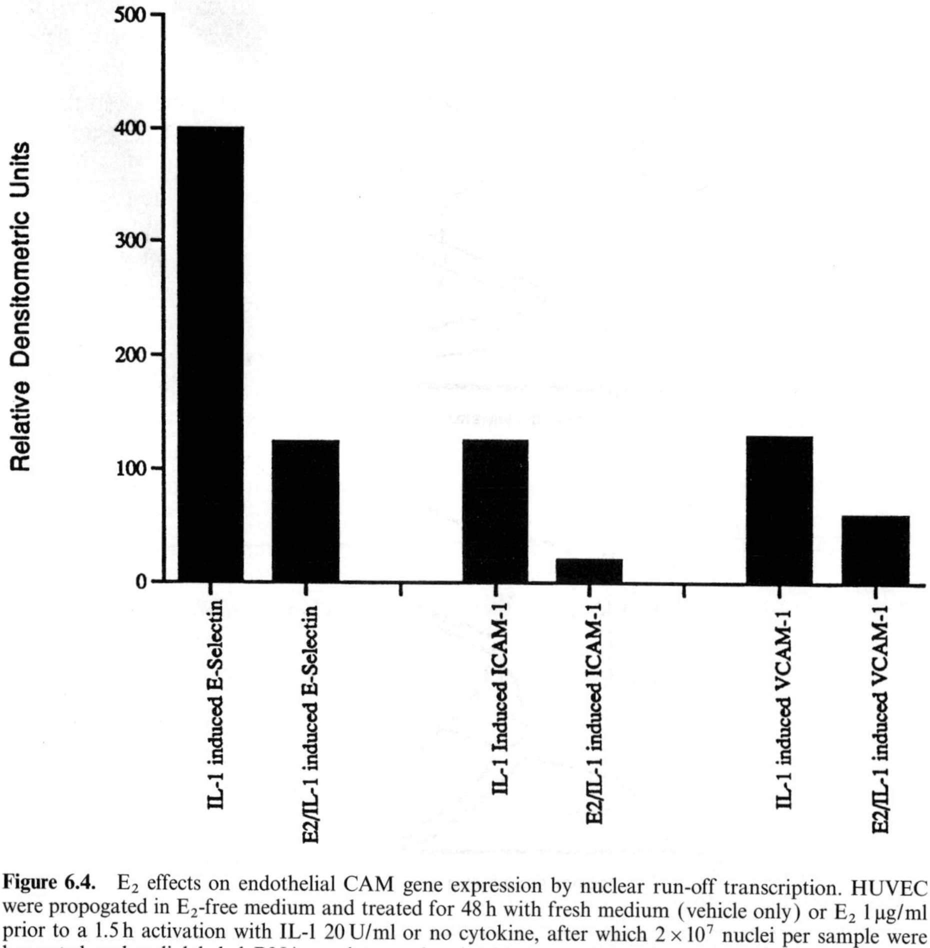


**Figure 6.4.** $E_2$ effects on endothelial CAM gene expression by nuclear run-off transcription. HUVEC were propogated in $E_2$-free medium and treated for 48 h with fresh medium (vehicle only) or $E_2$ 1 μg/ml prior to a 1.5 h activation with IL-1 20 U/ml or no cytokine, after which $2 \times 10^7$ nuclei per sample were harvested and radiolabeled RNA used to probe slot blots containing E-selectin, ICAM-1, VCAM-1, $\gamma$-actin or bluescript DNA. Histograms represent background (e.g. basal ICAM-1 signal)-subtracted, actin transcript-normalized densitometric analysis, comparing IL-1-induced nuclear transcripts in the absence or presence of $E_2$. Reproduced from *J. Clin. Invest.* 1996, **98**: 36–42 by copyright permission.

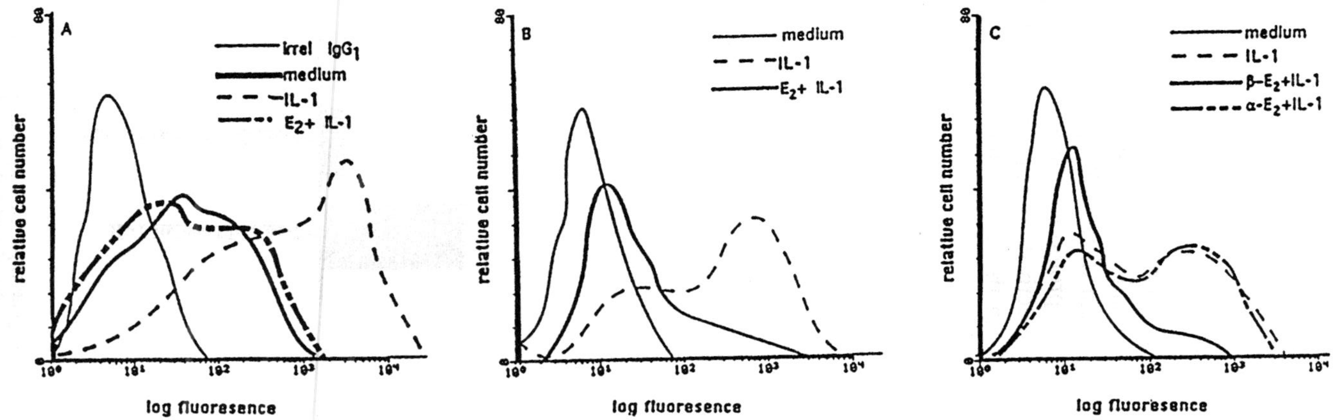


**Figure 6.5.** Flow cytometric analysis of $E_2$ effects on IL-1-mediated endothelial CAM induction. HUVEC were propogated in $E_2$-free medium and treated for 48 h with medium (vehicle), single dose $E_2$ 1000 ng/ml ($\beta$-$E_2$) or $\alpha$-estradiol 1000 ng/ml ($\alpha$-$E_2$), followed by a 4 h activation with IL-1 20 U/ml. Curves represent conditions as noted in each panel. Panel A represents ICAM-1 membrane expression, panels B and C represent E-selectin expression. 5000 events were analyzed per sample. Reproduced from *J. Clin. Invest.* 1996, **98**: 36–42 by copyright permission.

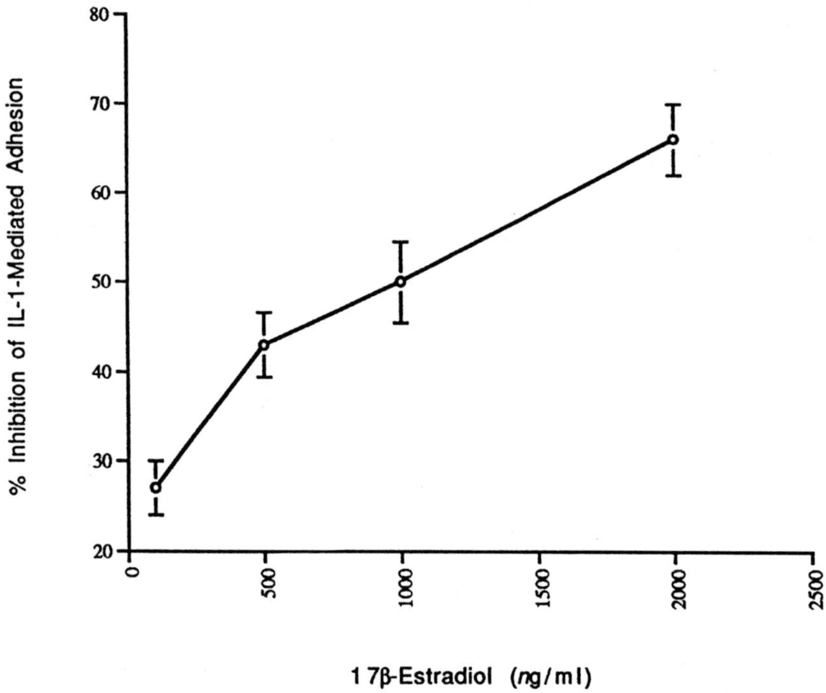


**Figure 6.6.** $E_2$ effect on IL-1 mediated augmentation of PBMC-EC adhesion. HUVEC propogated in gonadal hormone-free medium were pretreated (1 dose) × 48 h with the indicated concentrations of $E_2$, followed by a 16 h IL-1 (20 U/ml) activation. After washing out the $E_2$ and IL-1, $10^5$ $^{51}$Cr-labeled PBMCs (preactivated with PMA 50 nM × 30 min at 37°C) were added per well, and adhesion assays performed. IL-1 treatment (no $E_2$) resulted in a 150% increase in binding. Data are represented at each $E_2$ concentration as percent inhibition of IL-1-mediated adhesion, i.e., the 150% augmentation noted above. Samples were performed in quadruplicate. This experiment represents means ± SE for three seperate experiments. Reproduced from *J. Clin. Invest*. 1996, **98**: 36–42 by copyright permission.

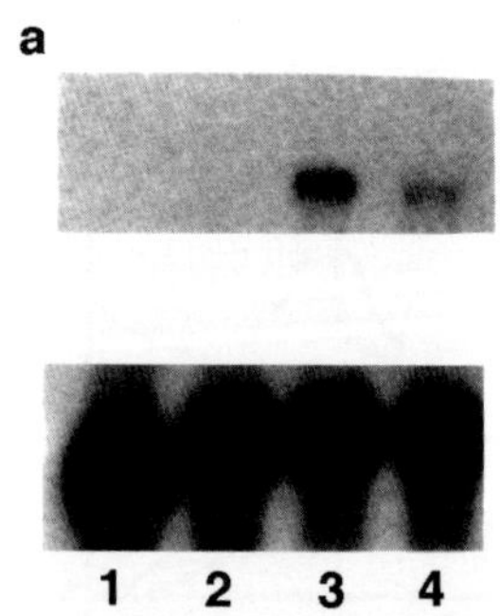


**Figure 6.7A.** Northern blot analysis of $E_2$ effects on EC E-selectin induction. HUVEC were propogated in $E_2$-free medium and treated for 48 h with fresh medium (containing vehicle) (lanes 1 and 3) or $E_2$ 1 μg/ml (lanes 2 and 4), prior to a 2 h activation with IL-1 20 U/ml (lanes 3 and 4) or no cytokine (lanes 1 and 2). 20 μg total RNA were hybridized per sample. GAPDH mRNA levels were utilized as internal controls. When densitometrically analyzed, lane 4 signal ($E_2$ + IL-1) represents a 64% decrease from lane 3 signal (IL-1). Reproduced from *J. Clin. Invest*. 1996, **98**: 36–42 by copyright permission.

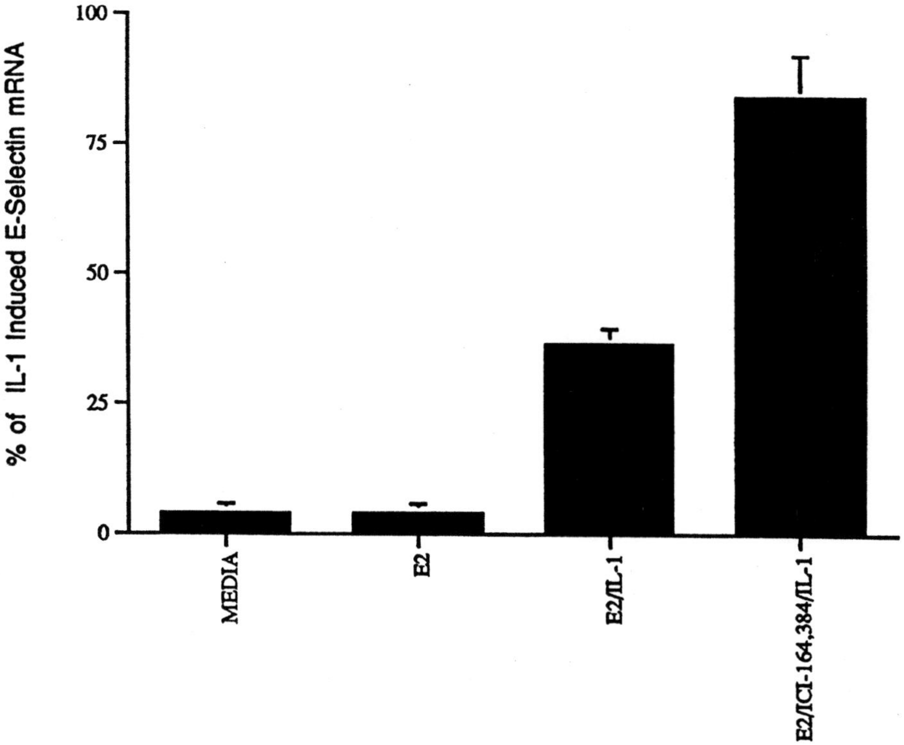


**Figure 6.7B.** Separate experiment with identical conditions as in Panel A, with the addition of ICI 164,184 ($E_2$ antagonist) in one of the experimental samples, as noted. Histograms represent phosphorimage-analyzed, GAPDH-normalized E-selectin mRNA levels by Northern blot. $E_2$ treatment results in a 78% decrease in IL-1-induced steady state E-selectin mRNA levels. The inhibition is largely abrogated by ICI 164,184. This experiment is representative of 4 separate experiments. Reproduced from *J. Clin. Invest.* 1996, **98**: 36–42 by copyright permission.

components of the NF-κB signaling pathway, negatively affecting transcription. $E_2$ could modulate human endothelial NF-κB nuclear translocation, DNA binding or transactivation, thereby affecting transcription of all three of these genes.

The antioxidant property of estradiol is potentially important for its non-genomic effect. Keaney *et al.* (1994) and Negre-Salvayre *et al.* (1993) have demonstrated that $E_2$ inhibits low-density lipoprotein oxidation *in vivo* and *in vitro* and this is thought to be one of estrogen's important cardioprotective effects. Haliwell and Grootveld (1987) have suggested that the antioxidant effect of $E_2$ can be attributed to its phenolic structure, which reacts with hydroxyl radicals to form hydroxylated products. Furthermore, antioxidants partially inhibit cytokine-mediated E-selectin and VCAM-1 gene activation, and diminish induced NF-κB binding activity on gel mobility shift assays (Mooradian, 1993).

The nature of the genes and the molecular mechanisms that are involved in the vascular genomic effects of estrogen are still uncertain and are currently being investigated.

## EFFECTS OF ESTROGEN ON NITRIC OXIDE RELEASE/ NITRIC OXIDE SYNTHASE ACTIVITY

Multiple experimental models have demonstrated that estrogen can modulate endothelial function by its effect on basal NO production and eNOS activity.

These effects of $E_2$ on eNOS activity and NO release have been investigated both *in vivo* and *in vitro*. Acute hemodynamic effects of estrogens in animal and human studies include increases in cardiac output and peripheral blood flow. Pregnancy has been associated with increased stimulated NO production. Weiner *et al.* (1989, 1991) demonstrated that acetylcholine induced NO-mediated relaxation of uterine and carotid arteries from guinea pig is increased with pregnancy. Keany *et al.* (1994) have shown that chronic estrogen treatment of oophorectomized hypercholesterolemic swine preserves the endothelium-dependent relaxation of coronary artery rings to bradykinin, whereas vessels from untreated animals exhibited impaired relaxation to the agonist. Chronic $E_2$ therapy increases relaxation to acetylcholine in aorta ring models from hypertensive rats. Collins *et al.* (1995) demonstrated that

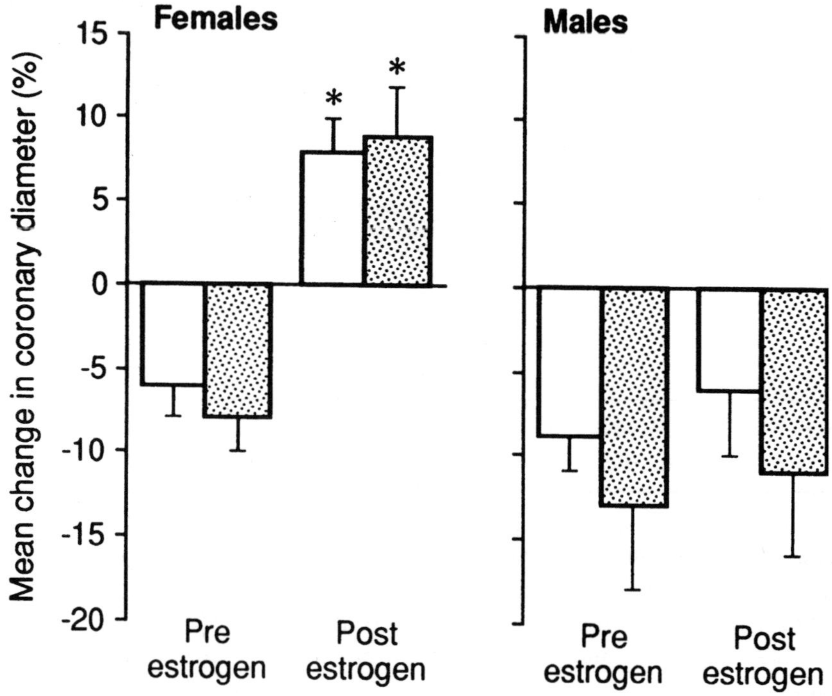


**Figure 6.8A.** Bar graphs showing change in the diameter of female coronary arteries (left) produced by an intracoronary infusion of acetylcholine (1.6 μmol/min, open bars; 16 μmol/min, stippled bars) before and after 2.5 μmol of intracoronary 17β-estradiol. Net constriction before estrogen was converted to dilation after estrogen (* indicates $p<0.01$ before vs after estrogen). Change in the diameter of male coronary arteries (right) produced by an intracoronary infusion of acetylcholine (1.6 μmol/min, open bars; 16 μmol/min, stippled bars) before and after 2.5 μmol of intracoronary 17β-estradiol ($p=$NS before vs after estrogen). Reproduced from *Circulation* 1995, **92**: 24–30 by copyright permission.

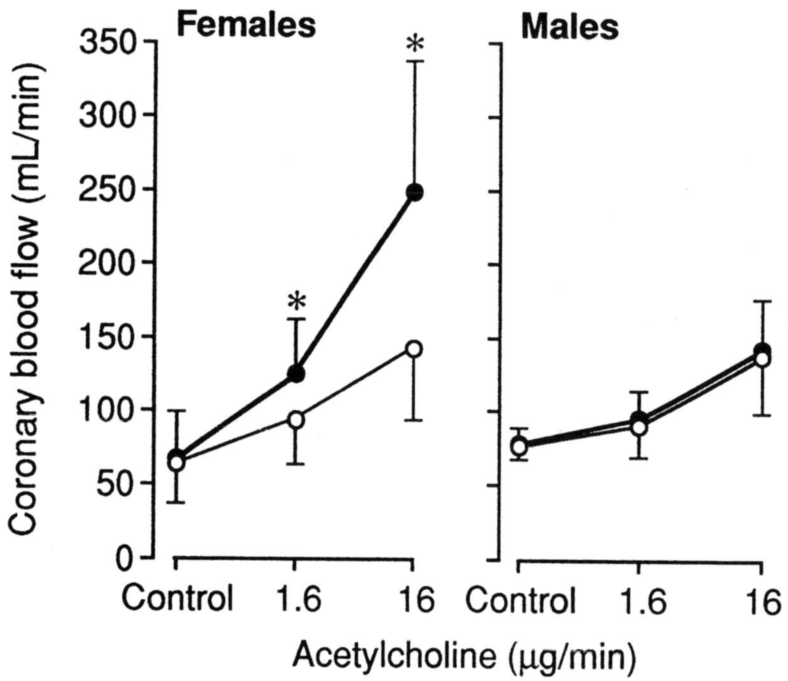


**Figure 6.8B.** Graphs showing increase in female coronary artery blood flow evoked by acetylcholine (left) before (open dot) and after (solid dot) estrogen. The dose-dependent increases in coronary blood flow induced by acetylcholine were significantly greater after estrogen (* indicates $p < 0.009$ before vs after estrogen, Wilcoxon test). Increase in male coronary blood flow evoked by acetylcholine (right) before (open dot) and after (closed dot) estrogen. The dose-dependent increases in coronary blood flow induced by acetylcholine were the same before and after estrogen ($p = \text{NS}$, Wilcoxon test). Reproduced from *Circulation* 1995, **92**: 24–30 by copyright permission.

estrogen can modulate acetylcholine-induced coronary vascular responses in females with CAD but not in men (Figures 6.8A and 6.8B). They propose that the augmentation of endothelium-dependent relaxation may be gender-specific. Improved receptor-mediated endothelium-dependent relaxation after short term $E_2$ administration has also been reported previously. Reis *et al.* (1994), have shown that 15 min post $E_2$ administration in postmenopausal women there is a decrease in vasomotor tone and attenuation of abnormal vasomotor response to acetylcholine. Preliminary data from our laboratory demonstrates that $E_2$ effectively augments basal NO release in an ER-dependent fashion. Therefore, data from several models suggest that $E_2$ can modulate eNOS activity and NO release but the exact mechanism of regulation has remained unresolved.

## SUMMARY

The effects of estrogen in the vasculature are quite complex and is regulated at various cellular levels. It appears that estrogen exerts its vascular protective effect, in part, by a direct action on the vessel wall. The exact cellular mechanisms underlying the effects of $E_2$ on the vasculature are still unclear and are being actively investigated. It is likely that the hormone mediates its protective effects through genomic as well as nongenomic mechanisms. Investigations into the cellular mechanisms involved in the hormone effect, the role of the estrogen receptor and/or other receptors in the induction of proteins involved in the hormone effect are needed to better define the action of $E_2$ on endothelial function, NO release and eNOS activation, and how the normal mechanisms are altered under states of vascular disease. Finally, it is also important to consider the effects of other hormones on $E_2$ action, such as progesterone, which is known to antagonize estrogen effects. Studies *in vivo* and *in vitro* of the combination of these two hormones as well as other hormones will be important for our understanding of the mechanisms of estrogen's action in the vasculature.

## REFERENCES

Adams, M.R., Clarkson, T.B., Kaplan, J.R. and Kortnik, D.R. (1990) Ovarian secretions and atherosclerosis. In: *Ovarian secretions and cardiovascular and neurological function*, edited by F. Naftolin, J.N. Gutmann, A.H. DeCherney, P.M. Sarrel, pp. 151–159. New York: Raven Press.

Aronica, S.M. and Katzenellenbogen, B.S. (1993) Stimulation of estrogen receptor-mediated transcription and alteration in the phosphorylation state of the rat uterine estrogen receptor by estrogen, cyclic adenosine monophosphate, and insulin-like growth factor-1. *Mol. Enocrinol.*, **7**, 743–752.

Awolesi, M.A., Widman, M.D., Sessa, W. and Sumpio, B.E. (1994) Cyclic strain increases endothelial nitric oxide synthase activity. *Surgery*, **116**, 439–445.

Bath, P.M., Hassal, W.D.G., Gladwin, A.M., Palmer, R.M.J. and Martin, J.F. (1991) Nitric oxide and prostacyclin: divergence of inhibitory effects on monocyte chemotaxis and adhesion to endothelium *in vitro*. *Arterioscler Thromb*, **11**, 254–260.

Caulin-Glaser, T., Sessa, W.C., Sarrel, P. and Bender, J.R. (1994) The effect of 17$\beta$-estradiol on human endothelial cell nitric oxide production. *Circulation*, **90**, 1–30.

Caulin-Glaser, T., Watson, C.A., Pardi, R. and Bender, J.R. (1996) Effects of 17$\beta$-estradiol on cytokine-induced endothelial cell adhesion molecule expression. *J. Clin. Invest.*, **98**, 36–42.

Cid, M.C., Kleinman, D.S., Grant, H.W., Schnaper, A.S., Fauci, A.S. and Hoffman, G.S. (1994) Estradiol enhances leukocyte binding to tumor necrosis factor (TNF) stimulated endothelial cells via an increase in TNF-induced adhesion molecules E-selectin, intercellular adhesion type 1, and vascular cell adhesion molecule type 1. *J. Clin. Invest.*, **93**, 17–25.

Collins, P., Rosano, G.M.C., Sarrel, P.M., Ulrich, L., Adamopoulos, S., Beale, C.M., McNeill, J.G. and Poole-Wilson, P.A. (1995) 17$\beta$-estradiol attenuates acetylcholine-induced coronary arterial constriction in women but not men with coronary heart disease. *Circulation*. **92**, 24–30.

Cotran, R. and Pober, J. (1990) Cytokine-endothelial interactions in inflammation, immunity and vascular injury. *J. Am. Soc. Nephrol.*, **1**, 225–235.

Cronstein, B., Kimmel, S., Levin, R., Martiniuk, F. and Weissmann, G. (1992) A mechanism for the antiinflammatory effects of corticosteroids: the glucocorticoid receptor regulates leukocyte adhesion to endothelial cells and expression of endothelial-leukocyte adhesion molecule 1 and intercellular adhesion molecule 1. *Proc. Natl. Acad. Sci. USA*, **89**, 9991–9995.

Degitz, K., Lian-Jie, L. and Caughman, S.W. (1991) Cloning and characterization of the 5' transcriptional regulatory region of the human intercellular adhesion molecule 1 gene. *J. Biol. Chem.*, **266**, 14024–14030.

García-Cardeña, G., Oh, P., Liu, J., Schnitzer, J.E. and Sessa, W.C. (1996) Targeting of nitric oxide synthase to endothelial cell caveolae via palmitoylation: Implications for nitric oxide signaling. *Proc. Natl. Acad. Sci. USA*, **93**, 6448–6453.

DeLuca, L., Johnson, D., Whitley, M., Collins, T. and Pober, J. (1994) cAMP and tumor necrosis factor competitively regulate transcriptional activation through and nuclear factor binding to the cAMP-responsive element/activating transcription factor element of the endothelial leukocyte adhesion molecule-1 (E-selectin) promoter. *J. Biol. Chem.*, **269**, 19193–19196.

Garg, U.C. and Hasid, A. (1989) Nitric oxide-generating vasodilators and 8-bromo-cyclic guanosine monophosphate inhibit mitogenesis and proliferation of cultured rat vascular smooth muscle cells. *J. Clin. Invest.*, **83**, 1774–1777.

Gilligan, D.M., Badar, D.M., Panza, J.A., Quyyumi, A.A. and Cannon, R.O. (1994) Acute vascular effects of estrogen in postmenopausal women. *Circulation*, **90**, 786–791.

Gossland, I.F., Wynn, V., Crook, D. and Miller, N.E. (1987) Sex, plasma lipoproteins and atherosclerosis: prevailing assumptions and outstanding questions. *Am. Heart. J.*, **114**, 1467–1503.

Haliwell, B. and Grootveld, M. (1987) The measurement of free radical reactions in humans. Some thoughts for future experimentation. *FEBS letters*, **213**, 9–16.

Hirata, K., Kuroda, R., Sakoda, T., Katayama M., Inoue, N., Suematsu, M., Kawashima, S. and Yokoyama, M. (1995) Inhibition of endothelial nitric oxide synthase activity by protein kinase C. *Hypertension*, **25**, 180–185.

Hou, J., Baichwal, V. and Cao, Z. (1994) Regulatory elements and transcription factors controlling basal and cytokine-induced expression of the gene encoding intercellular adhesion molecule 1. *Proc. Natl. Acad. Sci. USA*, **91**, 11641–11645.

Iadermarco, M., McQuillan, J., Rosen, G. and Dean, D. (1992) Characterization of the promoter for vascular cell adhesion molecule-1 (VCAM-1). *J. Biol. Chem.*, **267**, 16323–16329.

Ignarro, L.J., Buga, G.M., Wood, K.S., Byrns, R.E. and Chaudhuri, G.(1987) Endothelium-derived relaxing factor produced and released from artery and vein nitric oxide. *Proc. Natl. Acad. Sci. USA*, **84**, 9265–9269.

Ignar-Trowbridge, D.M., Teng, C.T., Ross, K.A., Parker, M.G., Korach, K.S. and McLachlan, J.A. (1993) Peptide growth factors elicit estrogen receptor-dependent transcriptional activation of an estrogen response element. *Mol. Endocrinol.*, **7**, 992–998.

Karas, R.H., Patterso, B.I. and Mendelsohn, M. (1994) Human vascular smooth muscle cells contain functional estrogen receptor. *Circulation*, **89**, 1943–1950.

Kaszubska, W., van Huijsduijnen, R.H., Ghers, P., DeRaemy-Schenk, A.M., Chen, B.P., Hai, J.F., DeLamarter, J.F. and Whelan, J. (1993) Cyclic AMP-independent ATF family members interact with NF-kB and function in the activation of the E-selectin promoter in response to cytokines. *Mol. Cell Biol.*, **13**, 7180–7190.

Keaney, J.F., Jr, Shwaery, G.T., Xu, A., Nicolosi, R.J., Loscalzo, J., Foxall, T.L. and Vita, J.A. (1994) 17 beta-estradiol preserves endothelial vasodilator function and limits low density lipoprotein oxidation in hypercholesterolemic swine. *Circulation*, **89**, 2251–2259.

Krishnan, V., Wanmg, X. and Safe, S. (1994) Estrogen receptor-Spl complexes mediate estrogen-induced cathepsin D gene expression in MCF-7 human breast tissue. *J. Biol. Chem.*, **269**, 15912–15917.

Kubes, P., Suzuki, M. and Granger, D.N. (1991) Nitric oxide: an endogenous modulator of leukocyte adhesion. *Proc. Natl. Acad. Sci. USA*, **88**, 4651–4655.

Kumar, V. and Chambon, P. (1988) The estrogen receptor binds tightly to its responsive element as a ligand-induced homodimer. *Cell*, **55**, 145 156.

Malinow, M.R., Moguilevsky, J.A., Lema, B. and Bur, G.E. (1963) Vascular and extravascular radioactivity after injection of estradiol-6.7 $H^3$ in the human being. *J. Clin. Endorinol. Metab.*, **23**, 306–310.

Marsden, P.A., Heng, H.H.Q., Scherer, S.W., Stewart, R.J., Hal, A.V., Shi, X.M., Tsui, L. and Schappert, K.T. (1993) Structure and chromasomal localization of the human constitutive endothelial nitric oxide synthase. *J. Biol. Chem.*, **268**, 17478–17488.

Michel, T., Li, G.K. and Busconi, L. (1993) Phosphorylation and subcellular translocation of endothelial nitric oxide synthase. *Proc. Natl. Acad. Sci. USA*, **90**, 6252–6257.

Miller, V.M. and Vanhoutte, P.M. (1988) Enhanced release of endothelium-derived relaxing factor by chronic increases in blood flow. *Am. J. Physiol.*, H446–H451.

Montgomery, K.F., Osborn, L., Hession, C., Tizard, R., Goff, D., Vassallo, C., Tarr, P.I., Bomsztyk, K., Lobb, R., Harlan, J.M. and Pohlman, T.H. (1991) Activation of endothelial-leukocyte adhesion molecule 1 (ELAM-1) gene transcription. *Proc. Natl. Acad. Sci. USA*, **88**, 6523–6527.

Mooradian, A.D. (1993) Antioxidant properties of steroids. *J. Steroid. Biochem. Mol. Biol.*, **45**, 509–511.

Nakai, K., Itoh, C., Hotta, K., Itoh, T., Yoshizumi, M. and Hiramori, K. (1994) Estradiol-17B regulates the induction of VCAM-1 mRNA expression by interleukin-1B in human umbilical vein endothelial cells. *Life Science*, **54**, 221–227.

Negre-Salvayre, A., Pierassi, M.T., Mobile, L. and Salvayre, R. (1993) Protective effects of 17 B-estradiol against the cytotoxicity if minimally oxidized LDL to cultured bovine aortic endothelial cells. *Atherosclerosis*, **99**, 207–217.

Neish, A.S., Williams, A.J., Palmer, H.J., Whitley, M.Z. and Collins, T. (1992) Functional analysis of the human vascular cell adhesion molecule 1 promoter. *J. Exp. Med.*, **176**, 1583–1593.

Nishida, K., Harrison, D.G., Navas, J.P., Fisher, A.A., Dockery, S.P., Uematsu, M., Nerem, R.M., Alexander, R.W. and Murphy, T.J. (1992) Molecular cloning and characterization of the constitutive bovine aortic endothelial cell nitric oxide synthase. *J. Clin. Invest.*, **90**, 2092–2096.

Palmer, R.M.J., Ferrige, A.G. and Moncada, S. (1987) Nitric oxide release accounts for the biological activity of endothelium-derived relaxing factor. *Nature*, **327**, 524–526.

Power, R.F., Mani, S.K., Codina, J., Conneely, O.M. and O'Malley, B. (1991) Dopaminergic and ligand-independent activation of steroid hormone receptors. *Science*, **254**, 1636–1639.

Ohara, Y., Sayegh, H.S., Yamin, J.J. and Harrison, D.G. (1995) Regulation of endothelial constitutive nitric oxide synthase by protein kinase C. *Hypertension*, **25**, 415–420.

Radomski, M.W., Palmer, R.M. and Moncada, S. (1990) An L-arginine/nitric oxide pathway present in human platelets regulates aggregation. *Proc. Natl. Acad. Sci. USA*, **87**, 5193–5197.

Radomski, M.W., Palmer, R.M.J., Moncada, S. (1987) Endogenous nitric oxide inhibits human platelet adhesion to vascular endothelium. *Lancet*, **2**, 1057–1058.

Ray, A., Prefontaine, K.E. and Ray, P. (1994) Down-modulation of interleukin-6 gene expression by 17B-estradiol by the estrogen receptor. *J. Biol. Chem.*, **269**, 12940–12946.

Read, M.A., Whitley, A.J., Williams, A.J. and Collins, T. (1994) NF-kB and IkBα: an inducible regulsatory system in endothelial activation. *J. Exp. Med.*, **179**, 503–512.

Rees, D.D., Palmer, R.M.J., Hodson, H.F. and Moncada, S. (1990) Characterization of three inhibitors of endothelial nitric oxide synthase *in vitro* and *in vivo*. *Br. J. Pharmacol.*, **101**, 746–752.

Rosano, G.M., Sarrel, P.M., Poole-Wilson, P.A. and Collins, P. (1993) Beneficial effect of oestrogen on exercise-induced myocardial ischaemia in women with coronary artery disease. *Lancet*, **342**, 133–136.

Rosenkranz-Weiss, P., Sessa, W.C., Milstien, S., Kaufman, S., Watson, C.A. and Pober, J.S. (1993) Regulation of nitric oxide synthesis by proinflammatory cytokines in human umbilical vein endothelial cells. *J. Clin. Invest.*, **93**, 2236–2243.

Ross, R. (1986) Medical Progress: The pathogenesis of atherosclerosis—an update. *NEJM*, **314**, 488–500.

Sarrel, P.M. (1988) Effects of ovarian hormones on the cardiovascular system. In *Circulation in the Female*. Edited by J. Ginsburg, pp. 117–141 Carsforth, United Kingdom: Parthenon Press.

Sessa, W.C.(1994) The nitric oxide synthase family of proteins. *J. Vasc. Res.*, **31**, 131–143.

Sessa, W.C., García-Cardeña, G., Liu, J., Keh, A., Pollock, J.S., Bradley, J., Thiru, S., Braverman, I.M. and Desa, K.M. (1995) The golgi association of endothelial nitric oxide synthase is necessary for the efficient synthesis of nitric oxide. *J. Biol. Chem.*, **270**, 17641–17644.

Sessa, W.C., Pritchard, K., Seyedi, N., Wang J. and Hintze, T.H. (1994) Chronic exercise in dogs increases coronary vascular nitric oxide production and endothelial cell nitric oxide synthase gene expression. *Circ. Res.*, **74**, 349–353.

Schule, R. and Evans, R.M. (1991) Cross-coupling of signal transduction pathways: Zinc finger meets leucine zipper. *Trends Genet*, **7**, 377–382.

Shyamala, G. and Guiot, M.C. (1992) Activation of kB-specific proteins by estradiol. *Proc. Natl. Acad. Sci. USA*, **89**, 10628–10632.

van der Wal, A., Das, P. and Tigges, A. (1992) Adhesion molecules on the endothelium and mononuclear cells in human atherosclerotic lesions. *Am. J. Pathol.*, **141**, 1427–1433.

von der Leyen, H.E., Gibbons, G.H., Lewis, N.P., Zhang, L., Nakajima, M., Kaneda, Y., Cooke, J.P. and Dza, V.J. (1995) Gene transfer inhibiting neointimal vascular lesion: *In Vivo* transfer of endothelial cell nitric oxide synthase gene. *Proc. Natl. Acad. Sci. USA*, **92**, 1137–1141.

Wakasugi, M., Noguchi, T., Kazama, Y.I., Kanemaru, Y. and Onaya, T. (1989) The effects of sex hormones on the synthesis of prostacyclin (PGI 2) by vascular tissue. *Prostaglandins*, **37**, 401–410.

Wenger, N.K. (1985) Coronary disease in women. *Annu. Rev. Med.*, **36**, 285–294.

Weiner, C., Martinez, E., Zhu, L.K., Ghodsi, A. and Chestnut, D. (1989) *In vitro* release of endothelium derived relaxing factor by acetylcholine is increased during the guinea pig pregnancy. *Am. J. Obstet. Gynecol.*, **161**, 1599–1605.

Weiner, C., Liu, K.Z., Thompson, L., Herrig, J. and Chestnut, D. (1991) Effect of pregnancy on endothelium and smooth muscle: Their role in reduced adrenergic sensitivity. *Am. J. Physiol.*, **261**, H1275–H1283.

Weisz, A. and Rosales, R. (1990) Identification of an estrogen response element upstream of the human c-fos gene that binds the estrogen receptor and AP-1 transcription factor. *Nucl. Acids Res.*, **18**, 5097–5106.

Williams, J.K., Adams, M.R. and Klopfenstein, H.S. (1990) Estrogen modulates responses of atherosclerotic coronary arteries. *Circulation*, **81**, 1680–1687.

Wolinsky, H. (1992) Effects of estrogen and progesterone treatment on the response of the aorta of male rats to hypertension:morphological and chemical studies. *Circ. Res.*, **30**, 341–349.

# 7 Non-Genomic Actions of Estrogens: Estrogens as Antioxidants *

**Carole L. Banka**

*The Scripps Research Institute, Department of Immunology,*
*10550 North Torrey Pines Road, La Jolla, CA 92037,*
*Tel.: 001 619 784 8235, Fax: 001 619 784 8480, E-mail: cbanka@scripps.edu.*
* *Grant support: NIH HL-50060 and NIH HL-55517.*

Recent research in humans and animal models suggests that estrogens protect women from heart disease through both receptor-mediated and nonreceptor-mediated mechanisms. The recently demonstrated efficacy of classical antioxidants such as vitamin E in reducing cardiovascular risk suggests that the nonreceptor-mediated mechanisms of estrogen protection may include their actions as antioxidants. Oxidative modification of low density lipoprotein (LDL), a process that contributes to the initiation and progression of atherosclerotic lesion formation, is inhibited by estrogens *in vitro*, but only at supraphysiologic concentrations. Paradoxically, estrogens appear to protect LDL *in vivo* at physiologic concentrations including those achieved during hormone replacement therapy (HRT). Several potential explanations for this discrepancy are proposed and discussed here. In addition, evidence is presented supporting the hypothesis that estrogens act preferentially as antioxidants for high density lipoprotein (HDL). This inhibition of HDL oxidative modification may enhance the ability of HDL to protect LDL from oxidation, a process considered to contribute to the inverse relationship between plasma HDL levels and cardiovascular risk. Data is also presented indicating that the antioxidant potential of specific estrogens for lipoproteins varies according to differences in molecular structure. Finally, the concept that estrogens may act as membrane antioxidants inhibiting oxidant-mediated cellular damage is reviewed.

KEYWORDS: Estrogen, antioxidant, lipoproteins, lipoprotein oxidation, hormone replacement therapy.

## INTRODUCTION

Premenopausal women have a lower risk of cardiovascular disease than age-matched males. Postmenopausal women exhibit a significant increase in risk, in part, due to aging and, in part, due to the loss of circulating ovarian steroid hormones (for review, see Barrett-Connor, 1997). Increasing evidence from clinical studies, including prospective trials, suggests that exogenous estrogen administered in hormone replacement regimens to postmenopausal women reduces both the risk factors for and the actual risk of cardiovascular disease (The Postmenopausal Estrogen/ Progestin Interventions Trial Writing Group, 1995; Grodstein *et al.*, 1996). The list of plausible mechanisms for the cardioprotective effects of estrogen has grown over the last decade. Favorable changes in lipoprotein levels, a decrease in LDL cholesterol and an increase in HDL cholesterol, are well-documented effects of unopposed estrogens and the increase in HDL cholesterol has been postulated to explain approximately half of the protection seen with estrogen administration (Bush *et al.*, 1987; Gruchow *et al.*, 1988). However, results from the Nurses Health Study and from extensive studies in primates have revealed cardioprotection in the absence of significant changes in plasma lipoproteins suggesting other important mechanisms (Grodstein *et al.*, 1996; Adams *et al.*, 1990).

Estrogens are thought to mediate a variety of physiologic processes through the classical interaction of a steroid/intracellular receptor complex with an estrogen response element (ERE) in the promoter region of responsive genes (Evans, 1988; Katzenellenbogen, 1996). However, there are a variety of estrogen responses in the vascular wall that cannot be explained by this classical pathway and may depend upon putative cell surface receptors, membrane effects or other non-genomic mechanisms (for review, see Farhat *et al.*, 1996). One such mechanism is the potential of estrogens to act as potent antioxidants, the subject of this chapter.

## ANTIOXIDANTS AND ATHEROSCLEROSIS

Interest and controversy concerning antioxidants as potential antiatherogenic therapeutic agents has grown considerably in the past decade primarily due to promising results from animal studies (Elinder and Walldius, 1994). Of the antioxidants examined, those shown to have atheroprotective effects are all lipophilic antioxidants with phenolic groups that endow the molecules with chain-breaking antioxidant potential. Probucol is a potent antioxidant with an additional capacity to lower cholesterol levels. Probucol and probucol analogues have been shown to inhibit atherosclerosis in rabbits (Carew *et al.*, 1987; Mao *et al.*, 1991; Baumann *et al.*, 1994) and nonhuman primates (Sasahara *et al.*, 1994). Two rabbit studies documented inhibition of atherosclerosis in the absence of changes in serum cholesterol levels suggesting that the atheroprotection resulted from the antioxidant component of the treatments (Carew *et al.*, 1987; Mao *et al.*, 1991). Similar effects have been seen in animal models with butylated hydroxytoluene (BHT) and *N,N'*-diphenyl 1,4-phenylenediamine (DPPD) (Bjorkham *et al.*, 1991; Sparrow *et al.*, 1992; Tangirala *et al.*, 1995). Results from treatments of animal atherosclerosis models with α-tocopherol have been mixed. Both inhibition of atherogenesis and regression of lesions have been documented in primates and hamsters (Verlangieri and Bush, 1992; Parker *et al.*, 1995) but not in rabbits (Kleinveld *et al.*, 1995; Fruebis *et al.*, 1995). These conflicting results may result from species differences or may reflect the requirement for threshold levels of antioxidants as suggested by Fruebis *et al.* (1995). Moreover, α-tocopherol treatment of rabbits reduced restenosis in femoral arteries after angioplasty suggesting that α-tocopherol inhibited lipoprotein oxidation or the proliferative effects of the oxidized lipoproteins. α-Tocopherol administration to human subjects resulted in protection of LDL from oxidative modification suggesting potential therapeutic value (Reaven *et al.*, 1993). The most dramatic effects of vitamin E on cardiovascular risk have come from two recent clinical studies. In a randomized 32-month prospective study of patients with documented cardiovascular disease, the Cambridge Heart Antioxidant Study (CHAOS), administration of vitamin E resulted in a 47% decrease in myocardial infarction (fatal and nonfatal combined) (Stephens *et al.*, 1996). In a seven-year study of cardiovascular disease-free postmenopausal women, mortality from heart disease was lower among women who consumed relatively high levels of vitamin E in their diets although consumption of vitamin E supplements was not associated with decreased risk (Kushi *et al.*, 1996). Taken together, these studies offer substantial evidence that antioxidants play a role in inhibiting atherogenic arterial lesion progression. Therefore, it is compelling to examine the antioxidant potential of estrogens as one non-genomic mechanism underlying cardiovascular protection by these steroid hormones.

## ESTROGENS AND *IN VITRO* LOW DENSITY LIPOPROTEIN OXIDATION

Oxidative modification of low density lipoprotein (LDL) is considered to be a seminal process in the initiation and progression of arterial atherosclerotic lesion formation (for review, see Steinberg, 1997). It is currently thought that oxidation of LDL *in vivo* is a cell-mediated event dependent on free radicals released by cells, i.e. superoxide anion' or generated intracellularly, for example by lipoxygenases' resulting in the transfer of lipoperoxides from cells to LDL. Similar oxidative modifications can be achieved *in vitro* by exposing LDL to iron or copper or to cultured cells (see Steinberg *et al.*, 1989). Oxidative modification of LDL is initiated by lipid peroxidation, a process of decomposition of lipid peroxides (LOOH) to peroxyl ($LO_2^{\cdot}$) and alkoxyl ($LO^{\cdot}$) radicals. A secondary effect of lipid peroxidation is the derivatization of lysine residues of apoprotein (apo) B, the major protein component of LDL, by the products of fatty acid degradation. The modified apo B appears to be recognized by several receptors for oxidized LDL (oxLDL) on macrophages (see Steinberg, 1997) and a recently described receptor restricted to endothelial cells (Sawamura *et al.*, 1997). In addition to mediating macrophage foam cell formation, a critical early step in lesion formation, oxLDL has been shown to inhibit macrophage migration and to induce synthesis and secretion of a variety of factors which may contribute to atherogenesis, including a macrophage chemotactic factor (for review, see Steinberg *et al.*, 1989). Products of LDL oxidation have been identified immunohistologically in atherosclerotic lesions from humans and LDL receptor-deficient rabbits (Yla-Herttuala *et al.*, 1989; Rosenfeld *et al.*, 1990). Finally, circulating autoantibodies to oxLDL have been identified in humans and experimental animals (Salonen *et al.*, 1992; Palinski *et al.*, 1995). Much of the cardioprotective effect of antioxidants in studies described above has been attributed to the prevention or inhibition of LDL oxidation. Therefore, it has been of interest to examine estrogens as potential antioxidants involved in LDL oxidation.

Prior to studies of lipoprotein oxidation, estrogens (estradiol, estriol and estrone) were shown to protect phospholipid from UV-radiation-mediated peroxidation and to inhibit lipid peroxidation in rat liver microsomes (Yagi and Komura, 1986). Subsequently, several studies documented that estrogens inhibit lipid peroxidation in

LDL at pharmacologic doses *in vitro*, and that this inhibition extends to oxidation of LDL by metal ions (copper), by free radical generators and by exposure of LDL to cultured cells (Huber *et al*., 1990; Maziere *et al*., 1991; Rifici and Khachadurian, 1992; Subbiah *et al*., 1993; Ayres *et al*., 1996). Inhibition of LDL oxidation by 17$\beta$-estradiol was comparable to inhibition by the lipophilic antioxidants, butylated hydroxytoluene (BHT) and probucol (Huber *et al*., 1990; Rifici and Khachadurian, 1992). Estrogen metabolites, specifically catecholestrogens, as well as the synthetic estrogen, 17$\alpha$-ethinylestradiol, and the equine estrogens are also potent inhibitors of LDL oxidation *in vitro* (Subbiah *et al*., 1993; Taniguchi *et al*., 1994; Parthasarathy *et al*., 1994). The antiestrogens tamoxifen and hydroxytamoxifen inhibit LDL oxidation (Wiseman *et al*., 1993). In protecting LDL from oxidation, 17$\beta$-estradiol also inhibits consequences of LDL oxidation, including loading of potential macrophage foam cells with cholesteryl esters (Huber *et al*., 1990) and cytotoxic effects on endothelial cells (Negre-Salvayre *et al*., 1993). Estrogenic compounds protect fatty acids, cholesterol and apoprotein B in LDL from oxidative modification (Rifici and Khachadurian, 1992; Wiseman *et al*., 1993; Schütze *et al*., 1993; Ayres *et al*., 1996).

The rank order of effectiveness of estrogens as antioxidants has been studied in several pure systems. In a system of phycoerythrin oxidation by a free radical generator, estrogen inhibition was demonstrated with estriol > 17$\beta$-estradiol > 17$\alpha$-estradiol, whereas estrone was ineffective (Mooradian, 1993). When compared with $\alpha$-tocopherol in reactions with a substituted phenoxyl radical in benzene, the rates of reaction were 2-hydroxyestrone > $\alpha$-tocopherol > 2-methoxyestrone > 17$\beta$-estradiol > estrone (Mukai *et al*., 1990). In the more complicated system of lipoprotein oxidation, rank order of effectiveness of estrogen compounds appears to parallel those seen in the pure systems. In summary, studies of lipoprotein oxidation *in vitro* suggest that 17$\beta$-estradiol is more effective than estriol; the equine estrogens, equilin and equilenin are more potent than 17$\beta$-estradiol; the catecholestrogens including 2-hydroxyestradiol, 2-hydroxyestrone, 2-hydroxyestriol and 4-hydroxyestrone are the most potent antioxidants (Rifici and Khachadurian, 1992; Subbiah *et al*., 1993; Taniguchi *et al*., 1994; Cirino *et al*., 1994; Tang *et al*., 1996). The potential role of estrone, the major circulating estrogen in postmenopausal women, as a lipoprotein antioxidant is not clear. Estrone has been shown to inhibit copper-mediated LDL oxidation (less effectively than 17$\beta$-estradiol) in some studies (Subbiah *et al*., 1993; Taniguchi *et al*., 1994) but not others (Rifici and Khachadurian, 1992; Tang *et al*., 1996). It is clear from all studies published to date that estrogens are the only steroids that serve as lipoprotein antioxidants; this action does not extend to androgens or progestins (Rifici and Khachadurian, 1992; Mooradian, 1993; Taniguchi *et al*., 1994; Banka and Curtiss, 1996).

The structure of estrogens suggests they may act as chelators and this is supported by the observed protection of LDL from copper-mediated oxidation. However, none of the studies published to date clearly establish chelation as the mechanism of estrogen antioxidant action. On the other hand, several lines of evidence suggest that estrogens act as chain-breaking antioxidants or free radical scavengers. The hydroxyphenolic A ring of 17$\beta$-estradiol and most estrogens can readily give up a hydrogen atom. Secondly, estrogens inhibit oxidation mediated by free radical generators such as 2,2'-azobis (2-amidinopropane)dihydrochloride (AAPH)

(Mooradian, 1993; Lacort *et al.*, 1995; Ayres *et al.*, 1996). Finally, chain-breaking antioxidants usually introduce or extend a lag phase in lipid peroxidation reactions (Halliwell, 1990), a characteristic of estrogens as antioxidants. Tamoxifen, which does not have a donatable hydrogen atom, may act through a different mechanism.

## ESTROGENS AND *IN VIVO* LOW DENSITY LIPOPROTEIN OXIDATION

As described above, estrogens protect LDL from oxidation *in vitro* only at pharmacologic concentrations. Therefore, little emphasis was placed on the antioxidant potential of estrogens as a mechanism of protection from atherosclerosis based on *in vitro* observations. However, recent investigations suggest that the antioxidant mechanism is operational *in vivo*. The possibility of estrogens protecting LDL *in vivo* is complicated by the interactions of estrogens with albumin and sex hormone binding globulin (reviewed by Hammond, 1990). Yet, Sack and his coworkers have seen an inhibition of oxidation of LDL *ex vivo* following the treatment of postmenopausal women with estrogens administered by transdermal patch for three weeks or by bolus infusion which resulted in mean plasma estradiol levels of 463 pmol/L and 1.6 nmol/L, respectively (Sack *et al.*, 1994). The antioxidant effect of the 17$\beta$-estradiol patch was reproduced in a second study (Guetta *et al.*, 1995a). However, Nenseter *et al.* (1996) found that a 12-week treatment of hypercholesterolemic postmenopausal women with hormone replacement therapy consisting of 2 mg oral 17$\beta$-estradiol sequentially combined with 1 mg norethisterone acetate(NETA) had no effect on susceptibility of LDL to oxidation. These discrepancies may result from differences in baseline plasma cholesterol levels (higher in the latter study) or differences in the route of estrogen administration. Alternatively, it is possible that the addition of an androgenic progestin (NETA) to the treatment regimen reverses the antioxidant effects of 17$\beta$-estradiol. 17$\beta$-Estradiol administered by implant to ovariectomized hypercholesterolemic swine rendered the LDL significantly less susceptible to *ex vivo* oxidation compared with LDL from the placebo control swine (Keaney *et al.*, 1994). Wander and coworkers (1996) documented a small but significant decrease in *ex vivo* LDL oxidation following oral hormone replacement therapy with 0.625 conjugated equine estrogens (Premarin) and 10 mg medroxyprogesterone. Tamoxifen administered in therapeutic doses to postmenopausal women increased the lag time of LDL oxidation by 20% in a study in which oral administration of 0.625 conjugated equine estrogens had no protective effect (Guetta *et al.*, 1995b). The most dramatic antioxidant effect of estrogens *in vivo* resulted from oral treatment of healthy postmenopausal women with individual constituents of conjugated equine estrogens (Premarin) (Wilcox *et al.*, 1997). Each of the three most prevalent components of Premarin, estrone sulfate, equilin sulfate and 17$\alpha$-dihydroequilin sulfate in doses comparable to those in 1.25 mg Premarin were administered separately to individual treatment populations in a prospective randomized cross-over study. Following a 30-day treatment period, the lag time of LDL oxidation was increased more than 2-fold in all treatment groups compared with that of LDL from the same subjects prior to treatment. Estrone sulfate was most effective followed by equilin sulfate and 17α-dihydroequilin sulfate (Wilcox *et al.*, 1997). Taken together, these studies suggest that one mechanism of the atheroprotective effects accompanying hormone replacement therapy is antioxidant protection of LDL.

How do we explain the dramatic differences in doses of estrogen required to inhibit LDL oxidation *in vitro* (μmol/L) and the effective plasma levels described for antioxidant effects *in vivo* (pmol–nmol/L)? Plausible explanations include metabolic alterations of the estrogens, estrogen-mediated changes in lipoprotein composition or indirect effects on other physiologic parameters. The following is a breakdown of four potential mechanisms underlying the antioxidant effects of estrogens *in vivo*.

(1) Estrogens may be metabolized to more potent antioxidant compounds. The results of *in vitro* studies summarized above implicate the catecholestrogens (compounds with an addition hydroxyl group on the A ring) as the most potent antioxidants. Yoshino and coworkers (1987) reported that intraperitoneal administration of 17β-estradiol to mice resulted in a decrease in serum and liver lipid peroxide levels that was much more dramatic in females than in males. However, administration of 2-hydroxyestradiol reduced lipid peroxides to the same extent in males and females. The authors concluded that the increased protection by 17β-estradiol in females resulted from hepatic metabolism of 17β-estradiol to the more potent catecholestrogen. Studies of lipid peroxidation in rat liver microsomes revealed that catecholestrogens were the most potent antioxidants and that antioxidant activity of other estrogens, including 17β-estradiol, estrone and estriol, increased markedly in the presence of intact microsomes and NADPH, conditions which favored metabolic conversion of the estrogens to catecholestrogens (Lacort *et al.*, 1995). Thus, metabolic conversion of estrogens to catecholestrogens is predicted to enhance antioxidant activity.

(2) Estrogen administration may alter the composition of lipoprotein particles rendering them less susceptible to deleterious oxidative modification. In addition to the well-established reduction in LDL cholesterol resulting from estrogen treatment, several clinical studies have documented an estrogen-dependent decrease in large, buoyant LDL (LDL1), an increase in small, dense LDL (LDL3) and a decrease in the cholesterol/triglyceride ratio in all LDL subclasses in postmenopausal women (Granfone *et al.*, 1992; Campos *et al.*, 1993; Tilly-Kiesi *et al.*, 1996). This pattern is more marked in women receiving oral estrogen than transdermal estrogen, is true in both normolipidemic and hypercholesterolemic women and is not blunted significantly by the addition of progestin to the hormone replacement regimen. Paradoxically, the small, dense LDL subfraction has been shown to be associated with increased risk of atherosclerosis (Austin *et al.*, 1988) and to be more susceptible to oxidative modification than larger, more buoyant LDL (Tribble *et al.*, 1992; De Graaf *et al.*, 1993). Thus, it remains to be established whether the small, dense LDL resulting from estrogen replacement therapy is similar to those particles seen in other subjects in susceptibility to oxidation. It is of interest that HRT alters the profile of fatty acids in plasma, another potential mechanism impacting on susceptibility of lipoproteins to oxidation (Wander *et al.*, 1996).

(3) Estrogen may induce or interact with other physiologic antioxidants. As described above, α-tocopherol is the antioxidant with greatest potential in reducing

risk of cardiovascular disease and protection of LDL from oxidation (Reaven *et al.*, 1993). $\alpha$-Tocopherol can be regenerated by other antioxidants such as ascorbate. Estrogens can also regenerate $\alpha$-tocopherol in a simple assay system by donating a hydrogen atom from the hydroxyl group on the aromatic A ring, and catecholestrogens are most effective in this regeneration (Mukai *et al.*, 1990). However, two clinical studies have failed to show synergistic effects of simultaneous administration of estrogen and $\alpha$-tocopherol to postmenopausal women although each compound was effective in inhibiting LDL oxidation when administered alone (Guetta *et al.*, 1995a; Wander *et al.*, 1996). These results suggest that, in a complex physiologic setting, estrogens are not exerting antioxidant effects by regenerating vitamin E. Whether estrogens induce synthesis of endogenous antioxidants or antioxidant enzymes remains to be determined .

(4) Estrogen antioxidant effects on LDL may be mediated indirectly through HDL. HDL is the most predictive risk factor of cardiovascular disease in women. The inverse correlation between HDL levels and heart disease has long been attributed to the role of HDL in reverse cholesterol transport, the process by which excess cholesterol in the periphery is transported to the liver for catabolism. Recently a new function has been identified for HDL, that of protecting LDL from oxidation (for review, see Navab *et al.*, 1996). What has received little attention is the possibility that in protecting LDL from oxidation, HDL may undergo oxidative damage that could impact its protective functions and accelerate its clearance from the circulation. If this is the case, any antioxidant that protects HDL would preserve its ability to protect LDL from oxidative damage. We now have evidence that estrogens preferentially protect HDL from lipid peroxidation and protein modification at physiologic concentrations *in vitro* (Banka and Curtiss, 1996; Banka, 1997). The differences between estrogen antioxidant effects on LDL *in vitro* and *in vivo* may be mediated indirectly by interactions of estrogen with HDL.

## ESTROGENS AND HIGH DENSITY LIPOPROTEIN OXIDATION

Substantial evidence now exists for the HDL antioxidant protection of LDL *in vitro* (for review, see Banka, 1996). The mechanism is thought to involve transfer of lipid peroxides from oxidized LDL to HDL and subsequent hydrolysis of the lipid peroxides by enzymes associated with HDL (see Banka, 1996). A single report has addressed the impact of HDL on oxidative parameters *in vivo*. Klimov and coworkers (1993) found a significant reduction in plasma conjugated dienes and trienes (transient rearrangements in unsaturated fatty acids that accompany lipid peroxidation) in hypercholesterolemic rabbits six hours after a bolus injection of human $HDL_3$. In the same report, the authors demonstrated a weak, but statistically significant, negative correlation between HDL cholesterol and the content of conjugated dienes in the plasma of healthy subjects and patients with heart disease. The possibility exists that HDL itself becomes oxidatively damaged in the process of protecting LDL. In fact, Parthasarathy and coworkers, who first demonstrated HDL protection of LDL, found that HDL was modified in the process (Parthasarathy *et al.*, 1990). Furthermore, using a non-exchangeable cholesteryl ester analogue,

Tribble and coworkers (1995) have documented oxidation of core lipids in HDL coincubated with LDL under oxidative conditions.

Oxidative modification impacts adversely on several of the protective functions of HDL including the promotion of cholesterol efflux from peripheral cells (Nagano *et al.*, 1991; Morel, 1994) and the activation of the plasma cholesterol esterifying enzyme, lecithin:cholesterol acyltransferase (LCAT) (McCall *et al.*, 1994). It is predicted that oxidative modification of HDL would also impact on its ability to protect LDL and that an antioxidant that protects HDL would indirectly protect LDL.

Prompted by the observations that selective interactions occur between 17$\beta$-estradiol and HDL, but not LDL (Leszczynski and Schafer, 1989; Leszczynski and Schafer, 1991), and that 17$\beta$-estradiol inhibits oxidation of LDL *in vitro* at high concentrations (see above), we compared the antioxidant effects of 17$\beta$-estradiol on HDL and LDL (Banka and Curtiss, 1996; Banka, 1997). When HDL and LDL isolated from a fasting male donor were normalized to phospholipid content and exposed to copper for four hours, we found the same degree of lipid peroxidation had occurred in the two lipoprotein fractions. Introduction of 17$\beta$-estradiol resulted in inhibition of LDL oxidation at μmol/L concentrations whereas HDL was protected in the presence of low nmol/L concentrations. Although the susceptibility to lipoprotein oxidation varied among donors, the selective effect of estradiol on HDL when compared with LDL from the same donor was a consistent finding. No protection of HDL or LDL from lipid peroxidation was seen with representative androgens or progestins. Furthermore, selective antioxidant protection of HDL was not seen with other antioxidants including vitamin C, Trolox (a water-soluble vitamin E analog), butylated hydroxytoluene (BHT) or probucol, that is, these antioxidants were equally effective at inhibiting lipid peroxidation in HDL and LDL. Only $\alpha$-tocopherol showed a similar preferential protection of HDL although the differences in concentrations required to protect HDL and LDL were less dramatic than the differences seen with estradiol. We have also shown that protection of HDL from lipid peroxidation by estradiol is accompanied by a protection of the protein component of HDL.

The potency of different estrogen compounds on HDL and LDL oxidation were compared using lipoproteins isolated from plasma of a fasting male donor (Figure 7.1). The lipoprotein fractions were normalized to phospholipid content (200 μg/ml) and dialyzed overnight against saline containing the estrogens at 2 μmol/L. The dialysis step served to introduce the estrogens and to remove aqueous antioxidants included during lipoprotein isolation. Following dialysis, the lipoproteins were oxidized by incubation for 4 hours at 37°C in the presence of 5 μmol/L. Oxidation reactions were stopped by addition of BHT (40 μmol/L) and chilling on ice. Lipid peroxidation was assessed by measuring the thiobarbituric acid reactive substances (TBARS) (Puhl *et al.*, 1994). This assay detects the aldehyde end products of fatty acid lipid peroxidation. TBARS values for LDL and HDL controls (oxidized in the absence of estrogen) were $30.94 \pm 0.51$ and $31.53 \pm 0.0$ respectively. Data in Figure 7.1 is expressed as percent of control and estrogens are shown in order of increasing effectiveness in inhibiting LDL oxidation. 17$\beta$-Estradiol (17$\beta$-E2) and 17$\alpha$-ethinylestradiol (ethinyl E2) were most effective in protecting LDL, inhibiting TBARS formation by more than 75%, followed by estrone. 17$\alpha$-Estradiol inhibited LDL oxidation by only 10%. In all cases the estrogens were dramatically

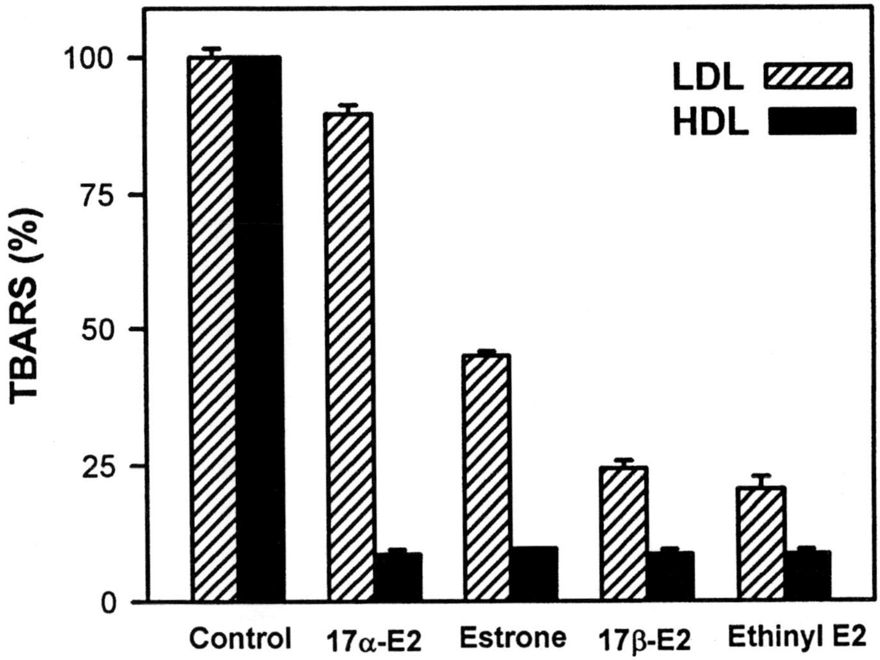


**Figure 7.1.** Impact of estrogen compounds on HDL and LDL lipid peroxidation. HDL and LDL were isolated from a fasting male donor by sequential ultracentrifugation. The lipoproteins (200 µg phospholipid/ml) were dialyzed for 16 h against 0.9% NaCl containing ethanol (0.1%, control) or estrogen (2 µmol/L) and then oxidized by exposure to copper acetate (5 µmol/L) for 4 h at 37°C. The reaction was stopped by addition of BHT (40 µmol/L). TBARS were measured against a standard curve of tetramethoxypropane (TMP). Bars represent the mean ± S.D. for each sample.

more effective at inhibiting lipid peroxidation in HDL. All compounds inhibited HDL oxidation by more than 90%. Previous experiments have demonstrated 17$\beta$-estradiol protection of HDL at concentrations as low as 25 nmol/L (Banka and Curtiss, 1996). Therefore, further titration experiments will be required to establish the rank order of effectiveness of these estrogens in inhibiting HDL oxidation.

These results indicate that the selective protection of HDL is a characteristic of the class of estrogenic steroids. Furthermore, the data illustrating protection of LDL suggest that structural alterations at the 3-position and the 17-position in estradiol can marked by effect antioxidant potential. As mentioned above, estrone has been effective in inhibiting LDL oxidation in some studies but not others. Estrone was effective in this experiment although less so than 17$\beta$-estradiol. Conjugation of the ethinyl group at the 17-position of the molecule had no effect on potency, yet, surprisingly, the $\alpha$-configuration of the hydroxyl group at the 17-position markedly reduced antioxidant potential for LDL. As described above, others have shown catecholestrogens to be dramatically more effective than estradiol as antioxidants. It is predicted, therefore, that the estradiol molecule can be structurally modified both to enhance the antioxidant potential and reduce or eliminate receptor binding capacity creating potentially powerful therapeutic agents.

Two other laboratories have examined the estrogen antioxidant effects on HDL. Taniguchi *et al.* (1994) found catecholestrogens to be more potent antioxidants than other estrogens in protecting LDL and HDL. Although their experiments were not designed to compare effects on HDL with those on LDL, their data reflects preferential protection of HDL. Ayres and coworkers (1996) added estradiol and $\alpha$-tocopherol (200 µmol/L) to human plasma, exposed the plasma for 1 hour at 37°C to AAPH (a free-radical generator) and then isolated the LDL and HDL fractions and measured TBARS. The addition of $\alpha$-tocopherol to plasma inhibited lipid peroxidation in both LDL and HDL; however, addition of 17$\beta$-estradiol protected HDL but had no effect on LDL oxidation.

The preferential protection of HDL in our experiments (Figure 7.1) as well as those of Ayers *et al.* can be explained by selective passive enrichment of the HDL particle with estrogen and this mechanism is supported by the observations of Leszczynski and Schafer described above (Leszczynski and Schafer, 1989; 1991). I have proposed that a second mechanism involving non-passive enrichment of HDL with estrogens operates *in vivo* (Banka, 1997). This proposed mechanism relies on the fatty acid esterification of estrogen by the plasma enzyme lecithin:cholesterol acyltransferase (LCAT). LCAT is the major cholesterol esterifying enzyme in plasma and acts primarily on HDL cholesterol and HDL phospholipid fatty acids due to the cofactor activity of apoprotein A-l, the major protein component of HDL (see Banka *et al.*, 1991). The fatty acid esterification of estradiol would result in a more hydrophobic molecule predicted to associate with higher affinity than estradiol with the HDL particle. Support for this hypothesis comes from several experimental approaches. The kinetic profiles reported for interaction of estradiol with HDL (Leszczynski and Schafer, 1989; 1991) suggest a rapid, passive interaction of the two components as well as a slower, unsaturable association of the estrogen with HDL. The authors suggest fatty acid esterification as a potential mechanism for the slower association. Hochberg and his colleagues have identified estradiol esterified at the 17-position in serum and tissues and found particularly high levels in follicular fluid (Janocko and Hochberg, 1983; Hochberg *et al.*, 1991; Larner *et al.*, 1992; 1993). Finally, a recent report by Shwaery and coworkers (1997) identified nonpolar forms of estradiol in LDL isolated from plasma incubated with physiologic concentrations of estradiol. The LDL was less susceptible to oxidation. The plasma-mediated association of the nonpolar estrogen compounds and subsequent inhibition of LDL oxidation was prevented by addition of an LCAT inhibitor to the plasma. These exciting results support the hypothesis that estradiol (and other estrogens with a hydroxyl group at the 17-position) are esterified by LCAT on HDL. Subsequent transfer of the estrogen fatty acid esters from HDL to LDL by cholesterol ester transfer protein (CETP) would result in enrichment of LDL with less polar species of estrogen that would confer antioxidant protection within the LDL particle. Thus, the interaction of estrogen with HDL would protect HDL directly preserving its antioxidant capacity for protection of LDL and, through LCAT-mediated conversion of estrogen to fatty acid esters, would contribute estrogen antioxidants to LDL. The results and hypothesis presented in this section suggest that the discrepancy between the high estrogen concentrations required to protect LDL *in vitro* and the low concentrations effective *in vivo* may be explained, in part, through indirect actions of the estrogens on HDL.

## NON-LIPOPROTEIN ANTIOXIDANT EFFECTS OF ESTROGENS

If estrogens are potent antioxidants for lipoproteins, then they most likely act as antioxidants for other lipid components in the body. The work of Yoshino *et al.* (1987) describing a decrease in serum and liver lipid peroxides following estrogen administration has been discussed above. The actions of estriol, estradiol and catecholestrogens as membrane phospholipid antioxidants have been described by others as have the membrane antioxidant effects of tamoxifen (Sugioka *et al.*, 1987; Wiseman, 1994; Lacort *et al.*, 1995). The chain-breaking antioxidant effects of estrogens have been implicated as the mechanism of membrane antioxidant protection (Lacort *et al.*, 1995). Wiseman has proposed that estrogens and tamoxifen act as membrane antioxidants through stabilization of membranes due to a decrease in membrane fluidity (Wiseman, 1994). Both mechanisms are likely to be important.

The antioxidant protection of cell membranes is likely to be an important factor in protecting the cells of the vascular wall from oxidative damage. This attribute may underlie the observation that preexposure of endothelial cells to estrogen protects the cells from the cytotoxic effects of oxidized LDL (Negre-Salvayre *et al.*, 1993). Wagner and colleagues (1991) have demonstrated that chronic estrogen and progesterone treatment in surgically postmenopausal primates reduces the accumulation of LDL in the artery wall and, in a subsequent study, that lipid peroxides, markers of oxidative stress, are also reduced in the artery wall following chronic equine estrogen ester treatment (Wagner *et al.*, 1996). We have shown that acute pre-exposure of carotid arteries from ovariectomized rats to 17$\beta$-estradiol reduces arterial accumulation of perfused oxidized LDL suggesting a direct antioxidant effect on the endothelium (Roberts *et al.*, 1996). Kim and coworkers (1996) examined the effect of two weeks of treatment with 17$\beta$-estradiol on myocardial and coronary endothelial function during reperfusion in dogs and found clear support for the action of estrogen in preventing or attenuating the incidence of ventricular arrhythmias, endothelial dysfunction and myocardial systolic dysfunction. Oxygen free radicals are generated during the early stages of reperfusion and this prompted the authors to measure exhaled *n-pentane*, a measure of *in vivo* lipid peroxidation during reperfusion in the dogs. Exhaled pentane levels in placebo-treated dogs were significantly elevated during reperfusion whereas pentane levels did not rise significantly in the estrogen-treated dogs, leading the authors to propose that some portion of the estrogen protection of endothelium and myocardium relies on antioxidant function. Finally, in the wake of observations suggesting that estrogen treatment may reduce the risk of Alzheimer's disease in women, two laboratories have demonstrated estrogen receptor-independent protection of cultured hippocampal neurons from oxidative stress-induced cell death caused by neurotoxins such as $FeSO_4$, amyloid $\beta$-peptide, hydrogen peroxide and glutamate (Behl *et al.*, 1995; Goodman *et al.*, 1996). Therefore, antioxidant protection of cell membranes may, in part, mediate the atheroprotective effects of estrogens.

Oxidative stress and associated free radicals are thought to accelerate a variety of disease states including atherosclerosis through induction of inflammatory genes (for review, see Sen and Packer, 1996). The list of genes responsive to and, therefore, reflective of oxidant stress grows annually and it is proposed that antioxidants are involved in expression of these genes (Suzuki *et al.*, 1994; Pinkus *et al.*, 1996). It remains to be seen whether estrogens can exert regulatory activities on genes through the antioxidant mechanism.

It is clear that oxidative stress plays some part in atherogenesis and it is compelling to include antioxidant effects as potential mechanisms of estrogen protection from atherogenesis. It is not necessary and may be misleading to attribute all antioxidant effects of estrogens to the protection of lipoproteins. In the near future we must identify those genes and/or physiologic processes that respond to oxidant stress, develop appropriate assays for the respondents *in vitro* and *in vivo*, and begin to assess the impact of estrogens with and without progestins on these markers of oxidant stress.

## REFERENCES

Adams, M.R., J.R. Kaplan, S.B. Manuck, D.R. Koritnik, J.S. Parks, M.S. Wolfe, and T.B. Clarkson (1990). Inhibition of coronary artery atherosclerosis by 17-beta estradiol in ovariectomized monkeys. Lack of effect of added progesterone. *Arteriosclerosis* **10**: 1051–1057.

Austin, M.A., J.L. Breslow, and C.H. Hennekens (1988). Low-density lipoprotein subclass patterns and risk of myocardial infarction. *JAMA* **260**: 1917–1921.

Ayres, S., M. Tang, and M.T.R. Subbiah (1996). Estradiol-17$\beta$ as an antioxidant: Some distinct features when compared with common fat-soluble antioxidants. *J. Lab. Clin. Med.* **128**: 367–375.

Banka, C.L. (1996). High density lipoprotein and lipoprotein oxidation. *Curr. Opin. Lipidol.* **7**: 139–142.

Banka, C.L. (1997). Antioxidant properties of estrogen: selective protection of high density lipoprotein. In Proceedings of the workshop on hormonal, metabolic and cellular influences on cardiovascular disease.T.M. Forte, editor. Futura Publishing, Inc. Mount Kisko, NY pp. 193–214.

Banka, C.L., D.J. Bonnet, A.S. Black, R.S. Smith, and L.K. Curtiss (1991). Localization of an apolipoprotein A-I epitope critical for activation of lecithin-cholesterol acyltransferase. *J. Biol. Chem.* **266**: 23886–23892.

Banka, C.L. and L.K. Curtiss (1996). Estrogens as antioxidants: specific protection of high density lipoprotein. *Circulation* **94**: I-152 (Abstr.).

Barrett-Connor, E (1997). Sex differences in coronary heart disease. *Circulation* **95**: 252–264.

Baumann, D.S., M. Doblas, G. Schonfeld, G.A. Sicard, and A. Daugherty (1994). Probucol reduces the cellularity of aortic intimal thickening at anastomotic regions adjacent to prosthetic grafts in cholesterol-fed rabbits. *Arterioscler. Thromb.* **14**: 162–167.

Behl, C., M. Widmann, T. Trapp, and F. Holsboer (1995). 17-$\beta$ Estradiol protects neurons from oxidative stress-induced cell death *in vitro. Biochem. Biophys. Res. Commun.* **216**: 473–482.

Bjorkham, I., A. Henricksson-Freyschuss, O. Breuer, U. Diczfalusy, L. Berglund, and P. Henrikson (1991). The antioxidant butylated hydroxytoluene protects against atherosclerosis. *Arterioscler. Thromb.* **11**: 15–22.

Bush, T.L., E. Barrett-Connor, L.D. Cowan, M.H. Criqui, R.B. Wallace, C.M. Suchindran, H.A. Tyroler, and B.M. Rifkind (1987). Cardiovascular mortality and noncontraceptive use of estrogen in women: Results from the Lipid Research Clinics' Program Follow-up study. *Circulation* **75**: 1102–1109.

Campos, H., F.M. Sacks, B.W. Walsh, I. Schiff, M.A. O'Hanesian, and R.M. Krauss (1993). Differential effects of estrogen on low-density lipoprotein subclasses in healthy postmenopausal women. *Metabolism* **42**: 1153–1158.

Carew, T.E., D.C. Schwenke, and D. Steinberg (1987). Antiatherogenic effect of probucol unrelated to its hypocholesterolemic effect: evidence that antioxidants *in vivo* can selectively inhibit low density lipoprotein degredation in macrophage-rich fatty streaks and slow the progression of atherosclerosis in the Watanabe heritable hyperlipidemic rabbit. *Proc. Natl. Acad. Sci. USA* **84**: 7725–7729.

Cirino, G., C. Cicala, L. Sorrentino, F.M. Maiello, and J.L. Browning (1994). Recombinant secreted nonpancreatic phospholipase $A_2$ induces a synovitis-like inflammation in the rat air pouch. *J. Rheumatol.* **21**: 824–829.

De Graaf, J., J.C. Hendriks, P.N. Demacker, and A.F. Stalenhoef (1993). Identification of multiple dense LDL subfractions with enhanced susceptibility to *in vitro* oxidation among hypertriglyceridemic subjects. *Arterioscler. Thromb.* **13**: 712–719.

Elinder, L.S. and G. Walldius (1994). Antioxidants and atherosclerosis progression: unresolved questions. *Curr. Opin. Lipidol.* **5**: 265–268.

Evans, R.M. (1988). The steroid and thyroid hormone receptor superfamily. *Science* **240**: 889–895.

Farhat, M.Y., S. Abi-Younes, and P.W. Ramwell (1996). Non-genomic effects of estrogen and the vascular wall. *Biochem. Pharmacol.* **51**: 571–576.

Fruebis, J., T.E. Carew, and W. Palinski (1995). Effect of vitamin E on atherogenesis in LDL receptor-deficient rabbits. *Atherosclerosis* **117**: 217–224.

Goodman, Y., A.J. Bruce, B. Cheng, and M.P. Mattson (1996). Estrogens attenuate and corticosterone exacerbates excitoxicity, oxidative injury, and amyloid $\beta$-peptide toxicity in hippocampal neurons. *J. Neurochem.* **66**: 1836–1844.

Granfone, A., H. Campos, J. McNamara, M.M. Schaefer, S. Lamon-Fava, J.M. Ordovas, and E.J. Schaefer (1992). Effects of estrogen replacement on plasma lipoproteins and apolipoproteins in postmenopausal, dyslipemic women. *Metabolism* **41**: 1193–1198.

Grodstein, F., M.J. Stampfer, J.E. Manson, G.A. Colditz, W.C. Willett, B. Rosner, F.E. Speizer, and C.H. Hennekens (1996). Postmenopausal estrogen and progestin use and the risk of cardiovascular disease. *N. Engl. J. Med.* **335**: 453–461.

Gruchow, H.W., A.J. Anderson, J.J. Barboriak, and V.A. Sobocinski (1988). Postmenopausal use of estrogen and occlusion of coronary arteries. *Am. Heart J.* **115**: 954–963.

Guetta, V., J.A. Panza, M.A. Waclawiw, and R.O. Cannon, III (1995a) Effect of combined 17$\beta$-estradiol and vitamin E on low-density lipoprotein oxidation in postmenopausal women. *Am. J. Cardiol.* **75**: 1274–1276.

Guetta, V., R.M. Lush, W.D. Figg, M.A. Waclawiw, and R.O. Cannon, III (1995b) Effects of the antiestrogen *tamoxifen* on low-density lipoprotein concentrations and oxidation in postmenopausal women. *Am. J. Cardiol.* **76**: 1072–1073.

Halliwell, B. (1990). How to characterize a biological antioxidant. *Free Rad. Res. Commun.* **9**: 1–32.

Hammond, G.L (1990). Molecular properties of corticosteroid binding globulin and the sex-steroid binding proteins. *Endocr. Rev.* **11**: 65–79.

Hochberg, R.B., S.L. Pahuja, J.E. Zielinski, and J.M. Larner (1991). Steroidal fatty acid esters. *J. Steroid Biochem. Molec. Biol.* **40**: 577–585.

Huber, L.A., E. Scheffler, T. Poll, R. Ziegler, and H.A. Dresel (1990). 17 beta-estradiol inhibits LDL oxidation and cholesteryl ester formation in cultured macrophages. *Free Rad. Res. Comms.* **8**: 167–173.

Janocko, L. and R.B. Hochberg (1983). Estradiol fatty acid esters occur naturally in human blood. *Science* **222**: 1334–1336.

Katzenellenbogen, B.S. (1996). Estrogen receptors: bioactivities and interactions with cell signaling pathways. *Biol. Reprod.* **54**: 287–293.

Keaney, J.F., Jr., G.T. Shwaery, A. Xu, R.J. Nicolosi, J. Loscalzo, T.L. Foxall, and J.A. Vita (1994). 17$\beta$-estradiol preserves endothelial vasodilator function and limits low-density lipoprotein oxidation in hypercholesterolemic swine. *Circulation* **89**: 2251–2259.

Kim, Y.D., B. Chen, J. Beauregard, P. Kouretas, G. Thomas, M.Y. Farhat, A.K. Myers, and D.E. Lees (1996). 17$\beta$-Estradiol prevents dysfunction of canine coronary endothelium and myocardium and reperfusion arrhythmias after brief ischemia/reperfusion. *Circulation* **94**: 2901–2908.

Kleinveld, H.A., H.L. Hak-Lemmers, M.P. Hectors, N.J. de Fouw, P.N. Demacker, and A.F. Stalenhoef (1995). Vitamin E and fatty acid intervention does not attentuate the progression of atherosclerosis in Watanabe heritable hyperlipidemic rabbits. *Arterioscler. Thromb. Vasc. Biol.* **15**: 290–297.

Klimov, A.N., V.S. Gurevich, A.A. Nikiforova, L.V. Shatilina, A.A. Kuzmin, S.L. Plavinsky, and N.P. Teryukova (1993). Antioxidative activity of high density lipoproteins *in vivo*. *Atherosclerosis* **100**: 13–18.

Kushi, L.H., A.R. Folsom, R.J. Prineas, P.J. Mink, Y. Wu, and R.M. Bostick (1996). Dietary antioxidant vitamins and death from coronary heart disease in postmenopausal women. *N. Engl. J. Med.* **334**: 1156–1162.

Lacort, M., A.M. Leal, M. Liza, C. Martin, R. Martinez, and M.B. Ruiz-Larrea (1995). Protective effect of estrogens and catecholestrogens against peroxidative membrane damage *in vitro*. *Lipids* **30**: 141–146.

Larner, J.M., C.H.L. Shackleton, E. Roitman, P.E. Schwartz, and R.B. Hochberg (1992). Measurement of estradiol-17-fatty acid esters in human tissues. *J. Clin. Endocrinol. Metab.* **75**: 195–200.

Larner, J.M., S.L. Pahuja, C.H. Shackelton, W.J. McMurray, G. Giordano, and R.B. Hochberg (1993). The isolation and characterization of estradiol-fatty acid esters in human ovarian follicular fluid. *J. Biol. Chem.* **268**: 13893–13899.

Leszczynski, D.E. and R.M. Schafer (1989). Characterization of steroid hormone association with human plasma lipoproteins. *Steroids* **54**: 37–53.

Leszczynski, D.E. and R.M. Schafer (1991). Metabolic conversion of six steroid hormones by human plasma high-density lipoprotein. *Biochim. Biophys. Acta* **1083**: 18–28.

Mao, S., M. Yates, R. Parker, E. Chi, and R. Jackson (1991). Attenuation of atherosclerosis in a modified strain of hypercholesterolemic Watanabe rabbits with use of a probucol analogue (MDL 29,311). that does not lower serum cholesterol. *Arterioscler. Thromb.* **11**: 1266–1275.

Maziere, C., M. Auclair, M. Ronveaux, S. Salmon, R. Santus, and J. Maziere (1991). Estrogens inhibit copper and cell-mediated modification of low density lipoprotein. *Atherosclerosis* **89**: 175–182.

McCall, M.R., J.J.M. van den Berg, F.A. Kuypers, D.L. Tribble, R.M. Krauss, L.J. Knoff, and T.M. Forte (1994). Modification of LCAT activity and HDL structure: New links between cigarette smoke and coronary heart disease risk. *Arterioscler. Thromb.* **14**: 248–253.

Mooradian, A.D. (1993). Antioxidant properties of steroids. *J. Steroid Biochem. Molec. Biol.* **45**: 509–511.

Morel, D.W. (1994). Reduced cholesterol efflux to mildly oxidized high density lipoprotein. *Biochem. Biophys. Res. Commun.* **200**: 408–416.

Mukai, K., K. Daifuku, S. Yokoyama, and M. Nakano (1990). Stopped-flow investigation of antioxidant activity of estrogens in solution. *Biochim. Biophys. Acta* **1035**: 348–352.

Nagano, Y., H. Arai, and T. Kita (1991). High density lipoprotein loses its effect to stimulate efflux of cholesterol from foam cells after oxidative modification. *Proc. Natl. Acad. Sci. USA* **88**: 6457–6461.

Navab, M., J.A. Berliner, A.D. Watson, S.Y. Hama, M.C. Territo, A.J. Lusis, D.M. Shih, B.J. Van Lenten, J.S. Frank, L.L. Demer, P.A. Edwards, and A.M. Fogelman (1996). The Yin and Yang of oxidation in the development of the fatty streak—A review based on the 1994 George Lyman Duff Memorial Lecture. *Arterioscler. Thromb. Vasc. Biol.* **16**: 831–842.

Negre-Salvayre, A., M.T. Pieraggi, L. Mabile, and R. Salvayre (1993). Protective effect of 17 beta-estradiol against cytotoxicity of minimally oxidized LDL to cultured bovine aortic endothelial cells. *Atherosclerosis* **99**: 207–217.

Nenseter, M.S., V. Volden, T. Berg, C.A. Drevon, L. Ose, and S. Tonstad (1996). Effect of hormone replacement therapy on the susceptibility of low-density lipoprotein to oxidation among post-menopausal hypercholesterolaemic women. *Eur. J. Clin. Invest.* **26**: 1062–1068.

Palinski, W., R.K. Tangirala, E. Miller, S.G. Young, and J.L. Witztum (1995). Increased autoantibody titers against epitopes of oxidized LDL in LDL receptor-deficient mice with increased atherosclerosis. *Arterioscler. Thromb. Vasc. Biol.* **15**: 1569–1576.

Parker, R.A., T. Sabrah, M. Cap, and B.T. Gill (1995). Relation of vascular oxidative stress, alpha-tocopherol, and hypercholesterolemia to early atherosclerosis in hamsters. *Arterioscler. Thromb. Vasc. Biol.* **15**: 349–358.

Parthasarathy, S., J. Barnett, and L.G. Fong (1990). High-density lipoprotein inhibits the oxidative modification of low-density lipoprotein. *Biochim. Biophys. Acta* **1044**: 275–283.

Parthasarathy, S., A.J. Morales, and A.A. Murphy (1994). Antioxidant: A new role for RU-486 and related compounds. *J. Clin. Invest.* **94**: 1990–1995.

Pinkus, R., L.M. Weiner, and V. Daniel (1996). Role of oxidants and antioxidants in the induction of AP-1, NF-kappaB, and glutathione S-transferase gene expression. *J. Biol. Chem.* **271**: 13422–13429.

Puhl, H., G. Waeg, and H. Esterbauer (1994). Methods to determine oxidation of low-density lipoproteins. *Methods Enzymol.* **233**: 425–441.

Reaven, P.D., A. Khouw, W.F. Beltz, S. Parthasarathy, and J.L. Witztum (1993). Effect of dietary antioxidant combinations in humans; protection of LDL by vitamin E but not $\beta$-carotene. *Arterioscler. Thromb.* **13**: 590–600.

Rifici, V.A. and A.K. Khachadurian (1992). The inhibition of low-density lipoprotein oxidation by 17-$\beta$ estradiol. *Metabolism* **41**: 1110–1114.

Roberts, K.A., B.A. Walsh, C.L. Banka, and J.C. Rutledge (1996). 17$\beta$-Estradiol reduces modified LDL accumulation in the artery wall. *FASEB J.* **10**: A617.

Rosenfeld, M.E., W. Palinski, S. Yla-Hertuala, G.C. Gurtner, S.A. Socher, S. Butler, and J.L. Witztum (1990). Distribution of oxidation specific lipid-protein adducts and apolipoprotein B in atherosclerotic lesions of varying severity from WHHL rabbits. *Arteriosclerosis* **10**: 336–349.

Sack, M.N., D.J. Rader, and R.O. Cannon, III (1994). Oestrogen and inhibition of oxidation of low-density lipoproteins in postmenopausal women. *Lancet* **343**: 269–270.

Salonen, J.T., S. Yla-Herttuala, R. Yamamoto, S. Butler, H. Korpela, R. Salonen, K. Nyyssonen, W. Palinski, and J.L. Witztum (1992). Autoantibody against oxidized LDL and progression of carotid atherosclerosis. *Lancet* **339**: 883–887.

Sasahara, M., E.W. Raines, A. Chait, T.E. Carew, D. Steinberg, P.W. Wahl, and R. Ross (1994). Inhibition of hypercholesterolemia-induced atherosclerosis in the nonhuman primate by probucol. I. Is the extent of atherosclerosis related to resistance of LDL to oxidation. *J. Clin. Invest.* **94**: 155–164.

Sawamura, T., N. Kume, T. Aoyama, H. Moriwaki, H. Hoshikawa, Y. Alba, T. Tanaka, S. Miwa, Y. Katsura, T. Kita, and T. Masaki (1997). An endothelial receptor for oxidized low-density lipoprotein. *Nature* **386**: 73–77.

Schütze, N., G. Vollmer, I. Tiemann, M. Geiger, and R. Knuppen (1993). Catecholestrogens are MCF-7 cell estrogen receptor agonists. *J. Steroid Biochem. Molec. Biol.* **46**: 781–789.

Sen, C.K. and L. Packer (1996). Antioxidant and redox regulation of gene transcription. *FASEB J.* **10**: 709–720.

Shwaery, G.T., J.A. Vita, and J.F. Keaney, Jr. (1997). Antioxidant protection of LDL by physiological concentrations of 17$\beta$-estradiol. Requirement for estradiol modification. *Circulation* **95**: 1378–1385.

Sparrow, C.P., T.W. Doebber, J. Olszewski, M.S. Wu, J. Ventre, K.A. Stevens, and Y.S. Chao (1992). Low density lipoprotein is protected from oxidation and the progression of atherosclerosis is slowed in cholesterol-fed rabbits by the antioxidant N,N'-diphenyl-phenylenediamine. *J. Clin. Invest.* **89**: 1885–1891.

Steinberg, D. (1997). Oxidative modification of LDL and atherogensis. *Circulation* **95**: 1062–1071.

Steinberg, D., S. Parthasarathy, T.E. Carew, J.C. Khoo, and J.L. Witztum (1989). Beyond cholesterol: Modifications of low-density lipoprotein that increase its atherogenicity. *N. Engl. J. Med.* **320**: 915–924.

Stephens, N.G., A. Parsons, P.M. Schofield, F. Kelly, K. Cheeseman, M.J. Mitchinson, and M.J. Brown (1996). Randomised controlled trial of vitamin E in patients with coronary disease: Cambridge Heart Antioxidant Study (CHAOS). *Lancet* **347**: 781–786.

Subbiah, M.T.R., B. Kessel, M. Agrawal, R. Rajan, W. Abplanalp, and Z. Rymaszewski (1993). Antioxidant potential of specific estrogens on lipid peroxidation. *J. Clin. Endocrinol. Metab.* **77**: 1095–1097.

Sugioka, K., Y. Shimosegawa, and M. Nakano (1987). Estrogens as natural antioxidants of membrane phospholipid peroxidation. *FEBS Lett* **210**: 37–39.

Suzuki, Y.J., M. Mizuno, and L. Packer (1994). Signal transduction for nuclear factor-kappaB activation: Proposed location of antioxidant-inhibitable step. *J. Immunol.* **153**: 5008–5015.

Tang, M., W. Abplanalp, S. Ayres, and M.T.R. Subbiah (1996). Superior and distinct antioxidant effects of selected estrogen metabolites on lipid peroxidation. *Metabolism* **45**: 411–414.

Tangirala, R.K., F. Casanada, E. Miller, J.L. Witztum, D. Steinberg, and W. Palinski. (1995). Effect of the antioxidant *N,N'*-diphenyl 1,4-phenylenediamine (DPPD) on atherosclerosis in apoE-deficient mice. *Arterioscler. Thromb. Vasc. Biol.* **15**: 1625–1630.

Taniguchi, S., T. Yanase, K. Kobayashi, R. Takayanagi, M. Haji, F. Umeda, and H. Nawata (1994). Catechol estrogens are more potent antioxidants than estrogens for the $Cu^{2+}$-catalyzed oxidation of low or high density lipoprotein: antioxidative effects of steroids on lipoproteins. *Endocrine J.* **41**: 605–611.

The Postmenopausal Estrogen/Progestin Interventions Trial Writing Group (1995). Effects of estrogen or estrogen/progestin regimens on heart disease risk factors in postmenopausal women. The Postmenopausal Estrogen/Progestin Interventions (PEPI) trial. *JAMA* **273**: 199–208.

Tilly-Kiesi, M., M. Lappi, J. Puolakka, H. Luotola, T. Pyorälä, and M.-R. Taskinen (1996). Different effects of continuous oestrogen-progestin and transdermal oestrogen with cyclic progestin regimens on low-density lipoprotein subclasses. *Eur. J. Clin. Invest.* **26**: 1125–1133.

Tribble, D.L., L.G. Holl, P.D. Wood, and R.M. Krauss (1992). Variations in oxidative susceptibility among six low density lipoprotein subfractions of differing density and particle size. *Atherosclerosis* **93**: 189–199.

Tribble, D.L., B.M. Chu, E.L. Gong, F. Van Venrooij, and A.V. Nichols (1995). HDL antioxidant effects as assessed using a nonexchangeable probe to monitor particle-specific peroxidative stress in LDL-HDL mixtures. *J. Lipid Res.* **36**: 2580–2589.

Verlangieri, A.J. and M.J. Bush (1992). Effects of d-alpha-tocopherol supplementation on experimentally induced primate atherosclerosis. *J. Am. Coll. Nutr.* **11**: 131–138.

Wagner, J.D., T.B. Clarkson, R.W. St.Clair, D.C. Schwenke, C.A. Shively, and M.R. Adams (1991). Estrogen and progesterone replacement therapy reduces low density lipoprotein accumulation in the coronary arteries of surgically postmenopausal cynomolgus monkeys. *J. Clin. Invest.* **88**: 1995–2002.

Wagner, J.D., L. Zhang, J.K. Williams, T.C. Register, D.M. Ackerman, B. Wiita, T.B. Clarckson, and M.R. Adams (1996). Esterified estrogens with and without methyltestosterone decrease arterial LDL metabolism in cynomolgus monkeys. *Arterioscler. Thromb. Vasc. Biol.* **16**: 1473–1480.

Wander, R.C., S.H. Du, S.O. Ketchum, and K.E. Rowe (1996). Effects of interaction of *RRR-α*-tocopheryl acetate and fish oil on low-density-lipoprotein oxidation in postmenopausal women with and without hormone-replacement therapy. *Am. J Clin. Nutr.* **63**: 184–193.

Wilcox, J.G., J. Hwang, H. Hodis, A. Sevanian, F.Z. Stanczyk, and R.A. Lobo (1997). Cardioprotective effects of individual conjugated equine estrogens through their possible modulation of insulin resistance and oxidation of low-density lipoprotein. *Fertil. Steril.* **67**: 57–62.

Wiseman, H., G. Paganga, C. Rice-Evans, and B. Halliwell (1993). Protective actions of tamoxifen and 4-hydroxytamoxifen against oxidative damage to human low-density lipoproteins: a mechanism accounting for the cardioprotective action of tamoxifen? *Biochem. J.* **292**: 635–638.

Wiseman, H. (1994). Tamoxifen and estrogens as membrane antioxidants: Comparison with cholesterol. *Methods Enzymol.* **234**: 590–602.

Yagi, K. and S. Komura (1986). Inhibitory effect of female hormones on lipid peroxidation. *Biochem. Int.* **13**: 1051–1055.

Yla-Herttuala, S., W. Palinski, M.E. Rosenfeld, S. Parthasarathy, T.E. Carew, S. Butler, J.L. Witztum, and D. Steinberg (1989). Evidence for the presence of oxidatively modified low density lipoprotein in atherosclerotic lesions of rabbit and man. *J. Clin. Invest.* **84**: 1086–1095.

Yoshino, K., S. Komuna, I. Watanabe, Y. Nakagawa, and K. Yagi (1987). Effect of estrogens on serum and liver peroxide levels in mice. *J. Clin. Biochem. Nutr.* **3**: 233–240.

# 8 Lipoprotein Metabolism in the Vessel Wall

**Janice D. Wagner [1,*] and Dawn C. Schwenke [2]**

*Departments of [1] Comparative Medicine and [2] Pathology, Bowman Gray School of Medicine of Wake Forest University, Winston-Salem, NC, USA.*
** Corresponding author: Janice D. Wagner, Department of Comparative Medicine, Bowman Gray School of Medicine, Medical Center Boulevard, Winston-Salem, NC 27157-1040, Tel.: 001 910 716 1630, Fax: 001 910 716 1515, E-mail: jwagner@cpm.bgsm.edu.*

Risk of coronary heart disease, the leading cause of death in women, is reduced by about 50% with estrogen replacement therapy. Consistent with these findings, estrogen replacement therapy also reduces atherosclerosis by about 50% in animal models. Both of these reductions are largely independent of plasma lipid concentrations, suggesting that estrogens may have direct effects on the artery wall. We have found that estrogens alone, or with other hormones, inhibit arterial degradation and accumulation of LDL in nonhuman primates. While this effect also was found to be largely independent of plasma lipid concentrations, there were changes in LDL heterogeneity and decreases in thiobarbituric acid-reactive substances, suggesting a decreased oxidative state with estrogens. In addition, the decreased arterial LDL metabolism appears to require a sufficient atherogenic stimulus, most likely recruitment of intimal macrophages, for treatment effects to be detected. These changes in LDLs and their metabolism may be partially responsible for the decreased atherogenesis and reduced subsequent atherosclerosis associated with estrogen replacement therapy.

KEYWORDS: Estrogens, atherogenesis, LDL metabolism, oxidation, nonhuman primates.

Coronary heart disease (CHD) remains the primary cause of death in women in Western societies. Estrogen replacement therapy (ERT) reduces the risk of CHD by about 50% in women (Stampfer and Colditz, 1991; Barrett-Connor and Bush, 1991). In addition, in experimental studies, ERT decreases arterial cholesterol content in rabbits (Hough and Zilversmit, 1986; Haarbo *et al.*, 1991; Sulistiyani *et al.*, 1995) and decreases coronary artery atherosclerosis (the leading cause of CHD) in nonhuman primates (Adams *et al.*, 1990; 1997).

Although some forms of estrogen replacement therapy have beneficial effects on plasma lipids and lipoproteins (Crook and Seed, 1990), estrogen's effects on both CHD risk and atherosclerosis extent are largely independent of plasma lipoprotein concentrations (Barrett-Connor and Bush, 1991; Haarbo *et al.*, 1991; Sulistiyani *et al.*, 1995; Adams *et al.*, 1990). This implies that estrogens may have direct effects on the artery or may affect risk factors other than plasma lipid concentrations. One mechanism by which estrogens may decrease atherosclerosis progression is by affecting arterial lipoprotein metabolism. This chapter will summarize studies on estrogens and arterial low density lipoprotein (LDL) metabolism and how their interaction influences atherogenesis.

## ARTERIAL LDL METABOLISM AND ATHEROSCLEROSIS

Plasma LDL concentrations are causally related to the development of atherosclerosis (Stewart *et al.*, 1994; Waters *et al.*, 1995), and most of the cholesteryl ester accumulating in atherosclerotic arteries derives from plasma lipoproteins (Newman and Zilversmit, 1962; Zilversmit, 1968). Intact LDL has been detected in the atherosclerotic lesions of several species, including human beings (Daugherty *et al.*, 1988; Smith and Ashall, 1984). LDL and other lipoproteins enter arteries at rates that are determined by arterial permeability, lipoprotein molecular size, plasma lipoprotein concentrations, and the extent of atherosclerosis (Stender and Zilversmit, 1981; Nordestgaard, Tybjaerghansen, and Lewis, 1992). Arterial permeability to LDL is not altered by hypercholesterolemia in areas free of atherosclerosis, but total influx of LDL is increased in proportion to the increase in LDL cholesterol (Schwenke and St. Clair, 1993; Schwenke and Carew, 1989a). However, arterial permeability to LDL and LDL influx are increased in atherosclerotic lesions compared with normal areas of artery (Schwenke and St. Clair, 1993). Permeability to LDL differs among arterial regions, but this heterogeneity only partially correlates with regional variation in susceptibility to atherosclerosis (Schwenke and St. Clair, 1993; Schwenke and Carew, 1989a; Nielsen *et al.*, 1992). LDL entering arteries may be retained in the extracellular space, degraded by the artery, or efflux from the artery (Smith and Ashall, 1984; Schwenke and St. Clair, 1993; Schwenke and Carew, 1989a,b; 1988; Carew, Schwenke, and Steinberg, 1987; Nordestgaard, Wooten, and Lewis, 1995); these removal processes, together with rates of LDL entry, determine arterial concentrations of LDL.

Several studies suggest that arterial accumulation and degradation of LDL are early events in atherogenesis. In normal rabbits, areas of artery that are susceptible to atherosclerosis demonstrate increased rates of LDL degradation and accumulate higher concentrations of intact LDL (Schwenke and Carew, 1988; 1989b). Short intervals of hypercholesterolemia exaggerated the focal increase in arterial LDL degradation rate and increased arterial accumulation of LDL by augmenting intra-arterial LDL retention (Schwenke and Carew, 1989a,b). These changes in arterial LDL metabolism occurred before significant numbers of foam cells accrued in the intima (Schwenke and Carew, 1989b) and before cholesterol began to accumulate in susceptible areas of artery (Schwenke, 1995). Such changes in arterial metabolism of LDL are further exaggerated after longer periods of hypercholesterolemia, when atherosclerotic lesions develop (Carew, Schwenke, and Steinberg, 1987; Schwenke and St. Clair, 1992). Thus, increased arterial metabolism of LDL appears to promote atherosclerosis.

Estrogens may inhibit atherosclerosis by influencing one or several aspects of arterial LDL metabolism. To date, the effect of estrogen on arterial permeability to LDL has received relatively little attention. However, the few available studies suggest that estrogen does not alter arterial permeability to LDL (Hough and Zilversmit, 1986; Haarbo *et al*., 1994; Schwenke, in press). This review will focus on other aspects of arterial metabolism of LDL for which more data are available.

## CYNOMOLGUS MONKEYS AND ARTERIAL LDL METABOLISM

The studies discussed below from our institution have used female cynomolgus macaques (*Macaca fascicularis*). These monkeys, when fed a moderately atherogenic diet that mimics the nutrient composition of the average American diet (usually 0.28 mg cholesterol/kcal), readily develop atherosclerosis. In addition, similar to human beings, premenopausal females develop half as much coronary artery atherosclerosis as males or surgically postmenopausal (ovariectomized) females (Clarkson *et al*., 1993). Subsequent studies by Adams *et al*. (1990, 1997) have shown that sex hormone replacement therapy (either parenteral estradiol therapy with or without progesterone, or oral conjugated equine estrogens) decreased atherosclerosis progression by at least 50%. Adjusting for plasma lipid concentrations did not affect these relationships, suggesting effects of estrogens on the artery wall.

In the series of studies described below, we assessed the effects of hormone treatment on arterial, hepatic, and whole body lipoprotein metabolism. To determine rates of LDL degradation and accumulation of LDL, we use a dual-labeling procedure originally described by Pittman *et al*. (1983) in which LDL is labeled directly with $^{131}I$ and also coupled to tyramine cellobiose (TC) labeled with $^{125}I$ ($^{125}I$-TC). By using these differently labeled LDLs, one can calculate separately the amount of undegraded LDL and products of LDL degradation (Schwenke and Carew, 1988; Carew *et al*., 1984). In our studies described below, labeled LDLs ($^{125}I$-TC and $^{131}I$) were injected via indwelling catheters and allowed to circulate for 24 hours. This allowed determination of plasma decay of labeled LDL as well as LDL metabolism in various arteries and tissues (i.e., rates of LDL degradation, amount of undegraded LDL, and total accumulation of $^{125}I$-TC representing both degraded and undegraded LDL).

As noted above, the presence and extent of atherosclerotic lesions influence several aspects of arterial LDL metabolism, including the rate of arterial LDL degradation. Thus, the initial studies were planned after a relatively short (16–18 weeks) period of dietary cholesterol exposure to stimulate intimal foam cell formation without affecting atherosclerosis extent.

## ESTROGEN REPLACEMENT THERAPY AND ARTERIAL LDL METABOLISM

### Study One: Parenteral Estrogen Replacement Therapy

In Study One, hormone replacement therapy (17$\beta$-estradiol and cyclic progesterone) was given in physiological concentrations to ovariectomized monkeys via subcutaneous Silastic$^{R}$ implants ($n = 9$), whereas the ovariectomized controls ($n = 8$) received no treatment. LDL were isolated from each monkey, labeled as described above, and re-injected 24 hours before necropsy, after which LDL uptake and metabolism were determined in various arteries and tissues.

The 18-week atherogenic stimulus caused only minimal changes in arterial morphology (i.e., small accumulations of subendothelial foam cells), with no effect of treatment. In contrast, hormone replacement therapy significantly decreased LDL degradation and accumulation in coronary arteries as well as other arteries (Wagner *et al.*, 1991; 1992) (Figure 8.1A). The rate of arterial LDL degradation was decreased by an average of 78% and total arterial LDL accumulation decreased by about 60%. Regional differences in indices of LDL metabolism among arterial sites were found. Accumulation of LDL, LDL degradation rates, and the concentrations of undegraded LDL were greatest in the coronary arteries and carotid bifurcations compared with the aorta, iliac and common carotid arteries, and least in cerebral arteries. Interestingly, the coronary arteries accumulated the most LDL yet had thinner intimas than other arterial sites, suggesting differences in inherent LDL metabolism among arterial sites. Sex hormone deficiency caused the same percent increase in arterial LDL degradation and accumulation in all arterial sites, but the absolute amount of the increase was greatest for the coronary arteries. Thus, it is possible that with the onset of menopause (surgical or natural), coronary arteries may be among the first arterial sites to increase metabolism and accumulation of LDL, contributing to the development of clinically significant atherosclerosis.

There was no effect of hormone therapy on hepatic or whole body LDL metabolism in this study. Also, no significant changes in plasma lipid, lipoprotein, or apolipoprotein concentrations were found. However, animals receiving hormone treatment had significantly more protein in the high density lipoprotein $(HDL)_{2b}$ subfraction and less in the $HDL_{3b}$ subfraction compared with controls. However, the combined $HDL_2$ and $HDL_3$ subfractions for controls (61% and 39%, respectively) and treatment group (65% and 35%, respectively) were not statistically different. LDL molecular weight was also reduced by hormone treatment, and this was the only variable consistently correlated with LDL degradation and accumulation, suggesting that sex hormone deficiency results in larger LDL particles that are more atherogenic. This is consistent with results of other studies in nonhuman primates, in which LDL molecular weight was highly associated with coronary artery atherosclerosis (Rudel, Bond, and Bullock, 1985).

The minimal effect of these hormones on plasma lipoprotein concentrations is consistent with reports of women given parenteral estrogen replacement therapy (Crook and Seed, 1990). The lack of effect of hormone therapy on hepatic and whole body LDL metabolism (i.e., fractional catabolic rate [FCR]) also may be due to the parenteral route of administration.

### Study Two: Oral Hormone Replacement Therapy

Estrogen/androgen combinations appear to be effective in ameliorating somatic complaints associated with estrogen deficiency, and also improve psychologic symptoms, libido, and sexual satisfaction (Burger *et al.*, 1984; Sherwin and Gelfand, 1985), yet may result in lowered plasma HDL cholesterol concentrations (Hickock, Toomey, and Speroff, 1993; Watts *et al.*, 1995). Thus, the objective of this study was to

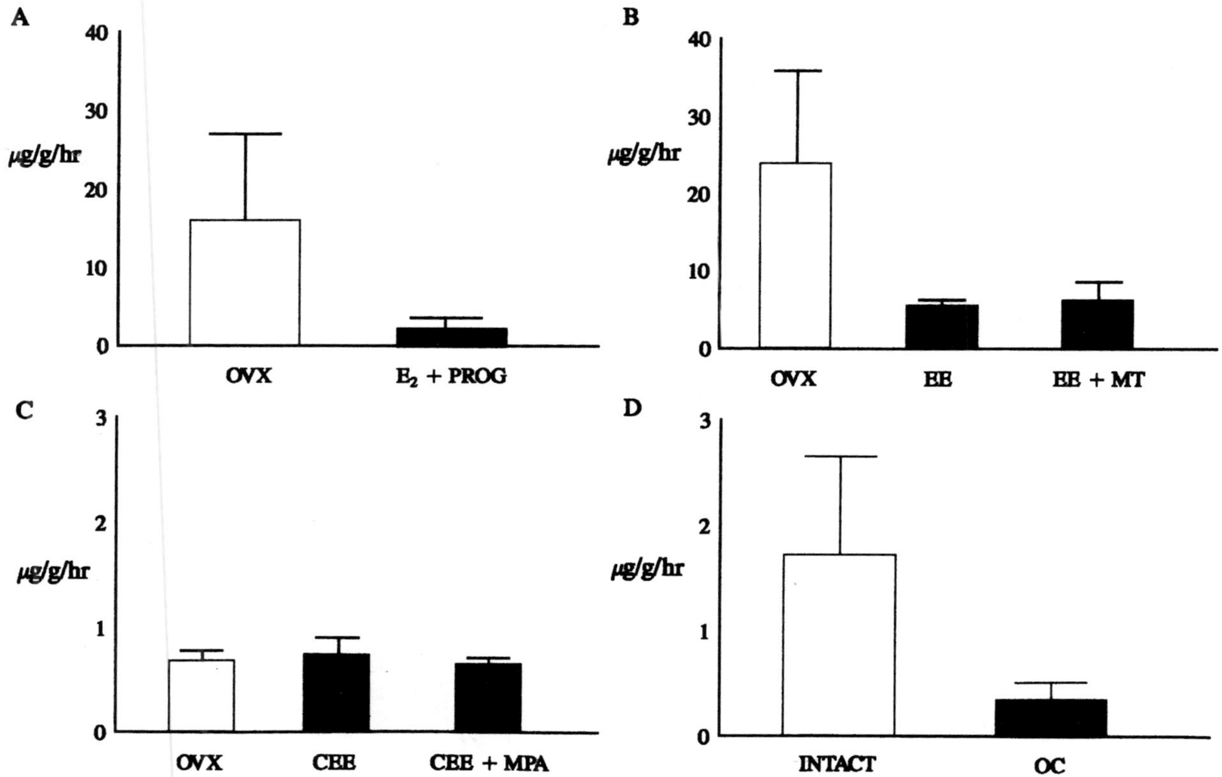


**Figure 8.1.** Effect of hormone treatments (black bars) on coronary artery LDL degradation rates of cynomolgus monkeys (μg/g/hr) compared with either ovariectomized control (OVX) or intact control (INTACT) groups. $E_2$ = estradiol; EE = esterified estrogen; MT = methyltestosterone; CEE = conjugated equine estrogens; MPA = medroxyprogesterone acetate; OC = oral contraceptives. See text for details of individual experiments.

examine the effects of oral esterified estrogens alone or in combination with an androgen, methyltestosterone, on LDL metabolism and early atherogenesis after 16 weeks of feeding an atherogenic diet. Three treatment groups of ovariectomized monkeys were studied: no drug treatment (controls, $n=11$), esterified estrogens (Estratab[R], Solvay Pharmaceuticals, Marietta, GA) (EE, $n=11$), and EE plus methyltestosterone (Estratest[R], Solvay Pharmaceuticals) (EE+MT, $n=10$). Drugs were given via a dosing solution calculated to be equivalent to a 60 kg woman receiving 2.5 mg EE and 5.0 mg of MT and resulted in estrone and equilin concentrations similar to women taking 1.25 mg EE.

Arterial LDL metabolism was again assessed using dual-labeled LDL as described above. LDL were isolated from pooled plasma from control animals, labeled, and injected 24 hours before necropsy. All indices of arterial LDL metabolism were decreased with both treatments (Wagner *et al*., 1996b). The hormone treatments had the greatest effect on the rate of arterial LDL degradation (Figure 8.1B), which was reduced an average of 73% with EE and 75% with EE+MT. This is consistent with our previous study of the effects of combined estradiol and progesterone, in which treatment decreased LDL degradation rate by 78% compared with ovariectomized controls. As above, LDL metabolism varied with arterial site but the treatments resulted in similar effects at all arterial sites. Furthermore, there were no significant differences in LDL degradation between EE alone and EE+MT. In addition, both treatments decreased cholesteryl ester content in the thoracic aorta.

Unlike the study using parenteral hormone replacement, treatment with EE alone significantly increased the whole body FCR of LDL ($p<0.05$), with a similar trend in the EE+MT group. Also, whereas parenteral hormone replacement did not influence plasma lipids, in this study total plasma cholesterol (TPC) concentrations were decreased by 30% in the EE group and 35% in the EE+MT group compared with the control group. Both treatments also reduced LDL cholesterol and combined very low density lipoprotein and intermediate density lipoprotein cholesterol concentrations. Although plasma HDL cholesterol concentrations did not differ significantly among treatment groups, triglyceride concentrations were increased in both EE and EE+MT groups compared with the control group.

In addition to decreasing lipoprotein concentrations, both treatments decreased LDL molecular weight ($p<0.01$). The decrease in apoB-containing lipoproteins may be due to the increased whole body LDL FCR, whereas the presence of smaller LDL particles may be due to higher triglyceride concentrations or a selective removal of larger particles (see below).

Although treatment did not significantly affect intimal area in the above two studies, EE treatment did change arterial morphology and significantly reduced cholesteryl ester content in the thoracic aorta. To assess the relative importance of macrophages in determining the effect of hormone treatment in arterial LDL metabolism, we next chose to study arterial LDL metabolism after only 12 weeks of atherogenic stimulus.

### Study Three: Short-Term Oral Hormone Replacement Therapy

In Study Three, ovariectomized monkeys were fed the atherogenic diet used previously with either (1) no treatment (control) ($n=10$), or the following hormones added to the diet: (2) conjugated equine estrogen (CEE) (Premarin[R], Wyeth-Ayerst, Radnor, PA) at a dose equivalent to 0.625 mg/day for women ($n=9$), (3) medroxy-progesterone acetate (MPA) at a dose equivalent to 2.5 mg (Cycrin[R], Wyeth-Ayerst) ($n=10$), or (4) combined CEE+MPA, at the above doses ($n=10$).

In this 12-week study, neither CEE nor MPA treatment had any effect on arterial LDL metabolism (Wagner *et al.*, 1997) (Figure 8.1C). We have recently completed a long-term (30 month) study of the same hormone treatments and found a 72% reduction in coronary artery atherosclerosis with CEE treatment compared with controls (Adams *et al.*, 1997). Thus, the lack of effect of CEE in this short-term study does not appear to be due to a lack of effect of CEE in decreasing atherosclerosis progression, but suggests that the 12-week period may have been too short to induce arterial changes required for the upregulation of arterial LDL degradation rate and subsequent effects of estrogens. In support of this, arterial degradation rates for postmenopausal monkeys in the control group at 12 weeks ranged from 4 to 14% of those for the corresponding arterial sites of postmenopausal monkeys in the control group of Study One examined after 18 weeks. Furthermore, cholesterol concentrations in the thoracic aorta at 12 weeks were 50% lower than those obtained after 16 weeks (Study Two). Thus, perhaps estrogen can only reduce arterial degradation rates that are elevated by early atherogenic changes, rather than normal basal values of arterial LDL metabolism. There may be other, unidentified reasons why CEE did not decrease arterial LDL accumulation at this stage of atherogenesis, since in another study we saw effects of tamoxifen, a mixed estrogen agonist/antagonist, after 12 weeks of treatment (Williams *et al.*, 1997).

Despite having no effect on arterial LDL metabolism, CEE treatment had a number of effects on plasma lipoproteins and on hepatic LDL metabolism. In general, CEE treatment resulted in slight decreases in the cholesterol content of lipoprotein fractions, but significantly reduced LDL molecular weight. In a subset of the monkeys from this study (Manning *et al.*, 1996), a detailed analysis of the LDL particles revealed that CEE treatment not only reduced LDL molecular weight, but also enriched LDL with protein and triglyceride while depleting LDL of cholesteryl ester and apolipoprotein E (apo E) compared with LDL from control monkeys. These changes in the LDL particles were most pronounced in the largest and lightest subfraction ($d=1.015$–$1.025$).

As in Study Two, CEE increased the whole body FCR by 40% ($p<0.10$) as well as hepatic LDL uptake ($p<0.01$) (Wagner *et al.*, 1997). Despite the increased removal of plasma LDL particles by the liver, CEE treatment substantially decreased hepatic lipid content ($p<0.001$). Our preliminary analyses (Colvin *et al.*, 1994) suggest that the decrease in hepatic cholesterol content is consistent with an increased biliary secretion of cholesterol as has been shown in women (Everson, McKinley, and Kern, 1991) and in baboons (Kushwaha and Born, 1991).

## ORAL CONTRACEPTIVES AND ARTERIAL LDL METABOLISM

Like estrogen replacement, oral contraceptives (OC) also attenuate arterial LDL metabolism (Wagner *et al.*, 1993). In a 16-week experiment, we studied intact female monkeys fed an atherogenic diet alone (controls, $n=7$) or with one of two OCs; a monophasic formulation (Ovral$^R$, Wyeth-Ayerst; equivalent to a human dose of 50 μg ethinyl estradiol and 500 μg of norgestrel per day) ($n=8$), or a triphasic formulation (Triphasil$^R$, Wyeth-Ayerst; equivalent to a human dose of 30–40 μg ethinyl estradiol and 50–125 μg of levonorgestrel per day) ($n=8$). Despite causing a slightly more atherogenic lipoprotein profile than diet alone (decreased HDL cholesterol and increased TPC : HDL cholesterol ratio), both OCs decreased arterial LDL degradation by over 80% (Figure 8.1D). As in the studies described previously, treatment also significantly reduced LDL molecular weights.

As with EE treatment, both OC treatments also reduced hepatic cholesterol content, which was independent of differences in LDL receptor mRNA and 3-hydroxy-3-methyl-glutaryl coenzyme A (HMG-CoA) reductase mRNA (Colvin *et al*., 1993). This occurred despite about a 30% increase in whole body LDL metabolism and a 50% increase in hepatic LDL uptake. As noted above, these findings are also consistent with an increased secretion of cholesterol into the bile acid biosynthetic pathway.

In all the studies described above, estrogen treatment consistently reduced LDL molecular weight. In addition, LDL molecular weight was highly correlated with measures of LDL degradation. To further explore the effects of LDL heterogeneity on arterial LDL metabolism, female monkeys were fed an atherogenic diet for 26 months with either no treatment (controls) or ethinyl estradiol (equivalent to 30–40 μg per day). Estrogen did not significantly influence plasma lipoprotein concentrations but again significantly reduced both LDL molecular weight and arterial LDL degradation (Wagner, Zhang, and Adams, 1996a). In half the animals, LDL were isolated from either control or treated monkeys and differentially labeled with either $^{125}$I-TC or $^{131}$I-TC. Labeled LDLs were then combined and re-injected into both groups. Although hepatic TC-LDL accumulation was significantly greater with LDL from control animals, there was minimal effect of LDL source on arterial TC accumulation. These data suggest that estrogen influences arterial LDL degradation via direct effects on the artery.

## ESTROGENS AND OXIDATION

Several lines of evidence suggest that oxidation of LDL contributes to atherogenesis (Parthasarathy, Steinberg, and Witztum, 1992). An additional mechanism for estrogen's cardioprotection and inhibitory effect on arterial LDL metabolism may be its antioxidant activity. The hormones 17$\beta$-estradiol (Huber *et al*., 1990), estradiol, estriol, and estrone (Maziere *et al*., 1991) reduce copper-mediated oxidation of LDL, oxidation of LDL by monocytes (Maziere *et al*., 1991), degradation of copper-treated LDL by macrophages (Maziere *et al*., 1991), and cholesterol accumulation in cultured macrophages incubated with copper-treated LDL (Huber *et al*., 1990). Thus, inhibition of intra-arterial oxidation of LDL may contribute to the anti-atherosclerotic effects of estrogen. However, the concentrations of estrogens required to inhibit oxidation *in vitro* are several orders of magnitude greater than physiological concentrations.

Some (Keaney *et al*., 1994; Guetta *et al*., 1995a; Sack, Rader, and Cannon, 1994) but not other (Guetta *et al*., 1995b) studies of *in vivo* estrogen treatment have observed decreased susceptibility of LDL to *in vitro* oxidation. There are several potential reasons for these divergent results. First, some studies that showed reduced susceptibility of LDL to *in vitro* oxidation after estrogen treatment used pharmacological supplements. For example, plasma estradiol concentrations reported by Sack *et al*. (1994) in postmenopausal women with intravenous or transdermal 17$\beta$-estradiol were 8–29 times greater than pretreatment concentrations and somewhat greater than normal midcycle estrogen concentrations of premenopausal women (Saaman and Crawford, 1995). Interestingly, both Keaney *et al*. (1994) and Sack *et al*. (1994) reported that resistance to oxidation was not correlated with estradiol concentrations during treatment, although the latter reported a positive correlation between lag times of LDL oxidation and pretreatment plasma estradiol concentrations.

Second, differences in type of estrogen used may have contributed to the contradictory findings. *In vitro* studies have indicated varying potencies of different forms of estrogen (Maziere *et al*., 1991; Matthys *et al*., 1996). Studies in which estrogen treatment was associated with reduced susceptibility of LDL to *in vitro* oxidation used 17$\beta$-estradiol (Keaney *et al*., 1994; Guetta *et al*., 1995a; Sack, Rader, and Cannon, 1994), whereas another study with conjugated equine estrogens (a common form of postmenopausal estrogen replacement) had no effect on lag time before oxidation (Guetta *et al*., 1995b). Third, route of delivery may also confound study results. Compared with endogenous estrogen or doses obtained via subcutaneous implants, concentrations of orally administered estrogen may well be lower on average when the entire day is considered.

Irrespective of whether physiological levels of estrogens affect oxidation of LDL in the plasma compartment, it is possible that estrogens may function as antioxidants within membranes of arterial cells. Antioxidants present in arterial membranes may have an influence on intra-arterial oxidation of LDL equal to or greater than that predicted by the susceptibility of LDL to *in vitro* oxidation. In this regard, it is relevant to note that pretreatment with estrogens inhibited monocytes' ability to oxidize LDL when tested in the absence of estrogens (Maziere *et al*., 1991). The study with macrophages demonstrates that membrancs of cells from other than reproductive tissues can be enriched with at least some estrogens. In addition, certain tissues, including the heart and liver, express the enzyme 2-hydroxylase (Hoffman, Paul, and Axelrod, 1980), which converts estradiol into the more potent antioxidant 2-hydroxyestradiol. The relationships between arterial concentrations of various estrogens and the presence or absence of 2-hydroxylating enzymes remain to be elucidated.

Consistent with estrogen's antioxidant ability, we have found that estrogens decrease the arterial content of thiobarbituric acid-reactive substances, a product of lipid peroxidation (Wagner *et al*., 1995; 1996b). Data from tissue culture experiments suggest that inhibition of intra-arterial oxidation would reduce degradation of LDL by macrophages (Parthasarathy *et al*., 1986). If reduction of intra-arterial oxidation of LDL contributes to estrogen's inhibition of atherosclerosis, estrogen should reduce metabolism of lipoproteins by arterial cells independent of effects on arterial permeability or intra-arterial concentrations of lipoproteins. As described above, estrogen appears not to influence arterial permeability (Hough and Zilversmit, 1986; Haarbo *et al.*, 1994; Schwenke, 1996). In addition, in the 16- to 18-week studies described above, estrogen treatment strikingly decreased the rate of arterial LDL degradation in all sites, despite small effects on arterial concentrations of undegraded LDL (Wagner *et al.*, 1991; 1992; 1993; 1996b). Also, an earlier study reported that arterial metabolism of lipoprotein cholesteryl ester was markedly decreased by estrogen treatment independent of differences in arterial permeability to lipoproteins (Hough and Zilversmit, 1986). Consistent with the idea that estrogens may reduce arterial degradation of LDL secondary to inhibiting intra-arterial oxidation, the degree of reduction in LDL degradation rate by estrogens was similar to that observed for atherosclerotic lesions of animals treated with the antioxidant probucol (Carew, Schwenke, and Steinberg, 1987). If in fact estrogen does inhibit intra-arterial oxidation of LDL, effects of estrogen on arterial metabolism might be less apparent when few macrophages are present in the arterial intima. This may explain why we failed to observe any effect of estrogen on arterial LDL degradation in our 12-week study, when few intimal macrophages were present.

## ESTROGENS AND LDL SIZE AND COMPOSITION

In previous studies, estrogens have consistently decreased LDL molecular weight (van der Mooren *et al.*, 1994; Campos *et al.*, 1988; 1993; Granfone *et al.*, 1992). Further, LDL molecular weight has been associated with coronary artery disease in both human (Crouse *et al.*, 1985) and nonhuman primates (Rudel, Bond, and Bullock, 1985). Reduced LDL molecular weight was associated with decreased arterial LDL degradation (see above) and decreased extent of coronary artery atherosclerosis in monkeys (Rudel, Bond, and Bullock, 1985). In contrast, some studies have found smaller LDLs associated with increased CHD risk in people (Crouse *et al.*, 1985) while others have not (Campos *et al.*, 1995). Smaller LDL are also reported to be more susceptible to oxidation (de Graaf *et al.*, 1991; Tribble *et al.*, 1992); however, this may be less important in the presence of an antioxidant, such as estrogen.

Decreased LDL size may be secondary to estrogen-induced increases in plasma triglyceride concentrations, which (upon lipolysis of LDL triglycerides) results in smaller LDL particles (Crouse *et al.*, 1985; Rudel *et al.*, 1986). This is consistent with the negative correlation between plasma triglycerides and LDL size in our studies (Wagner *et al.*, 1996b) and in humans (Crouse *et al.*, 1985; McNamara *et al.*, 1992). Since increased plasma triglyceride concentrations often are associated with insulin resistance, this may be responsible for the increased risk of CHD associated with small LDL in clinical trials. This is consistent with a recent study by Campos *et al.* (1995) where in normolipidemic men, those at greatest risk for CHD had predominantly larger LDL particles.

Another mechanism for the reduced size of LDL particles associated with estrogen treatment is selective removal of large LDL particles. Large LDL are more apo E-enriched (Stevenson, Sawyer, and Rudel, 1992) and may be removed selectively via either the apoB/E receptor or the lipoprotein-related protein receptor. The increased hepatic uptake of LDL seen with some estrogen therapies (together with a selective removal of large LDL) may be responsible for removal of larger, apoE-enriched particles, resulting in smaller LDL left in circulation. This may also account for the decreased plasma apoE concentrations associated with estrogen treatment (Manning *et al.*, 1996; Applebaum-Bowden *et al.*, 1989; Kushwaha *et al.*, 1991).

In conclusion, we have found that estrogens inhibit the arterial degradation and accumulation of LDL. We believe that this results in decreased progression of atherosclerosis and may explain the cardioprotective effects of estrogen. This effect appears to be independent of plasma lipid and lipoprotein concentrations; however, more subtle changes in lipoproteins may be affected. For example, LDL heterogeneity and oxidation are both affected by estrogens. Further studies are required to elucidate the relationships among these events.

## ACKNOWLEDGMENT

Research from our laboratory described in this review was supported in part by Grants P01 HL45666 from the National Heart, Lung and Blood Institute and RR0072 from the National Center for Research Resources, both from the National Institutes of Health, Bethesda, MD, USA.

## REFERENCES

Adams, M.R., Kaplan, J.R., Manuck, S.B., Koritnik, D.R., Parks, J.S., Wolfe, M.S. *et al.* (1990) Inhibition of coronary artery atherosclerosis by 17-beta estradiol in ovariectomized monkeys. Lack of an effect of an added progesterone. *Arteriosclerosis*, **10**, 1051–1057.

Adams, M.R., Register, T.C., Golden, D.L., Wagner, J.D. and Williams, J.K. (1997) Medroxyprogesterone acetate antagonizes inhibitory effects of conjugated equine estrogens on coronary artery atherosclerosis. *Arteriosclerosis, Thrombosis, and Vascular Biology*, **17**, 217–221.

Applebaum-Bowden, D., McLean, P., Steinmetz, P., Fontana, D., Matthys, C., Warnick, G.R. *et al.* (1989) Lipoprotein, apolipoprotein, and lipolytic enzyme changes following estrogen administration in postmenopausal women. *Journal of Lipid Research*, **30**, 1895–1906.

Barrett-Connor, E. and Bush, T.L. (1991) Estrogen and coronary heart disease in women. *Journal of the American Medical Association*, **265**, 1861–1867.

Burger, H.G., Hailes, J., Menelaus, M., Nelson, J., Hudson, B. and Balazs, N. (1984) The management of persistent menopausal symptoms with oestradiol-testosterone implants: Clinical, lipid and hormonal results. *Maturitas*, **6**, 351–358.

Campos, H., McNamara, J.R., Wilson, P.W.F., Ordovas, J.M. and Schaefer, E.J. (1988) Differences in low density lipoprotein subfractions and apolipoproteins in premenopausal and postmenopausal women. *Journal of Clinical Endocrinology and Metabolism*, **67**, 30–35.

Campos, H., Roederer, G.O., Lussier-Cacan, S., Davignon, J. and Krauss, R.M. (1995) Predominance of large LDL and reduced $HDL_2$ cholesterol in normolipidemic men with coronary artery disease. *Arteriosclerosis, Thrombosis, and Vascular Biology*, **15**, 1043–1048.

Campos, H., Sacks, F.M., Walsh, B.W., Schiff, I., O'Hanesian, M.A. and Krauss, R.M. (1993) Differential effects of estrogen on low-density lipoprotein subclasses in healthy postmenopausal women. *Metabolism*, **42**, 1153–1158.

Carew, T.C., Pittman, R.C., Marchand, E.R. and Steinberg, D. (1984) Measurement *in vivo* of irreversible degradation of low density lipoprotein in the rabbit aorta. *Arteriosclerosis*, **4**, 214–224.

Carew, T.E., Schwenke, D.C. and Steinberg, D. (1987) Antiatherogenic effect of probucol unrelated to its hypocholesterolemic effect: Evidence that antioxidants *in vivo* can selectively inhibit low density lipoprotein degradation in macrophage-rich fatty streaks and slow the progression of atherosclerosis in the Watanabe heritable hyperlipidemic rabbit. *Proceedings of the National Academy of Sciences of the United States of America*, **84**, 7725–7729.

Clarkson, T.B., Adams, M.R., Williams, J.K. and Wagner, J.D. (1993) Clinical implications of animal models of gender difference in heart disease. In *Cardiovascular Health and Disease in Women*, edited by P.S. Douglas, pp. 283–302. Philadelphia: W.B. Saunders Company.

Colvin, P.L., Jr., Wagner, J.D., Adams, M.R. and Sorci-Thomas, M.G. (1994) Sex steroid replacement increases hepatic 7 alpha-hydroxylase mRNA (abstract). *Circulation*, **90**, I-351.

Colvin, P.L., Wagner, J.D., Heuser, M.D. and Sorci-Thomas, M.G. (1993) Oral contraceptives decrease hepatic cholesterol independent of the LDL receptor in nonhuman primates. *Arteriosclerosis, Thrombosis, and Vascular Biology*, **13**, 1645–1649.

Crook, D. and Seed, M. (1990) Endocrine control of plasma lipoprotein metabolism: effects of gonadal steroids. *Baillière's Clinical Endocrinology and Metabolism*, **4**, 851–875.

Crouse, J.R., Parks, J.S., Schey, H.M. and Kahl, F.R. (1985) Studies of low density lipoprotein molecular weight in human beings with coronary artery disease. *Journal of Lipid Research*, **26**, 566–573.

Daugherty, A., Zweifel, B.S., Sobel, B.E. and Schonfeld, G. (1988) Isolation of low density lipoprotein from atherosclerotic vascular tissue of Watanabe heritable hyperlipidemic rabbits. *Arteriosclerosis*, **8**, 768–777.

de Graaf, J., Hak-Lemmers, H.L.M., Hectors, M.P.C., Demacker, P.N.M., Hendriks, J.C.M. and Stalenhoef, A.F.H. (1991) Enhanced susceptibility to *in vitro* oxidation of the dense low density lipoprotein subfraction in healthy subjects. *Arteriosclerosis and Thrombosis*, **11**, 298–306.

Everson, G.T., McKinley, C. and Kern, F., Jr. (1991) Mechanisms of gallstone formation in women. Effects of exogenous estrogen (Premarin) and dietary cholesterol on hepatic lipid metabolism. *Journal of Clinical Investigation*, **87**, 237–246.

Granfone, A., Campos, H., McNamara, J.R., Schaefer, M.M., Lamon-Fava, S., Ordovas, J.M. *et al.* (1992) Effects on estrogen replacement on plasma lipoproteins and apolipoproteins in post-menopausal, dyslipidemic women. *Metabolism*, **41**, 1193–1198.

Guetta, V., Lush, R.M., Figg, W.D., Waclawiw, M.A. and Cannon R.O. (1995b) Effects of the antiestrogen tamoxifen on low density lipoprotein concentrations and oxidation in postmenopausal women. *American Journal of Cardiology*, **76**, 1072–1073.

Guetta, V., Panza, J.A., Waclawiw, M.A. and Cannon, R.O. (1995a) Effect of combined 17 beta-estradiol and vitamin E on low-density lipoprotein oxidation in postmenopausal women. *American Journal of Cardiology*, **75**, 1274–1276.

Haarbo, J., Leth-Espensen, P., Stender, S. and Christiansen, C. (1991) Estrogen monotherapy and combined estrogen-progestogen replacement therapy attenuate aortic accumulation of cholesterol in ovariectomized cholesterol-fed rabbits. *Journal of Clinical Investigation*, **87**, 1274–1279.

Haarbo, J., Nielsen, L.B., Stender, S. and Christiansen, C. (1994) Aortic permeability to LDL during estrogen therapy. A study in normocholesterolemic rabbits. *Arteriosclerosis and Thrombosis*, **14**, 243–247.

Hickock, L.R., Toomey, C. and Speroff, L. (1993) A comparison of esterified estrogens with and without methyltestosterone: Effects on endometrial histology and serum lipoproteins in postmenopausal women. *Obstetrics and Gynecology*, **82**, 919–924.

Hoffman, A.R., Paul, S.M. and Axelrod, J. (1980) Estrogen 2-hydroxylase in the rat: Distribution and response to hormonal manipulation. *Biochemical Pharmacology*, **29**, 83–87.

Hough, J.L. and Zilversmit, D.B. (1986) Effect of 17-$\beta$ estradiol on cholesterol content and metabolism in cholesterol-fed rabbits. *Arteriosclerosis*, **6**, 57–63.

Huber, L.A., Scheffler, E., Poll, T., Ziegler, R. and Dresel, H.A. (1990) 17-beta-estradiol inhibits LDL oxidation and cholesteryl ester formation in cultured macrophages. *Free Radicals Research Communications*, **8**, 167–173.

Keaney, J.F., Shwaery, G.T., Xu, A.M., Nicolosi, R.J., Loscalzo, J., Foxall, T.L. *et al.* (1994) 17 Beta-estradiol preserves endothelial vasodilator function and limits low-density lipoprotein oxidation in hypercholesterolemic swine. *Circulation*, **89**, 2251–2259.

Kushwaha, R.S. and Born, K.M. (1991) Effect of estrogen and progesterone on the hepatic cholesterol 7$\alpha$-hydroxylase activity in ovariectomized baboons. *Biochimica et Biophysica Acta*, **1084**, 300–302.

Kushwaha, R.S., Foster, D.M., Barrett, P.H.R., Carey, K.D. and Bernard, M.G. (1991) Metabolic regulation of plasma apolipoprotein E by estrogen and progesterone in the baboon (*Papio* sp). *Metabolism*, **40**, 93–100.

Manning, J.M., Campos, G., Edwards, I.J., Wagner, W.D., Wagner, J.D., Adams, M.R. *et al.* (1996) Effects of hormone replacement modalities on low density lipoprotein composition and distribution in ovariectomized cynomolgus monkeys. *Atherosclerosis*, **121**, 217–229.

Matthys, K.E., Vanhove, C.E., Jorens, P.G., Rosseneu, M., Marescau, B., Herman, A.G. *et al.* (1996) Dual effects of oxidized low-density lipoprotein on immune-stimulated nitric oxide and prostaglandin release in macrophages. *European Journal of Pharmacology*, **298**, 97–103.

Maziere, C., Auclair, M., Ronveaux, M.-F., Salmon, S., Santus, R. and Maziere, J.C. (1991) Estrogens inhibit copper and cell-mediated modification of low density lipoprotein. *Atherosclerosis*, **89**, 175–182.

McNamara, J.R., Jenner, J.L., Zhengling, L., Wilson, P.W.F. and Schaefer, E.J. (1992) Change in LDL particle size is associated with change in plasma triglyceride concentration. *Arteriosclerosis and Thrombosis*, **12**, 1284–1290.

Newman, H.A.I. and Zilversmit, D.B. (1962) Quantitative aspects of cholesterol flux in rabbit atheromatous lesions. *Journal of Biological Chemistry*, **237**, 2078–2084.

Nielsen, L.B., Nordestgaard, B.G., Stender, S. and Kjeldsen, K. (1992) Aortic permeability to LDL as a predictor of aortic cholesterol accumulation in cholesterol-fed rabbits. *Arteriosclerosis and Thrombosis*, **12**, 1402–1409.

Nordestgaard, B.G., Tybjaerghansen, A. and Lewis, B. (1992) Influx *in vivo* of low density, intermediate density, and very low density lipoproteins into aortic intimas of genetically hyperlipidemic rabbits. Roles of plasma concentration, extent of aortic lesion, and lipoprotein particle size as determinants. *Arteriosclerosis and Thrombosis*, **12**, 6–18.

Nordestgaard, B.G., Wootton, R. and Lewis, B. (1995) Selective retention of VLDL, IDL, and LDL in the arterial intima of genetically hyperlipidemic rabbits *in vivo*. Molecular size as a determinant of fractional loss from the intima-inner media. *Arteriosclerosis, Thrombosis, and Vascular Biology*, **15**, 534–542.

Parthasarathy, S., Steinberg, D. and Witztum, J.L. (1992) The role of oxidized low-density lipoproteins in the pathogenesis of atherosclerosis. *Annual Review of Medicine*, **43**, 219–225.

Parthasarathy, S., Young, S.G., Witztum, J.L., Pittman, R.C. and Steinberg, D. (1986) Probucol inhibits oxidative modification of low density lipoprotein. *Journal of Clinical Investigation*, **77**, 641–644.

Pittman, R.C., Carew, T.E., Glass, C.K., Green, S.R., Taylor, C.A. Jr. and Attie, A.D. (1983) A radioiodinated, intracellularly trapped ligand for determining the sites of plasma protein degradation *in vivo*. *Biochemical Journal*, **212**, 791–800.

Rudel, L.L., Bond, M.G. and Bullock, B.C. (1985) LDL heterogeneity and atherosclerosis in nonhuman primates. *Annals of the New York Academy of Sciences*, **454**, 248–253.

Rudel, L.L., Parks, J.S., Johnson, F.L. and Babiak, J. (1986) Low density lipoproteins in atherosclerosis. *Journal of Lipid Research*, **27**, 465–474.

Samaan, S.A. and Crawford, M.H. (1995) Estrogen and cardiovascular function after menopause. *Journal of the American College of Cardiology*, **26**, 1403–1410.

Sack, M.N., Rader, D.J. and Cannon, R.O. (1994) Oestrogen and inhibition of oxidation of low-density lipoproteins in postmenopausal women. *Lancet*, **343**, 269–270.

Sherwin, B.B. and Gelfand, M.M. (1985) Differential symptom response to parenteral estrogen and/or androgen administration in the surgical menopause. *American Journal of Obstetrics and Gynecology*, **151**, 153–160.

Schwenke, D.C. (1995) Selective increase in cholesterol at atherosclerosis-susceptible aortic sites after short-term cholesterol feeding. *Arteriosclerosis, Thrombosis, and Vascular Biology*, **15**, 1928–1937.

Schwenke, D.C. (in press) Gender differences in aortic permeability to low density lipoprotein at atherosclerosis-prone aortic sites in rabbits: lack of effect of 17$\beta$ estradiol. *Arteriosclerosis, Thrombosis, and Vascular Biology*.

Schwenke, D.C. and St. Clair, R.W. (1992) Accumulation of 125-I-tyramine cellobiose-labeled low density lipoprotein is greater in the atherosclerosis-susceptible region of White Carneau pigeon aorta and further enhanced once atherosclerotic lesions develop. *Arteriosclerosis and Thrombosis*, **12**, 446–460.

Schwenke, D.C. and St. Clair, R.W. (1993) Influx, efflux, and accumulation of LDL in normal arterial areas and atherosclerotic lesions of White Carneau pigeons with naturally occurring and cholesterol-aggravated aortic atherosclerosis. *Arteriosclerosis and Thrombosis*, **13**, 1368–1381.

Schwenke, D.C. and Carew, T.E. (1988) Quantification *in vivo* of increased LDL content and rate of LDL degradation in normal rabbit aorta occurring at sites susceptible to early atherosclerotic lesions. *Circulation Research*, **62**, 699–710.

Schwenke, D.C. and Carew, T.E. (1989a) Initiation of atherosclerotic lesions in cholesterol-fed rabbits. II. Selective retention of LDL versus selective increases in LDL permeability in susceptible sites of arteries. *Arteriosclerosis*, **9**, 908–918.

Schwenke, D.C. and Carew T.E. (1989b) Initiation of atherosclerotic lesions in cholesterol-fed rabbits. I. Focal increases in arterial LDL concentration precede development of fatty streak lesions. *Arteriosclerosis*, **9**, 895–907.

Schwenke, D.C. and St. Clair, R.W. (1993) Influx, efflux, and accumulation of LDL in normal arterial areas and atherosclerotic lesions of White Carneau pigeons with naturally occurring and cholesterol-aggravated aortic atherosclerosis. *Arteriosclerosis and Thrombosis*, **13**, 1368–1381.

Smith, E.B. and Ashall, C. (1984) Compartmentalization of water in human atherosclerotic lesions: Changes in distribution and exclusion volumes for plasma macromolecules. *Arteriosclerosis*, **4**, 21–27.

Stampfer, M.J. and Colditz, G.A. (1991) Estrogen replacement therapy and coronary heart disease: A quantitative assessment of the epidemiologic evidence. *Preventive Medicine*, **20**, 47–63.

Stender, S. and Zilversmit, D.B. (1981) Transfer of plasma lipoprotein components and of plasma proteins into aortas of cholesterol-fed rabbits. Molecular size as a determinant of plasma lipoprotein influx. *Arteriosclerosis*, **1**, 38–49.

Stevenson, S.C., Sawyer, J.K. and Rudel, L.L. (1992) Role of apolipoprotein E on cholesteryl ester-enriched low density lipoprotein particles in coronary artery atherosclerosis of hypercholesterolemic nonhuman primates. *Arteriosclerosis and Thrombosis*, **12**, 28–40.

Stewart, B.F., Brown, B.G., Zhao, X.Q., Hillger, L.A., Sniderman, A.D., Dowdy, A. *et al.* (1994) Benefits of lipid-lowering therapy in men with elevated apolipoprotein B are not confined to those with very high low density lipoprotein cholesterol. *Journal of the American College of Cardiology*, **23**, 899–906.

Sulistiyani, Adelman, S.J., Chandrasekaran, A., Jayo, J. and St. Clair, R.W. (1995) Effect of 17α-dihydroequilin sulfate, a conjugated equine estrogen, and ethynylestradiol on atherosclerosis in cholesterol-fed rabbits. *Arteriosclerosis, Thrombosis and Vascular Biology*, **15**, 837–846.

Tribble, D.L., Holl, L.G., Wood, P.D. and Krauss, R.M. (1992) Variations in oxidative susceptibility among six low-density lipoprotein subfractions of differing density and particle size. *Atherosclerosis*, **93**, 189–199.

van der Mooren, M.J., de Graaf, J., Demacker, P.N., de Haan, A.F. and Rolland, R. (1994) Changes in the low-density lipoprotein profile during 17β-estradiol-dydrogesterone therapy in postmenopausal women. *Metabolism Clinical and Experimental*, **43**, 799–802.

Wagner, J.D., Adams, M.R., Schwenke, D.C. and Clarkson, T.B. (1993) Oral contraceptive treatment decreases arterial LDL degradation in female cynomolgus monkeys. *Circulation Research*, **72**, 1300–1307.

Wagner, J.D., Clarkson, T.B., St. Clair, R.W., Schwenke, D.C., Shively, C.A. and Adams, M.R. (1991) Estrogen and progesterone replacement therapy reduces LDL accumulation in the coronary arteries of surgically postmenopausal cynomolgus monkeys. *Journal of Clinical Investigation*, **88**, 1995–2002.

Wagner, J.D., Schwenke, D.C., Zhang, L., Applebaum-Bowden, D., Bagdade, J.D. and Adams, M.R. (1997) Effects of short-term hormone replacement therapies on low density lipoprotein metabolism in cynomolgus monkeys. *Arteriosclerosis, Thrombosis, and Vascular Biology*, **17**, 1128–1134.

Wagner, J.D., St. Clair, R.W., Schwenke, D.C., Shively, C.A., Adams, M.R. and Clarkson, T.B. (1992) Regional differences in arterial low density lipoprotein metabolism in surgically postmenopausal cynomolgus monkeys: Effects of estrogen and progesterone therapy. *Arteriosclerosis and Thrombosis*, **12**, 717–726.

Wagner, J.D., Washburn, S.A., Zhang, L., Honore, E.K., Williams, J.K., Schwenke, D.C. *et al.* (1995) A non-feminizing conjugated equine estrogen decreases arterial LDL degradation and improves vascular reactivity in male rhesus monkeys (abstract). *Circulation*, **92**, I–627.

Wagner, J.D., Zhang, L. and Adams, M.R. (1996a) Ethinyl estradiol inhibits arterial LDL metabolism independent of LDL size and composition (abstract). *Circulation*, **94**, I–703.

Wagner, J.D., Zhang, L., Williams, J.K., Register, T.C., Ackerman, D.M., Wiita, B. *et al.* (1996b) Esterified estrogens with and without methyltestosterone decrease arterial LDL metabolism in cynomolgus monkeys. *Arteriosclerosis, Thrombosis, and Vascular Biology*, **16**, 1473–1480.

Waters, D., Higginson, L., Gladstone, P., Boccuzzi, S.J., Cook, T. and Lesperance, J. (1995) Effects of cholesterol lowering on the progression of coronary atherosclerosis in women: A Canadian Coronary Atherosclerosis Intervention Trial (CCAIT) substudy. *Circulation*, **92**, 2404–2410.

Watts, N.B., Notelovitz, M., Timmons, M.C., Addison, W.A., Wiita, B. and Downey, L.J. (1995) Comparison of oral estrogens and estrogens plus androgen on bone mineral density, menopausal symptoms, and lipid-lipoprotein profiles in surgical menopause. *Obstetrics and Gynecology*, **85**, 529–537.

Williams, J.K., Wagner, J.D., Zhang, L., Golden, D.L. and Adams, M.R. (1997) Tamoxifen inhibits arterial accumulation of LDL degradation products and progression of coronary artery atherosclerosis in monkeys. *Arteriosclerosis, Thrombosis, and Vascular Biology*, **17**, 403–408.

Zilversmit, D.B. (1968) Cholesterol flux in the atherosclerotic plaque. *Annals of the New York Academy of Sciences*, **149**, 710–724.

# 9 Regulation of Lp(a) by Estrogen

**Katalin Kauser * and Gabor M. Rubanyi**

*Cardiovascular Department, Berlex Biosciences,*
*15049 San Pablo Ave, Richmond, CA 94806-0099, USA.*
**Corresponding author, Tel.: 001 510 669 4282, Fax: 001 510 669 4246.*

Elevated levels of plasma lipoprotein(a) (Lp(a)) is considered an independent risk factor for coronary heart disease. Lp(a) consists of two apolipoproteins, such as the LDL particle apoB-100 and apo(a), which is unique for Lp(a). Circulating levels of Lp(a) is determined by the synthetic rate of apo(a), rather than its degradation. Sexual steroid hormones have been suggested to be potent regulators of circulating Lp(a), which is highly resistant to conventional lipid lowering therapies. Vasculoprotection by estrogen possibly includes the effect of the hormone on plasma lipids. Lp(a) lowering effect of estrogen has been demonstrated in response to endogenous or exogenously applied hormones. In recent studies from our laboratory we have investigated the mechanism of action of estrogen on apo(a) levels utilizing the YAC apo(a) transgenic mice, in which liver expression of apo(a) is driven by the human promoter. We found that hormonal changes during the estrus cycle of these mice affected circulating apo(a) levels, suggesting that this model is useful to study the regulation of apo(a). Our results with exogenously applied estrogens suggest that the Lp(a) lowering effect of the hormone is likely mediated by decreased mRNA expression of apo(a) in the liver of the transgenic animals.

KEYWORDS: Lipoprotein(a), atherosclerosis, vasculoprotection, estrogen, gene regulation.

## INTRODUCTION

Lp(a) is considered as an independent risk factor for premature coronary heart disease (Simon *et al.*, 1993; Kraft *et al.*, 1996). Epidemiological studies revealed strong correlation between high levels of circulating Lp(a) and atherosclerotic cardiovascular disease (Ridker, 1995). Studies with apo(a) transgenic mice showed that these animals, on high lipid diet, are susceptible to develop fatty streak lesions representing early stage atherosclerosis (Lawn *et al.*, 1992). Plasma levels of Lp(a) are determined via the synthesis of apo(a), one of the major constituent of Lp(a), by the liver (Rader *et al.*, 1994). Formation of Lp(a) is highly resistant to treatment with lipid lowering agents or to alteration of diet (Bostom *et al.*, 1994). Gender difference and endogenous or exogenous sexual steroid hormones seem to be the most effective factors regulating Lp(a) levels (Julius *et al.*, 1994; Shewmon *et al.*, 1994;

Caruso *et al.*, 1994). The Lp(a) lowering effect of estrogen is likely to contribute to the hormone's beneficial effect on cardiovascular diseases. This chapter will review current opinion about the role of Lp(a) in atherosclerosis and provide a summary of recent data investigating hormonal regulation of Lp(a) in apo(a) transgenic mice.

## CHARACTERISTICS OF LP(a)

The protein moiety of Lp(a) consists of two covalently bound apoproteins, such as apoB-100, the main constituent of low density lipoprotein (LDL) and apo(a), which is a unique component of Lp(a). Apo(a) is a high molecular weight glycoprotein, synthesized by the liver, although expression of apo(a) has also been found in the testes and brain (McLean *et al.*, 1987). ApoB-100 is also synthesized by the liver, but the assembly of the Lp(a) molecule takes place outside of the hepatocytes in the plasma. Circulating levels of Lp(a) is determined by the biosynthetic rate of apo(a). Individual differences can reach three orders of magnitude (Utermann, 1989).

The apo(a) gene is highly homologous to the plasminogen gene (McLean *et al.*, 1987). It is found in humans and in some sub-human primates, as well as the European hedgehog but in which it is encoded by a different gene (Lawn, NAVBO Meeting, Seattle, 1996). There is genetic polymorphism in the size of apo(a) resulting from different numbers of the plasminogen-like kringle IV sequence repeats (McLean *et al.*, 1987). The variation in the kringle IV repeats between individuals ranges from 11–50 (Kraft *et al.*, 1992; Lackner *et al.*, 1993). Some of these alleles can be expressed below the detection limit contributing to some of the discrepancies in the literature (Ridker, 1995; Kraft *et al.*, 1996). There is a strong inverse correlation of apo(a) size with the circulating levels of Lp(a) (Lackner *et al.*, 1993; Rader *et al.*, 1994; Gaw *et al.*, 1994).

It has been postulated that different apo(a) isoforms bear with different predictability of risk for coronary heart disease (Gaubatz *et al.*, 1990; Sandholzer *et al.*, 1992; Molgaard *et al.*, 1992). A detailed analysis of apo(a) gene variation in association with coronary heart disease is provided by a recent publication (Kraft *et al.*, 1996). This study using apo(a) phenotyping of genomic DNA revealed that the low kringle repeat number was associated with significantly elevated risk for coronary heart disease, myocardial infarction, stroke or peripheral vascular disease (Kraft *et al.*, 1996).

## POTENTIAL ATHEROGENIC ROLE OF LP(a)

Besides the strong association between high Lp(a) levels and premature coronary heart disease, the exact mechanism of how Lp(a) participates in the pathomechanism of atherosclerosis is not known. Strongest evidence for its role in atherogenesis was provided by the studies with the apo(a) transgenic mice, demonstrating fatty streak formation in the aorta of the lipid fed transgenics compared to control mice on the same diet (Lawn *et al.*, 1992).

The presence of apo(a) and apoB-100 has been demonstrated in the intima of the arterial wall from human specimens (Niendorf *et al.*, 1990). Marked accumulation of Lp(a) has been observed in lesions of human coronary arteries (Hajjar *et al.*, 1989) as well as in saphenous vein bypass grafts (Cushing *et al.*, 1989). The way how Lp(a) gets into the arterial wall is not entirely understood, diffusion as well as uptake by macrophages have been suggested (Niendorf *et al.*, 1990). Accumulation of Lp(a) in macrophage-foam cells is probably enhanced by malonaldehyde (Haberland *et al.*, 1992) or oxidative (Hoff *et al.*, 1993) modification of Lp(a) in the vessel wall.

Due to the similarity to plasminogen, Lp(a) has been suggested to contribute to the thrombogenicity of the plaque material in arterial lesions. Indeed, it has been demonstrated *in vitro* that Lp(a) inhibited binding of plasminogen to fibrin (Harpel *et al.*, 1989). Lp(a) has also thought to bind competitively to plasminogen binding sites on the vascular endothelium reducing its antithrombotic properties (Nachman *et al.*, 1991). Another site, where Lp(a) can interfere with plasminogen activation, involves inhibition of tissue plasminogen activator (tPA) binding to fibrin on the surface of the thrombi (Harpel *et al.*, 1989; Loscalzo, 1990). There are recent *in vivo* evidence supporting the inhibitory effect of Lp(a) on fibrinolysis, such as development of occlusive thrombi in balloon injured monkey carotid arteries in animals with high circulating levels of Lp(a) compared to monkeys having low plasma Lp(a) concentrations (William *et al.*, 1993).

Lp(a) has also shown to alleviate the vasodilator function of endothelium (Sorensen *et al.*, 1994). Flow induced endothelium-dependent relaxation was significantly attenuated in hypercholesterolemic children with significant inverse relation to Lp(a) levels, but not to total cholesterol level (Sorensen *et al.*, 1994).

## CONTROL OF CIRCULATING LP(a) LEVELS

Unlike other circulating lipids, Lp(a) levels are refractory to alterations in diet or treatment with most lipid lowering agents, such as resins, fibrates, HMG-CoA reductase inhibitors or probucol (Bostom *et al.*, 1994). Only niacin has been reported to evoke reductions in high plasma Lp(a) (Gurakar *et al.*, 1985; Carlson *et al.*, 1989). Significant Lp(a) regulating effects has been reported by hormonal treatments, such as increases in Lp(a) concentration by growth hormone (Eden *et al.*, 1993; Olivecrona *et al.*, 1993) or reduction by sexual steroid hormones: estrogen or androgen (Henriksson *et al.*, 1992; Soma *et al.*, 1993; Sacks *et al.*, 1994). More recently Lp(a) lowering effect of insulin-like growth factor-I (IGF-I) has been demonstrated independent of estrogen or growth hormone (Olivecrona *et al.*, 1995).

Reduction of cardiovascular mortality and morbidity of postmenopausal women on estrogen replacement therapy is well documented (Stampfer *et al.*, 1991; Rosano and Sarrel, 1994; Cacciabaudo and August, 1995). The benefit of estrogen replacement therapy on coronary heart disease is related, at least in part, to changes in lipid metabolism. In addition to favorable alteration of LDL and HDL levels, estrogen also reported to lower plasma Lp(a) concentration (Bostom *et al.*, 1994). Besides exogenous administration of estrogen, plasma lipids and Lp(a) have been shown to be regulated by cyclical variation of the endogenous hormones during the menstrual cycle (Tonolo *et al.*, 1995). Significant increases in Lp(a) levels have been reported after menopause (Jenner *et al.*, 1993; Brown *et al.*, 1993).

Orchidectomy of prostatic patients slightly increased plasma concentration of Lp(a), while estrogen therapy resulted in a 50% reduction in Lp(a) levels suggesting that the sexual steroid hormones are effective regulators of Lp(a) in men (Henriksson *et al.*, 1992).

Due to the limited phylogenetic distribution of Lp(a), the development of transgenic mice was an important milestone to our understanding of the role of apo(a) in atherogenesis (Lawn *et al.*, 1992). In order to study the regulation of the apo(a) gene a new transgenic mice was developed, which unlike the earlier version under the control of the transferrin promoter (Lawn *et al.*, 1992), contained an intact apo(a) gene, with 12 kringle IV-like repeats, in a 270-kb yeast artificial chromosome (YAC) (Frazer *et al.*, 1995). Expression of apo(a) was only found in the liver of the transgenic mice and gave rise of significantly elevated plasma apo(a) levels. Murine apoB-100 is unable to form a disulfite bond to apo(a) and therefore there is no Lp(a) formed in the apo(a) transgenic animals. First studies with the YAC apo(a) transgenic mice revealed the apo(a) lowering effect of testosterone in orchidectomized mice (Frazer *et al.*, 1995).

## EFFECT OF ESTROGENS IN APO(a) TRANSGENIC MICE

Studies in our laboratory were designed to assess the effect of endogenous and exogenous ovarian sexual steroid hormones on circulating apo(a) and on apo(a) mRNA expression in the liver of YAC apo(a) transgenic mice.

In order to investigate the effect of endogenous sexual steroid hormones on circulating apo(a) levels estrus cycle of intact female YAC apo(a) transgenic mice was monitored by daily microscopical analysis of the vaginal smears. Appearance of specifically shaped epithelial cells in the smear is characteristic for the different stages of the 4 day estrus cycle. Animals were followed throughout several cycles before daily blood samples were taken for plasma apo(a) determination. Only mice exhibiting consistent 4-day cycles were used for the study. Blood samples were taken by tail bleeding into a $K^+$EDTA containing tube (SARSTEDT, Numbrecht, Germany) and after centrifugation the plasma was assayed by the Macra ELISA Kit (Strategic Diagnostic, Newark, DE) for apo(a) content. Plasma apo(a) levels on the different days of the estrus cycle (day 1 = diestrus, day 2 = proestrus, day 3 = estrus, day 4 = metestrus) were compared to published levels of circulating 17$\beta$-estradiol on the corresponding days (Walmer *et al.*, 1992). We found an inverse correlation between apo(a) and serum estrogen levels throughout the estrus cycle of mice (Figure 9.1).

To study the effect of exogenously applied female steroid hormones female apo(a) transgenic mice were ovariectomized. Two weeks after surgery mice were treated with daily subcutaneous injection of 17alpha-ethinyl-estradiol (5 μg/day) to study the effect of estrogens on apo(a) regulation. Other mice were treated with progesterone (3 mg/day) or with the combination of 17alpha-ethinyl-estradiol (5 μg/day) and

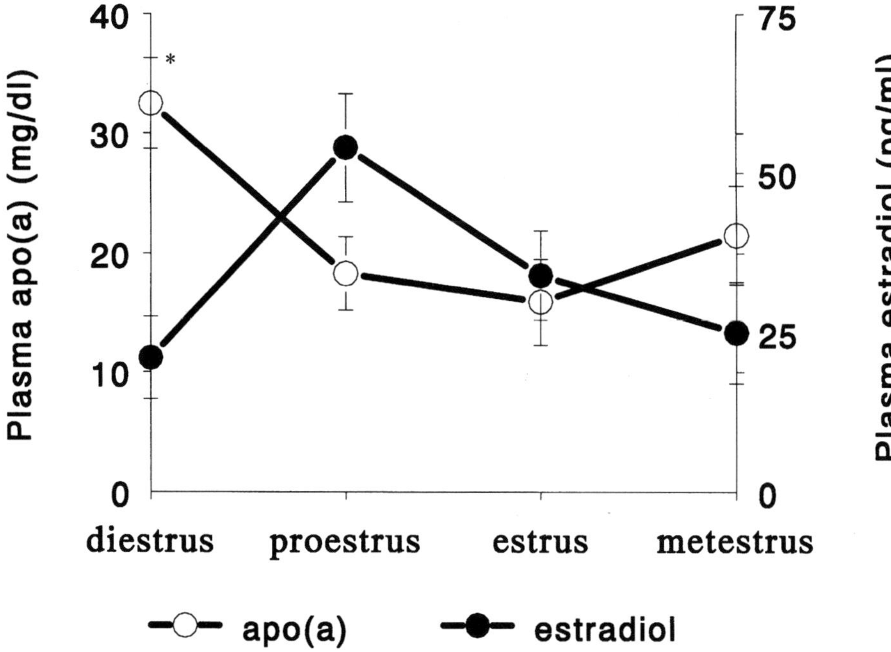


**Figure 9.1.** Changes in plasma apo(a) and 17$\beta$-estradiol level during the estrus cycle in mature female transgenic mice ($n=6$). Values represent mean $\pm$ SEM of plasma apo(a) and 17$\beta$-estradiol levels. Days of the estrus cycle were determined by microscopical analysis of the vaginal smear. 17$\beta$-estradiol levels throughout the estrus cycle: day 1 = diestrus: 21 $\pm$ 6.5 pg/mL; day 2 = proestrus: 54 $\pm$ 8.5 pg/mL; day 3 = estrus: 34 $\pm$ 7 pg/mL; day 4 = metestrus: 25 $\pm$ 8 pg/mL, according to Walmer *et al.* (1992). Asterisk (*$p<0.05$) represents statistically significant difference in plasma apo(a) level from that of in proestrus.

progesterone (3 mg/day) to evaluate the effect of hormonal combination on apo(a). Changes in plasma apo(a) levels were followed weekly by tail bleeding. Animals were sacrificed after two weeks treatment, when livers were collected in liquid nitrogen and changes in apo(a) mRNA were determined by Northern blot analysis. 17alpha-ethinyl-estradiol treatment resulted in significant decrease in circulating plasma apo(a) levels with a concomitant significant reduction in mRNA expression of apo(a) in the liver (Figure 9.2). The effect of the hormone treatment was confirmed by the measurement of uterine dry weight of the mice (Figure 9.3). Progesterone had no effect on apo(a) levels and mRNA expression when it was administered alone, and it has also not affected the responses to estrogen (Figure 9.2). It did however evoked a significant increase in uterine weight of the treated animals (Figure 9.3).

Our data indicate that the mechanism underlying estrogen-induced lowering of plasma apo(a) levels involves regulation of hepatic apo(a) mRNA. The exact mechanism by which estrogen suppresses apo(a) gene expression remains to be elucidated.

To evoke their biological actions steroid hormones, such as estrogen, bind to their intracellular receptor, forming an active complex. The activated steroid receptor

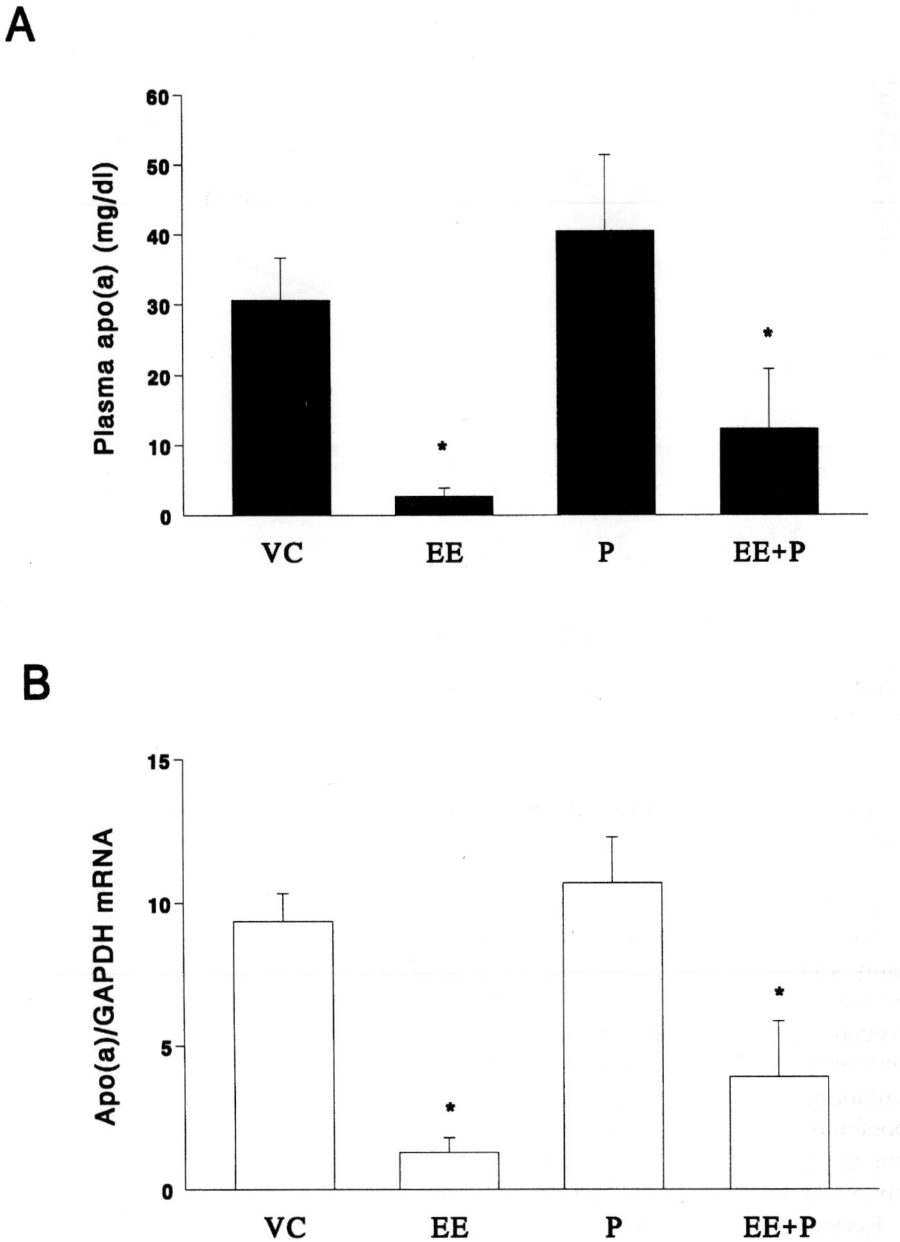


**Figure 9.2.** Effect of two weeks treatment with vehicle (VC; benzylbenzoate:castoroil/1:9; $n=7$), 17alpha-ethinyl-estradiol (EE; 5 μg/animal/day sc.; $n=7$), progesterone (P; 3 mg/animal/day sc., $n=4$) and combination of 17alpha-ethinyl-estradiol (5 μg) and progesterone (3 mg) (EE+P; $n=4$) on plasma apo(a) levels (A) and apo(a) gene expression in the liver (B). Bars represent mean ± SEM of $n=3$–7 animals. Asterisks illustrate statistically significant differences between vehicle control and treatment groups. 17alpha-ethinyl-estradiol significantly ($p<0.05$) decreased plasma apo(a) level, which correlated well with decreased expression of apo(a) gene in the liver. Progesterone alone had no effect and when given in combination did not influence the effect of 17alpha-ethinyl-estradiol.

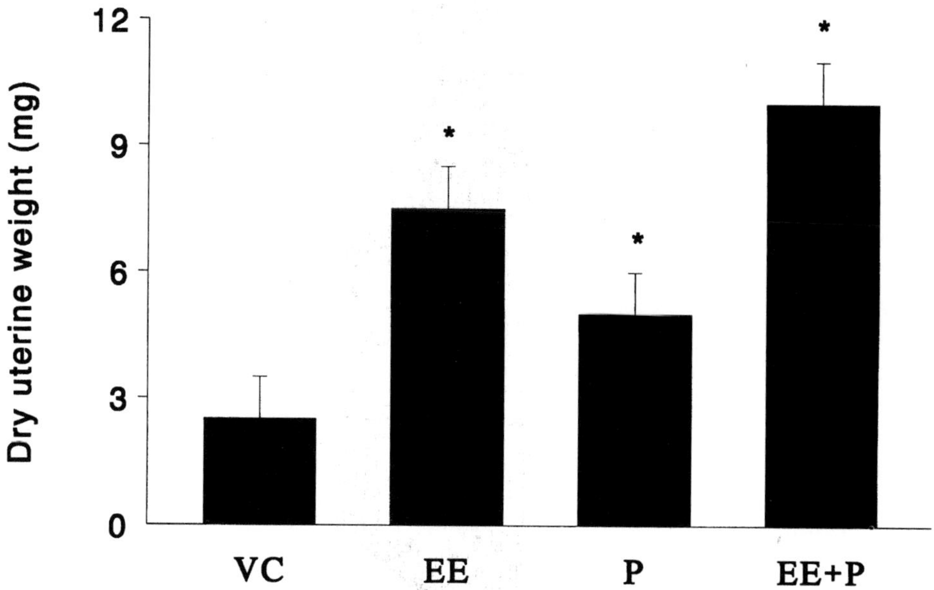


**Figure 9.3.** Dry uterine weight was determined at the end of the study to assess the estrogenic activity of the different compounds. Uteri were dissected, cleaned of adherent connective tissue and after 24 h drying in an oven, weight was measured. Values represent mean ± SEM of mg weight of the dried organs. Asterisk ($*p<0.05$) represents statistically significant difference from ovariectomized vehicle treated controls.

is able to act as a transcriptional enhancer binding to specific steroid hormone response elements on the DNA or acting as a repressor of gene expression (Beato, 1989).

It is possible that the effect of estrogen is mediated by the production of another factor. It has been proposed that insulin-like growth factor-1 (IGF-1) could potentially serve as a mediator of the effect of estrogen, since its level is decreased by estrogen treatment and shown to be increased by growth hormone, which is an upregulator of Lp(a) (Olivecrona *et al.*, 1995). Direct treatment with IGF-1, however, showed that IGF-1 is an inhibitor of Lp(a) rather than an upregulator, therefore it cannot mediate the effect of estrogen on Lp(a) (Olivecrona *et al.*, 1995). While the possibility, that estrogen exerts its effect by an indirect mechanism cannot be ruled out by our results, its mechanism involves decrease in the expression of apo(a) mRNA in the liver of the apo(a) transgenic mice.

Recent *in vitro* studies have demonstrated that inhibition of IL-6 gene expression in HeLa and murine bone marrow cells by estrogen is mediated through the estrogen receptor in the absence of high affinity DNA binding (Ray *et al.*, 1994; Pottratz *et al.*, 1994). This inhibition is mediated by a protein-protein interaction between the activated estrogen receptor and the transcription factors nuclear factor-IL6 (NF-IL6) or CCAAT/enhancer-binding protein (C/EBPb) (Stein and Yang, 1995; Galien *et al.*, 1996). This cross talk provides a new mechanism of estrogen mediated gene repression, which may play a role in apo(a) regulation by estrogen.

Our study also showed, that the other ovarian steroid hormone, progesterone, had no effect on apo(a) gene expression or plasma levels of apo(a) in transgenic mice. Progesterone also did not affect the apo(a) lowering action of estrogen, when the two were given in combination. These findings suggest, that the progestin component of combined hormone replacement therapy (HRT) preparations may not influence the beneficial effect of estrogen on plasma Lp(a) levels in postmenopausal women.

## SUMMARY

In conclusion, both epidemiological and experimental evidence supports the role of estrogen in the regulation of Lp(a) levels. Since elevated levels of Lp(a) are associated with increased risk for cardiovascular diseases, it is likely that Lp(a) lowering effect of estrogen contributes to its well-documented cardioprotective role. Although the exact mechanism of action whereby estrogen evokes its inhibition on the circulating levels of Lp(a) is not entirely understood, studies with the apo(a) transgenic animals demonstrate that its mechanism involves the suppression of apo(a) expression in the liver.

## REFERENCES

Bostom, A.G., Gagnon, D.R., Cupples, L.A., Wilson, P.W.F., Jenner, J.L., Ordovas, J.M., Schaefer, E.J. and Castelli, W.P. (1994) A prospective investigation of elevated lipoprotein(a) detected by electrophoresis and cardiovascular disease in women. *Circulation*, **90**, 1688–1695.

Brown, S.A., Morrisett, J.D., Boerwinkle, E., Hutchinson, R. and Patsch, W. (1993) The relation of lipoprotein(a) concentrations and apolipoprotein(a) phenotypes with asymptomatic atherosclerosis in subjects of the Atherosclerosis Risk in Communities (ARIC) Study. *Arteriosclerosis and Thrombosis*, **13**, 1558–1566.

Cacciabaudo, J.M. and August, P. (1995) Female sex hormones and cardiovascular disease: implications for therapy. In *Hypertension: Pathophysiology, Diagnosis, and Management, Second Edition*, edited by J.H. Laragh and B.M. Brenner, pp. 2391–2406. New York: Raven Press, Ltd.

Carlson, L.A., Hamsten, A. and Asplund, A. (1989) Pronounced lowering of lipoprotein Lp(a) in hyperlipidemic subjects treated with nicotinic acid. *Journal of Internal Medicine*, **226**, 271–276.

Caruso, M.G., Berloco, P., Notarnicola, M. and Di Leo, A. (1994) Lipoprotein(a) serum levels in post-menopausal women treated with oral estrogens administered at different times. *Hormone Metabolism Research*, **26**, 379–382.

Cushing, G.L., Gaubatz, J.W., Nava, M.L. *et al.* (1989) Quantification and localization of apolipoprotein(a) and B in coronary bypass grafts resected at reoperation. *Arteriosclerosis*, **9**, 593–603.

Edén, S., Wiklund, O., Oscarsson, J., Rosén, T. and Bengtsson, B.-Å. (1993) Growth hormone treatment of growth hormone-deficient adults results in a marked increase in Lp(a) and HDL cholesterol concentrations. *Arteriosclerosis and Thrombosis*, **13**, 296–301.

Frazer, K.A., Narla, G., Zhang, J.L. and Rubin, E.M. (1995) The apolipoprotein(a) gene is regulated by sex hormones and acute-phase inducers in YAC transgenic mice. *Nature*, **9**, 424–431.

Gaubatz, J.W., Ghanem, K.I., Guevara, J., Jr., Nava, M.L., Patsch, W. and Morrisett, J.D. (1990) Polymorphic forms of human apolipoprotein(a): inheritance and relationship of their molecular weights to plasma levels of lipoprotein(a). *Journal of Lipid Research*, **31**, 603–613.

Gaw, A., Boerwinkle, E., Cohen, J.C. and Hobbes, H.H. (1994) Comparative analysis of the apo(a) gene, apo(a) glycoprotein, and plasma concentrations of Lp(a) in three ethnic groups: evidence for no common Lp(a) allele at the apo(a) locus. *Journal of Clinical Investigation*, **93**, 2526–2534.

Gurakar, A., Hoeg, J.M., Kostner, G., Papadopoulos, N.M. and Brewer, H.B. (1985) Levels of lipoprotein Lp(a) decline with neomycin and niacin treatment. *Athersclerosis*, **57**, 293–301.

Haberland, M.E., Fless, G.M., Scanu, A.M. and Fogelman, A.M. (1992) Malondialdehyde modification of lipoprotein(a) produces avid uptake by human monocyte-macrophages. *Journal of Biological Chemistry*, **267**, 4143–4151.

Hajjar, K.A., Gavish, D., Breslow, J.L. and Nachman, R.L. (1989) Lipoprotein(a) modulation of endothelial cell surface fibrinolysis and its potential role in atherosclerosis. *Nature*, **339**, 303–305.

Harpel, P.C., Gordon, B.R. and Parker, T.S. (1989) Plasmin catalyzes binding of lipoprotein(a) to immobilized fibrinogen and fibrin. *Proceedings of the National Academy of Sciences USA*, **86**, 3847–3851.

Henriksson, P., Angelin, B. and Berglund, L. (1992) Hormonal regulation of serum Lp(a) levels: opposite effects after estrogen treatment and orchidectomy in males with prostatic carcinoma. *Journal of Clinical Investigation*, **89**, 1166–1171.

Hoff, H.F., O'Neil, J. and Yashiro, A. (1993) Partial characterization of lipoproteins containing apo[a] in human atherosclerotic lesions. *Journal of Lipid Research*, **34**(5), 789–798.

Jenner, J.L., Ordovas, J.M., Lamon-Fava, S., Schaefer, M.M., Wilson, P.W., Castelli, W.P. and Schaefer, E.J. (1993) Effects of age, sex, and menopausal status on plasma lipoprotein(a) levels: The Framingham Offspring Study. *Circulation*, **87**, 1135–1141.

Julius, U., Fritsch, W., Fritsch, W., Rehak, E., Fücker, K., Leonhardt, W. and Hanefeld, M. (1994) Impact of hormone replacement therapy on postprandial lipoproteins and lipoprotein(a) in normolipidemic postmenopausal women. *Clinical Investigations*, **72**, 502–507.

Kraft, H.G., Köchl, S., Menzel, H.J., Sandholzer, C. and Utermann, G. (1992) The apolipoprotein(a) gene: a transcribed hypervariable locus controlling plasma lipoprotein(a) concentration. *Human Genetics*, **90**, 220–230.

Kraft, H.G., Lingenhel, A., Köchl, S., Hoppichler, F., Kronenberg, F., Abe, A., Mühlberger, V., Schönitzer, D. and Utermann, G. (1996) Apolipoprotein(a) Kringle IV repeat number predicts risk for coronary heart disease. *Arteriosclerosis, Thrombosis, and Vascular Biology*, **16**, 713–719.

Lackner, C., Cohen, J.C. and Hobbs, H.H. (1993) Molecular definition of the extreme size polymorphism in apolipoprotein(a). *Human Molecular Genetics*, **2**, 933–940.

Lawn, R.M., Wade, D.P., Hammer, R.E., Chiesa, G., Verstuyft, J.G. and Rubin, E.M. (1992) Atherogenesis in transgenic mice expressing human apolipoprotein(a). *Nature*, **360**, 670–672.

Loscalzo, J., Weinfeld, M., Fless, G.M. and Scanu, A.M. (1990) Lipoprotein(a), fibrin binding, and plasminogen activation. *Arteriosclerosis*, **10**, 240–245.

McLean, J.W., Tomlinson, J.E., Kuang, W., Eaton, D.L., Chen, E.Y., Fless, G.M., Scanu, A.M. and Lawn, R.M. (1987) cDNA sequence of human apolipoprotein(a) is homologous to plasminogen. *Nature*, **300**, 132–137.

Molgaard, J., Klausen, I.C., Lassvik, C., Faergeman, O., Gerdes, L.U. and Olsson, A.G. (1992) Significant association between low-molecular-weight apolipoprotein(a) isoforms and intermittent claudication. *Arteriosclerosis and Thrombosis*, **12**, 895–901.

Nachman, R.L., Gavish, D., Azrolan, N. and Clarkson, T.B. (1991) Lipoprotein(a) in diet-induced atherosclerosis in nonhuman primates. *Arteriosclerosis and Thrombosis*, **11**, 32–38.

Niendorf, A., Rath, M., Wolf, K., Peters, S., Arps, H., Beisiegel, U. and Dietel, M. (1990) Morphological detection and quantification of lipoprotein(a) deposition in atheromatous lesions of human aorta and coronary arteries. *Virchows Archiv A Pathological Anatomy and Histopathology*, **417**, 105–111.

Olivecrona, H., Ericsson, S., Berglund, L. and Angelin, B. (1993) Increased concentrations of serum lipoprotein(a) in response to growth hormone treatment. *British Medical Journal*, **306**, 1726–1727.

Olivecrona, H., Johansson, A.G., Lindh, E., Ljunghall, S., Berglund, L. and Angelin, B. (1995) Hormonal regulation of serum lipoprotein(a) levels: contrasting effects of growth hormone and insulin-like growth factor-I. *Arteriosclerosis, Thrombosis, and Vascular Biology*, **15**, 847–849.

Pottratz, T.S., Bellido, T., Mocharia, H., Crabb, D. and Monolagas, S.C. (1994) 17β-Estradiol inhibits expression of human interleukin-6 promoter-reporter constructs by a receptor-dependent mechanism. *Journal of Clinical Investigation*, **93**, 944–950.

Rader, D.J., Cain, W., Ikewaki, K., Talley, G., Zech, L.A., Usher, D. and Brewer, H.B., Jr. (1994) The inverse association of plasma lipoprotein(a) concentrations with apolipoprotein(a) isoform size is not due to differences in Lp(a) catabolism but to differences in production rate. *Journal of Clinical Investigation*, **93**, 2758–2763.

Ray, A., Prefontaine, K.E. and Ray, P. (1994) Down-modulation of interleukin-6 gene expression by 17β-Estradiol in the absence of high affinity DNA binding by the estrogen receptor. *Journal of Biological Chemistry*, **269**, 12940–12946.

Ridker, P.M. (1995) An epidemiologic reassessment of lipoprotein(a) and atherothrombotic risk. *Trends in Cardiovascular Medicine*, **5**, 225–229.

Rosano, G.M.C. and Sarrel, P.M. (1994) Ovarian hormones and the cardiovascular system: recent findings. *Cardiologia*, **39**, 275–279.

Sacks, F.M., McPherson, R. and Walsh, B.W. (1994) Effect of postmenopausal estrogen replacement on plasma Lp(a) lipoprotein concentrations. *Archives of Internal Medicine*, **154**, 1106–1110.

Sandholzer, C., Saha, N., Kark, J.D., Rees, A., Jaross, W., Dieplinger, H., Hoppichler, F., Boerwinkle, E. and Utermann, G. (1992) Apo(a) isoforms predict risk for coronary heart disease. A study in six populations. *Atherosclerosis and Thrombosis*, **12**, 1214–1226.

Shewmon, D.A., Stock, J.L., Rosen, C.J., Heiniluoma, K.M., Hogue, M.M., Morrison, A., Doyle, E.M., Ukena, T., Weale, V. and Baker, S. (1994) Tamoxifen and estrogen lower circulating lipoprotein(a) concentrations in healthy postmenopausal women. *Arteriosclerosis and Thrombosis*, **14**, 1586–1593.

Simon, D.I., Ezratty, A.M. and Loscalzo, J. (1993) Lipoprotein(a) and atherothrombosis. *Current Opinion in Cardiology*, **8**, 814–820.

Soma, M.R., Osnago-Gadda, I., Paoletti, R., Fumagalli, R., Morrisett, J.D., Meschia, M. and Crosignani, P. (1993) The lowering of lipoprotein(a) induced by estrogen plus progesterone replacement therapy in postmenopausal women. *Archives of Internal Medicine*, **153**, 1462–1468.

Sorensen, K.E., Celermajer, D.S., Georgakopoulos, D., Hatcher, G., Betteridge, D.J. and Deanfield, J.E. (1994) Impairment of endothelium-dependent dilation is an early event in children with familiar hypercholesterolemia and is related to the lipoprotein(a) level. *Journal of Clinical Investigation*, **93**, 50–55.

Stampfer, M.J., Colditz, G.A., Willett, W.C., Manson, J.E., Rosner, B., Speizer, F.E. and Hennekens, C.H. (1991) Postmenopausal estrogen therapy and cardiovascular disease: ten-year follow-up from the Nurses' Health Study. *New England Journal of Medicine*, **325**, 756–762.

Stein, B. and Yang, M. (1995) Repression of the interleukin-6 promoter by estrogen receptor is mediated by NF-k$\beta$ and C/EBP$\beta$. *Molecular and Cellular Biology*, **15**, 4971–4979.

Tonolo, G., Ciccarese, M., Brizzi, P., Milia, S., Dessole, S., Puddu, L., Secchi, G. and Maioli, M. (1995) Cyclical variation of plasma lipids, apolipoproteins, and lipoprotein(a) during menstrual cycle of normal women. *American Journal of Physiology*, **269**, E1101–E1105.

Utermann, G. (1989) The mysteries of lipoprotein(a). *Science*, **246**, 904–910.

Walmer, D.K., Wrona, M.A., Hughes, C.L. and Nelson, K.G. (1992) Lactoferrin expression in the mouse reproductive tract during the natural estrous cycle: correlation with circulating estradiol and progesterone. *Endocrinology*, **131(3)**, 1458–1466.

Williams, J.K., Bellinger, D.A., Nichols, T.C., Griggs, T.R., Bumol, T.F., Fouts, R.L. and Clarkson, T.B. (1993) Occlusive arterial thrombosis in cynomolgus monkeys with varying plasma concentrations of lipoprotein(a). *Arteriosclerosis and Thrombosis*, **13**, 548–554.

# 10 Effects of Estrogen on Hemostatic and Leukocyte Function in Women

**Bjarne Østerud [1,*] and Pål Øian [2]**

[1] *Department of Biochemistry, Institute of Medical Biology and* [2] *Department of Obstetrics and Gynecology, University of Tromsø, Tromsø 9037, Norway.*
**Corresponding author, Tel.: 0047 776 44730, Fax: 0047 776 45350, E-mail: bjarne@fagmed.uit.no.*

This chapter describes the differences between men and women in hemostatic parameters and the reactivity of blood cells as well as the effect of exogenous estrogens on these parameters. Premenopausal women have higher levels of fibrinogen and factor VIIc, two of the risk factors for ischemic heart disease, as well as prothrombin factor 1 + 2. They are therefore more hypercoagulable than men, which is counterbalanced by a significantly more active fibrinolytic system than that seen in men. Thus, the most important regulator of fibrinolytic activity, PAI-1, is markedly lower in premenopausal women as compared to men and postmenopausal women. Postmenopausal women have a significantly higher level of fibrinogen and factor VIIc, which is greater than in men of the same age. Platelet adherence to subendothelium is lower in premenopausal women than in men, and men have about 30% more P-selectin (from activated platelets and endothelial cells) in plasma than premenopausal women.

In contrast to the alkylated estrogen in oral contraceptive pills, the estrogen used in hormone replacement therapy (HRT) appears to have beneficial effects on the coagulation- and fibrinolytic systems. Factor VIIc and fibrinogen are reduced, whereas PAI-1 returns to the premenopausal level.

HRT reduce cellular activation of monocytes and platelets as measured by the concentration of tissue factor, TNF$\alpha$ and TxB2 in stimulated blood. This down-regulation of the reactivity of blood cells may be important in explaining the beneficial effect of HRT on the risk of cardiovascular disease.

KEYWORDS: Coagulation, estrogen, fibrinolysis, monocytes, platelets, atherosclerosis.

## BACKGROUND

### The Hemostatic Mechanism and Its Regulation

The initiation of the hemostatic plug starts with the accumulation of platelets at the site of vessel wall injury, as the result of platelets sticking to collagen in exposed subendothelial structure. In this reaction the platelet receptor, glycoprotein Ib (GP Ib) is important as it links with von Willebrand factor to form a bridge between the platelets and collagen (Clemetson, 1995; Weiss, 1995). This platelet adhesion to the subendothelium is followed by cohesion or aggregation of the platelets whereby they bind to each other in a fibrinogen dependent reaction. This interaction is mediated by the binding of fibrinogen to GPIIb/IIIa receptor on the surface of activated platelets (Clemetson, 1995; Weiss, 1995). Collagen and the first traces of thrombin formed at the injury sites stimulate the platelet aggregation.

## The Initiation of Blood Coagulation

A primary platelet plug as described above is unstable and easily swept away. The stabilization of the platelet plug is reinforced by the formation of fibrin as a result of increased thrombin generation at the injury site where tissue factor (TF) is exposed (Drake *et al.*, 1989; Rapaport, 1995). The tissue injury disrupts an endothelial cell barrier that normally separates the circulating blood from tissue factor (TF) with the resultant binding of plasma factor VII to TF. Tissue factor is the cellular receptor and cofactor for plasma factor VIIa which initiates the coagulation protease cascade leading ultimately to the generation of thrombin and fibrin. This activation mechanism involves a series of reactions as shown in Figure 10.1. The limiting factor in the initiation of the coagulation activation is probably the rate of factor VII activation. Although not yet proven, a few initial factor VIIa/TF complexes may form after tissue injury that expose blood to the extravascular tissue factor sites because circulating plasma normally contains a small concentration of factor VIIa (Morrissey *et al.*, 1993). This could explain how the extrinsic system is triggered before any factor Xa is formed. When factor Xa is generated, it very efficiently

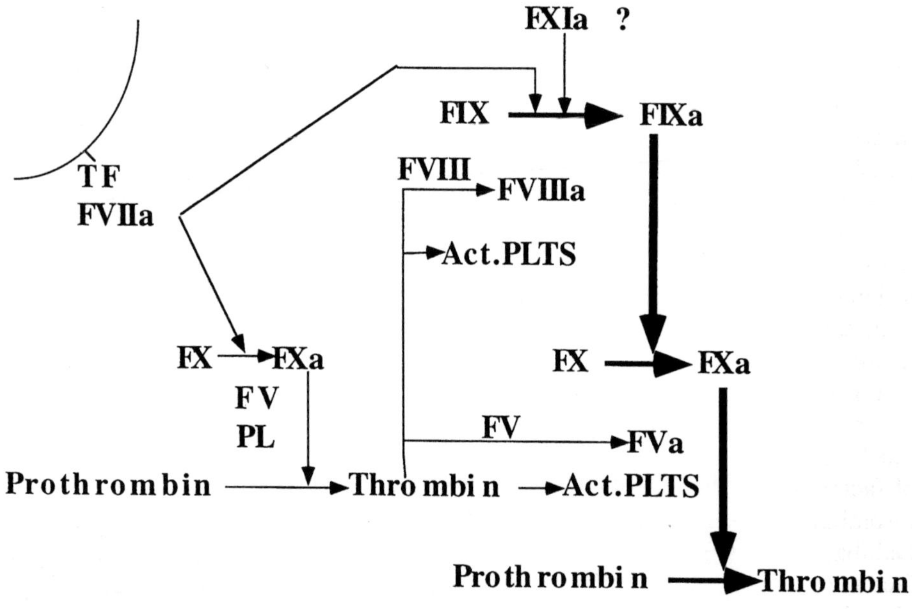


**Figure 10.1.** A simplified scheme of thrombin formation in the blood coagulation system. Tissue factor (TF) in the upper left on the figure is located on TF bearing cells, whereas the reactions to the right take place on activated platelets.

mediates the single peptide-bond cleavage that converts native factor VII to the enzyme, factor VIIa, when factor VII is bound to TF (Rao and Rapaport, 1988).

The initial factor Xa formed on the TF-bearing cells may be crucial for the production of the first traces of thrombin (Hoffman *et al.*, 1995; 1996). Why this is so important can be explained by thrombin's positive feedback reactions, including the activation of factor VIII and factor V in addition to the potent activation of platelets. However, the major thrombin generation may not be associated with the TF bearing cells. This hypothesis (Hoffman *et al.*, 1996) is based on the relatively rapid inactivation of factor Xa when factor Xa, formed on TF bearing cells is released into the plasma. In contrast, activated factor IX, factor IXa, which is also generated by the TF/VIIa complex (Østerud and Rapaport, 1977) and TF bearing cells (Rao *et al.*, 1992) may be transported from these cells to, for example, activated platelets without inactivation by plasma inhibitors since factor IXa is more slowly inactivated than factor Xa (Pieters *et al.*, 1988). The binding of factor IXa to factor VIIIa on thrombin-activated platelets (Ahmad *et al.*, 1995) mediates a significant production of factor Xa which propagates a substantially higher thrombin generation and thereby provides the necessary fibrin clot formation.

## The Regulation of the Coagulation System

The two-cell model system for thrombin generation also explains why the extrinsic activation of factor X alone is not sufficient for normal hemostasis but also requires the participation of the hemophilia factors IX and VIII. This becomes even more clear when we know that the TF initiated activation of factor IX and factor X is rapidly turned off by the tissue factor pathway inhibitor (TFPI) when the first traces of factor Xa are formed (Rapaport and Rao, 1992; Broze, 1995). Thus, factor IXa, which is formed by VIIa/TF activation of factor IX, and then released and transferred to activated platelets, is thought to take over the activation of factor X in association with thrombin activated factor VIII (VIIIa) on the surface of the platelets. Since we do not know whether factor XI is activated during hemostasis, most probably factor IXa may continue to activate factor X, and the resultant factor Xa, which, when bound to activated platelets, is protected from inactivation by plasma inhibitors (Miletich *et al.*, 1978), can convert more prothrombin to thrombin as long as factor Va is not inactivated.

In addition to TFPI, which regulates the activation of factor IX and factor X by factor VIIa/TF, thrombin generation is regulated by antithrombin that inhibits factor IXa, factor Xa, thrombin and also factor VIIa bound to TF. However, another inhibitory system appears to be quite important for the prevention of too much thrombin formation. Thus, activated protein C downregulates the activation of factor X and the convertion of prothrombin to thrombin by its inactivation of thrombin activated factor VIII (VIIIa) and activated factor V (Va) (for review see Dahlback, 1995), both essential cofactors for the enzyme complex of factor IXa/factor VIIIa and phospholipid and the enzyme complex of factor Xa, factor Va and phospholipid, respectively. Activation of protein C is mediated by the binding of thrombin to thrombomodulin on the endothelial cells of the vessel wall followed by the cleavage and conversion of protein C to protein Ca by thrombin. This proteolytic cleavage is accelerated by Protein S and probably by factor V, as well as by phospholipids serving as a template for the enzyme complex. This system in particular appears to be affected by the sex hormones in women. A recent discovery has clarified that a frequent mutation in factor V leads to failure of protein Ca to inactivate factor Va (Dahlback and Hildebrand, 1994; Bertina *et al.*, 1994). This mutation, together with changes associated with the use of alkylated estrogens, appears to be a strong risk factor in thrombotic events (Bloemenkamp *et al.*, 1995).

## The Fibrinolytic System

The fibrinolytic system is the counterpart of the blood coagulation system. Apparently there is a fine balance between these two systems and a diversion from the equilibrium will result either in bleeding or thrombosis. It is not the intention here to give a review on the fibrinolytic system, and we will refer to recent review articles (Collen and Lijnen, 1995; Verstraete, 1995). However, a short summary of the fibrinolytic system is illustrated in Figure 10.2.

The regulation of fibrinolysis has become more and more in focus for its role in a series of pathological reactions. Thus strong evidence has been put forward for a major role of the plasminogen activator inhibitor (PAI-1) in cardiovascular diseases, particularly coronary heart disease (Francis *et al.*, 1988; Wiman, 1995).

Tissue plasminogen activator (t-PA) is released from the endothelial cells by e.g. stress reactions. Although high levels of t-PA favour the dissolution of fibrin, high t-PA was found to be associated with an increased risk of cardiovascular disease.

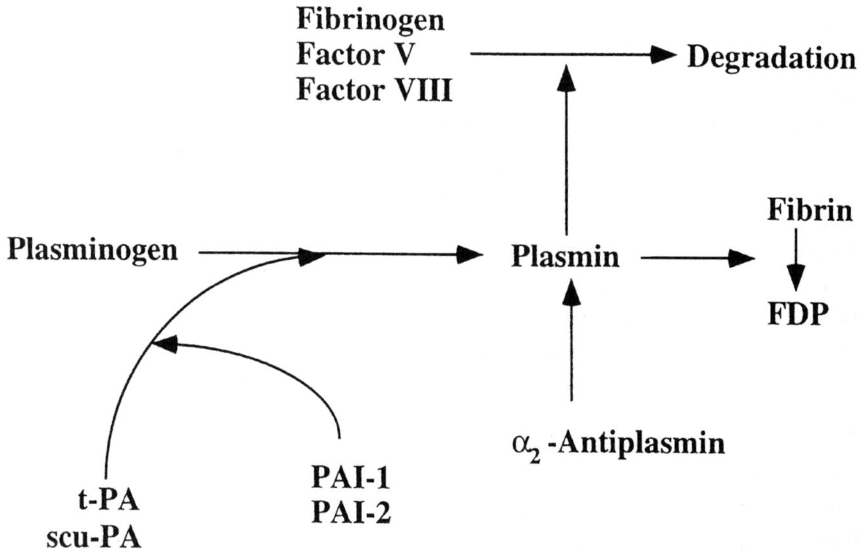


**Figure 10.2.** A brief summary of the fibrinolytic system. Plasmin associated with fibrin degrades fibrin, whereas the free plasmin degrades factors V, VIII and fibrinogen. Free plasmin is inhibited by α-antiplasmin.

High levels of t-PA may be caused by two different mechanisms:

(a) t-PA is constitutively released as a result of stress on the endothelium and a higher sensitivity for provocation of endothelium cells to release t-PA. The higher sensitivity may be associated with reduced nitrogen oxygen (NO) production. NO, which is also called the endothelial derived relaxing factor, seems to be a good parameter of the functional condition of the endothelial cells of the vessel wall.
(b) t-PA is rapidly inactivated by the formation of an inactive complex between t-PA and plasminogen activator inhibitor (PAI-1). PAI-1 is also released from the endothelial cells. When the level of PAI-1 is high, this is associated with an increase in t-PA in complex with PAI-1, which means that t-PA is blocked in its ability to activate plasminogen to plasmin.

t-PA and PAI-1 are released independently. Based on studies on the effect of strenuous physical activity, t-PA is released immediately, whereas PAI-1 is more dependent on mechanically injury-mediated release (Hansen *et al.*, 1990). It is interesting to note that well-trained athletes experience fewer changes in their t-PA and PAI-1 levels after hard physical exercise than non-trained athletes. Furthermore, the base levels of t-PA and PAI-1 are lower in well-trained athletes, as compared with non-trained individuals (Szymanski *et al.*, 1994; Kvernmo and Østerud, unpublished).

Another aspect of the fibrinolytic system is its relation to the lipid composition in blood. It is well established that high levels of triacylglycerol (TG) are associated with high levels of PAI-1 (Hamsten *et al.*, 1985; Mehta *et al.*, 1987; Juhan-Vague *et al.*, 1987). Indeed, the high levels of TG as a risk factor in myocardial infarction (MI), may rather be due to its association with high PAI-1 levels. Thus high levels of PAI-1 promote coronary heart disease through the promotion of fibrin deposits that are important in thrombotic diseases as well as in atherosclerosis.

Although PAI-2, which is located in and released from activated platelets, has not been thought to play an important role in the fibrinolytic system, we have seen that some individuals with highly reactive platelets, have a significant prolongation of whole blood clot lysis time after strenuous physical exercise (Hansen *et al.*, 1990), which could stem from platelet activation *in vivo*.

The estrogen effect on the fibrinolytic system is outlined later in this chapter, but already now it can be said that part of the beneficial effect of estrogens regarding cardiovascular diseases is clearly mediated through effects on the endothelium, as well as blood cells involved in the balance of t-PA and PAI-1 release.

## Thrombosis

The most frequent incidence of thrombosis is associated with a defect in either the activation of protein C or the failure of this system to inactivate factor Va (APC resistance) (for review see Dahlback, 1995). Deficiency of antithrombin leads to thrombosis, but this is a relatively rare disease (Lane *et al.*, 1992).

The role of platelets in thrombosis is not well established, but we have observed that individuals with large platelets (big middle platelet volume (MPV)) are more prone to develop thrombosis after surgery. Similarly, subjects with highly reactive monocytes that express TF are more susceptible to thrombosis after surgery (Østerud, unpublished).

The blood coagulation system is counterbalanced by the fibrinolytic system which is responsible for dissolving fibrin. This is important in the hemostatic mechanism whereby fibrinolysis dissolves fibrin and confines the clot to the injured vessel wall. This system is also essential in the prevention of thrombosis in the blood circulation under normal as well as pathological conditions.

## Atherosclerosis

The development of atherosclerosis is a series of complex long lasting reactions that probably start at the site of an injured vessel wall, e.g. at the bifurcations of the vessel. The flow of blood has then stressed the endothelium and made it less functional, as exemplified by suppressed production of NO, prostacyclin etc. and inreased endothelin (Myers *et al.*, 1991; Tanner *et al.*, 1993; Cardipan *et al.*, 1996). It is established that platelets adhere to the injury sites, and most probably, the adherence of monocytes to the endothelium at these sites is a requirement for the initiation of the atherosclerotic process. Whether thrombin formation is one of the early events in monocyte recruitment and the subsequent movement through the endothelium and into the intima is not known. It is quite possible that thrombin, through its interaction with platelets and monocytes may exert a vital role in the early events of atherosclerosis.

In recent studies, it has been shown in knock out mice that the adherence of white cells to the vessel wall is dependent on adhesion molecules expressed on the surface of the endothelium as well as on platelets (Frenette *et al.*, 1995). Adhesion molecules may therefore also be crucial in the reactions leading to the accumulation of monocytes in the intima and subsequent foam-cell formation.

### *Hyperreactive blood cells*

Our recent finding that several families with a high incidence of MI appear to have highly reactive monocytes (Østerud, unpublished) may be part of the explanation as to why many individuals with relatively normal cholesterol levels generate atherosclerosis and subsequent MI. Thus, the interaction between granulocytes, platelets and monocytes in the generation of TF (Halvorsen *et al.*, 1993; Sissener Engstad *et al.*, 1995) may be an early event in monocyte associated thrombin production, as well as monocyte transformation into macrophages when they pass into the intima. As highly reactive monocytes generate more TNFa and IL-6 than normal (Østerud, unpublished), the inflammatory reactions taking place as part of the development of atherosclerosis are enhanced by the more reactive cells.

### *Platelets in atherosclerosis*

Although some have ruled out platelets as participants in the early phase of atherosclerosis, platelets are well known to be part of the atherogenic process through their release of platelet-derived growth factor (PDGF). PDGF, together with monocyte-colony stimulating factor (M-CSF) and heparin-binding EGF-like growth factor (HB-EGF), propagate the proliferation of smooth muscle cells, and probably even macrophages in the lesions (Ross, 1995). However, it is our opinion that activated platelets may also play an important role in the early phase of monocyte recruitment to the endothelium and the resultant macrophage/foam cell formation in the intima. In a recent study, we found a high incidence of MI in the families of subjects with large MPV (Østerud, unpublished). Further support of this hypothesis was recently reported (Wehyrich *et al.*, 1996). It was shown that thrombin-activated platelets induced IL-8 production and monocytic chemoattractant protein 1 (MCP-1), which is thought to play an important role in the mobility of monocytes through the endothelium into the intima. In this regard, thrombin generation, which is quite dependent on platelet activation, may exert the trigger of the atherosclerotic process, although the final proof for this hypothesis is still warranted. In the long run, hyperactive platelets that have large MPV may efficiently contribute at several steps in the development of atherosclerosis. Another aspect where the reactivity of platelets may be very important, is the original adhesion of platelets to the vessel wall at the sites where the endothelium has been stressed to become injured as described above. Whatever mechanism, large and hyperactive platelets may indeed be a serious risk factor for coronary heart disease (Martin *et al.*, 1991), and it is well established that hyperactive large platelets promote the development of restenosis (Smyth *et al.*, 1993).

### *LDL-oxidation*

Studies in hypercholesterolemic animal models indicate that oxidation of low density lipoprotein (LDL) is likely to play an important role in atherogenesis (Yla-Herttuala *et al.*, 1989; for review see Holvoet and Collen, 1994). Oxidized LDL may accelerate the progression of this process by enhancing monocyte adhesion to the endothelium and macrophage foam cell generation, inducing smooth muscle cell migration and proliferation, foam cell formation; enhancing platelet adhesion and aggregation, and impairing vasodilation which results in increased shear stress. The formation of foam cells has been suggested to result from the uptake of oxidized LDL through a specific oxidized LDL receptor (scavenger). Whether LDL is oxidized in the circulation or after it has been transferred into the arterial wall is still debated, but as plasma contains potent antioxidants, LDL oxidation has been thought to take place in the intima (for review see Nielsen, 1996). However, it is quite possible that LDL may be oxidized in the close microenvironment around e.g. the granulocytes since this space may be protected from antioxidants (Weitz, 1987). The uptake of oxidized LDL in the macrophages is most likely associated with the induction and release of several products such as cytokines, growth factors, and chemoattractants. All of these factors are important for recruitment of more monocytes, proliferation of macrophages and smooth muscle cells, and further oxidation or enzymatic modification of LDL. Aggregates of lipid-rich macrophages are a central part of the earliest recognizable lesion, the so-called fatty streak, which also contains T-lymphocytes within the innermost layer of the artery wall, the intima

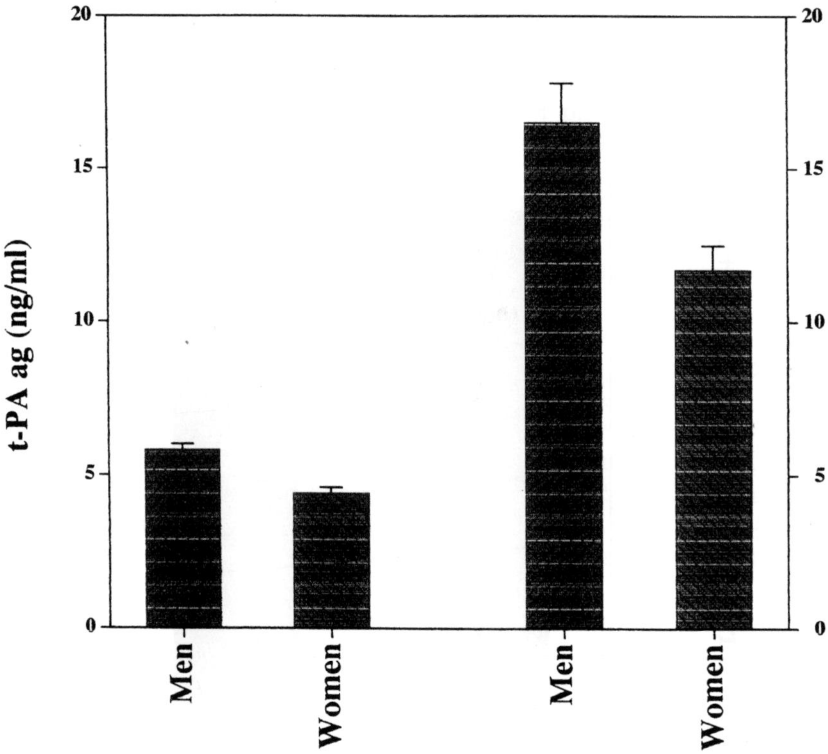


**Figure 10.4.** t-PA and PAI-1 antigen in plasma of 214 men (mean age 35.7) and 186 women (mean age 34.3). Blood was drawn between 8.00 a.m. and 10.00 a.m. Difference between men and women for both t-PA and PAI-1: $p < 0.0001$.

by a similar rise in the fibrinolytic activity. However, this is not the case, as PAI-1 is greatly enhanced in postmenopausal as compared to premenopausal women. Indeed, the PAI-1 level rises to the same level as for men of the same age. In a recent study, $14.8 \pm 0.6$ ng/ml PAI-1 was found in premenopausal women and $20.3 \pm 0.8$ ng/ml PAI-1 in men of the same age, whereas postmenopausal women had $19.6 \pm 0.7$ ng/ml and men of the same age had $21.7 \pm 0.7$ ng/ml (Gebara *et al.*, 1995). Fortunately for both men and women there is a rise in the levels of protein C and protein S at older ages (Henkens *et al.*, 1995). TFPI, which probably is a very important inhibitor of the TF triggered pathway, is also enhanced in postmenopausal as compared to premenopausal women (Ariens *et al.*, 1995). Since only about 20% of TFPI is in the circulation, and 80% is associated with the endothelium, it is difficult to interpret the significance of this.

### Sex Differences in Platelets

Although differences in the biological activities of platelets in men and women are less documented, sex differences in platelet function have been reported. In one study it was shown that platelet adherence to subendothelium was significantly less in premenopausal women than in men of the same age (Lawrence *et al.*, 1995).

*P-selectin*

The evidence for less reactive platelets in women than in men was recently further supported by our own study, where there was found to be 29% more P-selectin in the plasma of men ($n=142$) as compared to premenopausal women ($n=124$) of the same age (Østerud, unpublished). The free form of P-selectin in plasma probably reflects P-selectin released from activated platelets and endothelial cells. This observation may be very important, as it indicates a significant difference between men and women in exposure to one of the most important adhesion molecules thought to be crucial in the recruitment of white cells to the vessel wall. Interestingly it has been observed that the level of the soluble form of P-selectin in plasma was markedly higher in patients with acute myocardial infarction (Ikeda *et al.*, 1994).

## EXOGENOUS ESTROGENS

Estrogens can be classified into two main types: natural and synthetic. The "natural" estrogens are currently manufactured synthetically, but are still referred to as "natural" since they are chemically indistinguishable from the endogenous estrogens. A better and more important classification of estrogens is according to potency:

1. High-potency estrogens
2. Medium-potency estrogens
3. Low-potency estrogens

The potency of a hormone is dependent upon dose, method of administration, and the number of receptors in different organs. The high-potency estrogens are found in oral contraceptives, usually as ethinylestradiol. The medium potency estrogens are used in HRT and are entirely different from the administration of high-potency agent with respect to their effects on the coagulation system.

In the USA, conjugated equine estrogens produced from the urine of pregnant mares, are widely used for hormone replacement. In Europe, 17$\beta$-estradiol is more widely used.

Estradiol can be given orally or transdermally. Oral administration means a marked first-pass effect on hepatic metabolism and thus important for the coagulation system. Low-potency estrogens such as estriol are used for local treatment of urogenital problems. They have no systemic effects and are not dealt with in this review.

### Oral Contraceptives (OC)

Although oral contraceptive pills have been improved in regard to their pro thrombotic effects, they still result in increased coagulability. This is probably due to the alkylated estrogens increase of procoagulant activity in a dose-dependent manner (Winkler *et al.*, 1995). A large proportion of orally administered estrogen is, therefore, confined to the enterohepatic circulation, the concentration in the hepatic portal vein is six to seven times higher that in the peripheral circulation. The incidence of the venous thromboembolism with OC is in the order of 4 per 10,000 women-years. Recent epidemiological studies have reported an increased odds ratio for these complications in third—as compared to second—generation OC's (World Health Organization, 1995; Jick *et al.*, 1995; Spitzer *et al.*, 1996 ). However, these studies do not indicate any increase in the absolute risk of venous thromboembolism. Instead, data show a decrease in the incidence for second-generation OC's over time. The observed differences are probably explained by prescribing bias and the so-called healthy-user effect, i.e. that women susceptible for thromboembolism have left the group of second-generation OC users, either because they have stopped using OC's or have switched over to third-generation pills.

The most consistent effects of these alkylated estrogens on coagulation proteins are elevations of fibrinogen, prothrombin, factors VII, IX, X and XII; protein C and plasminogen (Wessler, 1992). Although these elevated levels have been attributed to the estrogenic component, the progestogen component cannot be ruled out. Despite a rise in protein C, this inhibitory system is probably impaired in oral contraceptive users as the level of protein S, the cofactor of activated protein C, is reduced (Malm *et al.*, 1988). Even more important, the free protein S is significantly reduced in OC users. The recent discovery that a mutant of factor V (the Leiden mutant) is not inactivated by activated protein C (protein C resistance) may account for the majority of thrombotic events among OC users (Bloemenkamp *et al.*, 1995).

When fibrin deposition on vascular endothelium was studied, it was shown that oral contraceptives with a low estrogen content (20 µg ethinylestradiol and 150 µg desogestrel per day), caused no enhancement, whereas a high dose estrogen (50 µg ethinylestradiol and 125 µg desogestrel per day) enhanced fibrin deposition. The subendothelial deposition of platelets and platelet thrombi was not changed by contraceptive treatment (Inauen *et al.*, 1991).

The effects of OC on platelets have been contradictory. We believe that some of the thrombotic effects, and perhaps be the most important effect imposed by the OC, is the enhanced activity of platelets . This is supported by our observation that whole blood clotting time, which is performed on non anticoagulated blood in plastic tubes, was significantly shortened by two triphasic oral contraceptive pills containing ethinyl estradiol plus levonorgestrel or gestodene (Omsjø *et al.*, 1989). Since the availability of phosphatidylserin in the inner membrane of platelets, which becomes exposed on the surface of activated platelets, is the limiting factor in this clotting assay (Østerud and Brox, 1983), we proposed that the shortening of the whole blood clotting time reflected increased sensitivity of the platelets to become activated. In the same study on oral contraceptive users, no significant changes were observed in fibrinogen, antithrombin, factors V, VII or VIII after 3 and 6 months use of OC. The fibrinolytic system was also not significantly altered based on an assay utilizing diluted whole blood clot lysis time (Omsjø *et al.*, 1989). However, in a couple of individuals, a drastic prolongation of the lysis time was seen after use of oral contraceptive pills.

**Hormone Replacement Therapy (HRT)**

It has been interesting to observe that the changes in the hemostatic and fibrinolytic parameters associated with the menopause and the reduced endogenous hormone production may be fully reversed by HRT.

*Effects on the coagulation system*

In contrast to the estrogen in oral contraceptive pills, the estrogen used in HRT, has in almost every study revealed a beneficial effect. A significant reduction in fibrinogen appears to be estrogen-induced (Folsom *et al.*, 1991; Meilahn *et al.*, 1992; Scarabin *et al.*, 1993), whereas the evidence for a lowering in the level of factor VII is contradictory, as both a reduction and a rise in factor VII have been observed (Sporrong *et al.*, 1990; Nabulsi *et al.*, 1993; Winkler *et al.*, 1995). For example, perorally administered conjugated estrogens and transdermally administered estradiol led to increases in the levels of factor VIIc, factor VII antigen and prothrombin fragment 1+2 in one study (Kroon *et al.*, 1994). Interestingly, there was a reduction in the antithrombin-thrombin complex, which conflicts with the rise in prothrombin fragment 1+2, since the antithrombin–thrombin complex is the most sensitive parameter of thrombin generation *in vivo*.

*Effects on the fibrinolytic system*

Probably the most important antithrombotic effect associated with HRT, is the strikingly reduction in PAI-1, a finding that seems to be well established in all studies (Meilahn *et al.*, 1992; Scarabin *et al.*, 1993; Winkler *et al.*, 1995; Gilabert *et al.*, 1995; Gebara *et al.*, 1995). The return of PAI-1 to about the same levels as in premenopausal women may be one of the most important effect of HRT, since this may prevent thrombosis in the microcirculation, an event associated with the aging process. Furthermore, prevention of fibrin depositions may in the long run delay the development of atherosclerosis (Smith, 1993). Although the reduction in PAI-1 is followed by a reduction in t-PA, the fibrinolytic potential is greatly enhanced during HRT.

## AN *EX VIVO* MODEL FOR ANALYZING THE REACTIVITY OF MONOCYTES IN WHOLE BLOOD

We have for more than 10 years used an *ex vivo* model to explore the functional properties of white cells and platelets in their normal environment of whole blood (for review see Østerud, 1995). This has been accomplished by exposing whole blood, anticoagulated with either heparin or hirudin, to a low dose of lipopolysaccharides (LPS). In our opinion, it is important to use a model system like this in order to search for the reactivity of cells in the circulation, since many reactions in blood are dependent on interactions between blood cells. Using isolated cells one fails to detect such interactions, in addition to the problem that the cells are altered and many times already activated during the isolation procedure. Both endothelial cells and monocytes in cell cultures behave quite differently than in their natural environment (Solberg *et al.*, 1990; Halvorsen *et al.*, 1993; Østerud, unpublished). LPS can be claimed to be physiological since it occurs naturally in the gut, and may leak to the circulatory system and thereby provoke cellular activation. In addition, numerous studies have shown an association between atherosclerosis and both viral and bacterial infections (for review see Lopes-Virella, 1993).

Although monocytes can be induced to produce a whole series of important reactants, such as cytokines (TNF$\alpha$, IL-1$\beta$, IL-6, IL-8, etc), chemoattractants (MCP-1), proteolytic enzymes (elastase, cathepsins), oxygen radicals, prostaglandins, leukotrienes, etc., we have used the above system particularly to evaluate the potency of monocytes of whole blood to generate TF and TNF$\alpha$ when stimulated. TF is essential for normal hemostasis, but may also be mandatory in the development of atherosclerosis.

Interestingly, this system revealed a fifty-fold difference in the expression of TF activity in monocytes in the blood of different individuals (Østerud, 1995). Since it was observed that those who generate very high TF activity in monocytes (high responders) appeared to belong to families with a high incidence of cardiovascular disease (especially coronary heart disease), an intensive search for the explanation for this phenomenon has been performed. Recently, we also found that the high responder phenomenon is associated with a high generation of TNF$\alpha$ and IL-6 (Østerud, unpublished), two of the most important mediators of inflammatory reactions in humans.

At least part of the high response regarding LPS-induced TF activity was found to be associated with an interaction of platelets and monocytes. Even before it was recognized that TF was produced in monocytes, it was shown that platelets enhanced TF activity in leukocytes (Niemetz and Marcus, 1974). In our study we found that when partially platelet depleted white cells from high responders were crosscombined with platelets (platelet rich plasma-PRP) from a low responder and then stimulated with LPS, this led to a marked drop (60–70%) in TF activity generation compared to the values obtained with autologous platelets (Østerud *et al.*, 1990). The role of platelets in LPS-induced TF activity was further established by stimulating white blood cells recombined with either platelet poor plasma (PPP) or PRP. Only about one third of TF activity was induced in the absence of platelets as compared to cells recombined with PRP (Halvorsen *et al.*, 1993).

A mandatory role of granulocytes in provoking the platelet enhancement of LPS-induced TF activity in monocytes was demonstrated using a plasma system with physiological concentrations of recombined mononuclear cells, granulocytes and platelets (Halvorsen *et al.*, 1993). This interaction was totally blocked by antibody against P-selectin (Sissener Engstad *et al.*, 1995). We believe that the role of granulocytes in this reaction is to make P-selectin maximally available on the surface of the platelets. This may be mediated by the release of platelet factor 4 (PF4) which appeared to promote the granulocyte effect in the platelet enhancement of LPS-induced TF activity of the monocytes (Sissener Engstad *et al.*, 1995). Thus, any downregulation of the reactivity of platelets, monocytes and granulocytes would probably be quite beneficial with regard to risk of cardiovascular disease.

**The Effect of Endogenously Produced Estrogen and Hormone Replacement Therapy on the Activation of Blood Cells**

In 5 different ways we have obtained results implying a very important role of estrogens in the downregulation of the reactivity in monocytes in whole blood:

1. LPS-induced TF activity is significantly lower in premenopausal women than in men of the same age when the blood system was fortified with liposomes that amplify the lipoxygenase pathway (Hansen *et al.*, 1989). This pathway is most likely crucial for the production of inflammatory products in monocytes.
2. In a study on the effects of dietary supplementation with marine oils, the LPS-induced TNF$\alpha$ and IL-6 in whole blood was significantly lower in premenopausal women ($n=124$) than in men of the same age ($n=142$) (Figure 10.5).
3. It was shown that during normal pregnancy, LPS-induced TF activity is reduced to 40% of that in the non-pregnant state (Figure 10.6) (Øian *et al.*, 1985).
4. OC users had a much lower LPS-induced TF activity in the monocytes as compared to non-users (Figure 10.7) (Omsjø *et al.*, 1986).
5. Recently it was found that after 12 months of HRT (both oral and transdermal treatment), the TF activity was significantly reduced both in unstimulated (73%) and stimulated monocytes (56%) (Figure 10.8) (Aune *et al.*, 1995). Furthermore, the production of TNF$\alpha$ was also reduced by 30%. In contrast, treatment associated with *in vitro* fertilization, with supraphysiological estradiol levels, induced and enhanced TF activity in stimulated monocytes (Aune *et al.*, 1993).

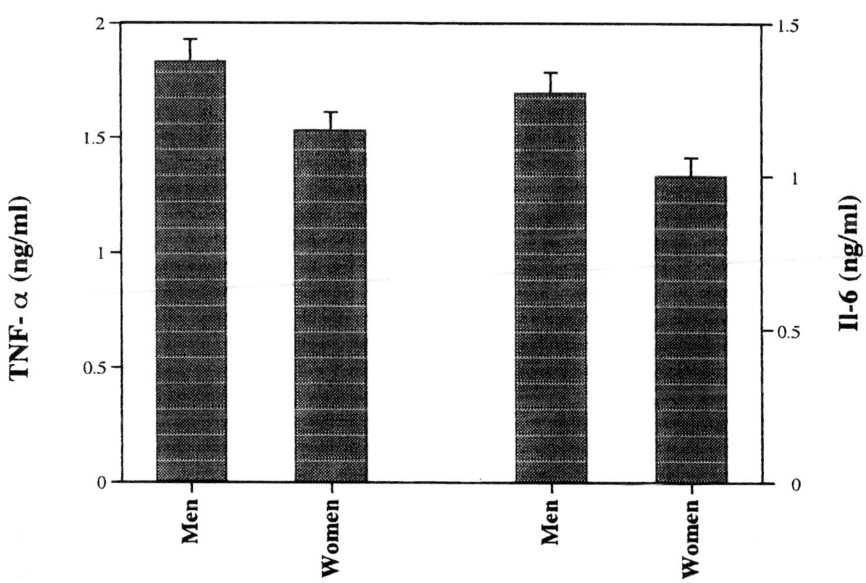


**Figure 10.5.** LPS-induced TNF$\alpha$ and IL-6 in whole blood (5 ngLPS/ml) of 142 men (mean age 37.9) and 124 women (mean age 34.3). Differences between men and women for both TNF$\alpha$ and IL-6: $p<0.0001$.

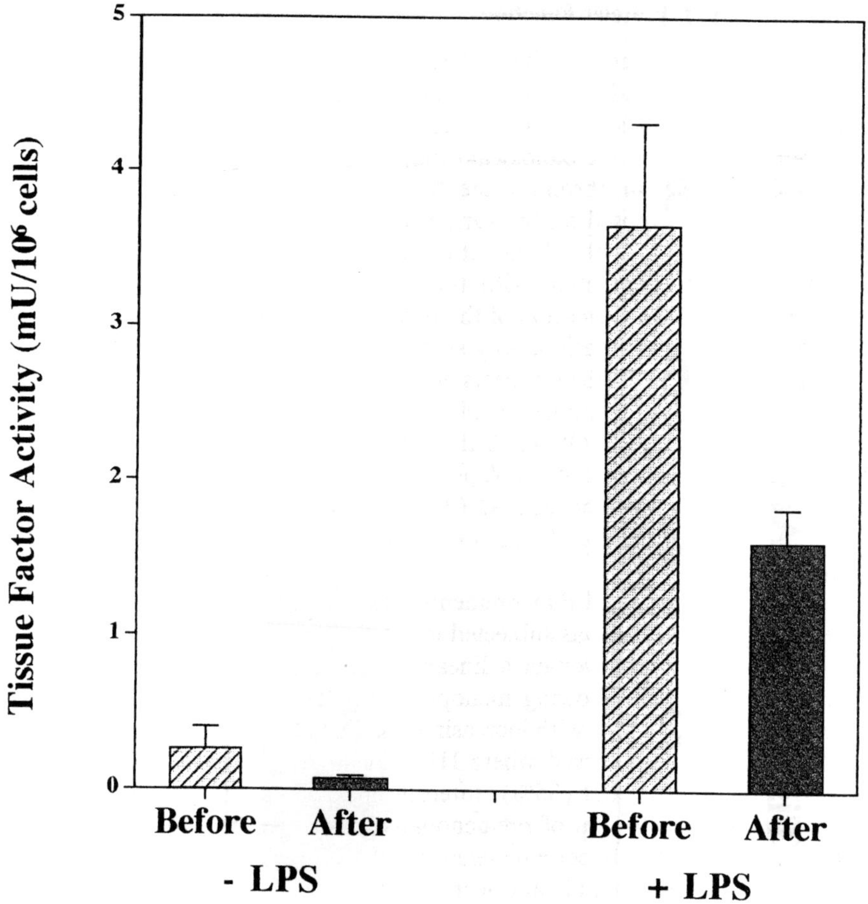


**Figure 10.8.** LPS-induced TF activity in monocytes of whole blood before and after hormone replacement therapy ($n=25$). −LPS represents induced TF activity in monocytes of blood samples incubated in the absence of LPS. From Aune *et al.* (1995), *Am. J. Obstet. Gynecol.*, **173**, 1816–1820.

reacts to low concentrations of stimuli. This is the background for the hypothesis that the atherosclerotic process today is viewed as a "defense system gone awry" (Ross, 1993). In this process TNF$\alpha$ and IL-1$\beta$ are probably two of the most important mediators of the inflammatory response and it has been suggested that TNF$\alpha$ in particular may be an etiologic factor in the progression of atheroma from intima thickening to endothelial disruption (Barath *et al.*, 1990).

An inflammatory reaction also includes the lymphocytes. In accordance with the observations of the estrogen effect on monocytes, estrogen also appeared to suppress natural killer cell cytotoxicity which was suggested to be associated with an increased concentration of Ig-producing cells (Nilsson and Carlsten, 1994).

### Effect of Estrogen on Platelet Function

The importance of platelets in atherosclerosis as well as in hemostasis and thrombosis has already been eluded to. In our opinion, the platelets may be more essential in cardiovascular diseases, particularly coronary heart disease, than is currently accepted. This is based on the biological products associated with activated platelets, such as PDGF, P-selectin, thromboxane A2, and negatively charged phospholipids as phosphatidylserin exposed on the surface of activated platelets, that are mandatory in, respectively, growth of smooth muscle cells, adherence of platelets to leukocytes and their interaction with the endothelium, platelet aggregation and vasoconstriction, and the formation of the two major coagulation-enzyme complexes required for thrombin generation. Any suppression of platelet reactivity will therefore be associated with beneficial aspects of pathological reactions. In addition to the above mentioned platelet components, platelets release the following mediators upon activation: platelet factor 4 (PF4), b-thromboglobulin, serotonin (5-hydroxytryptamine), ADP, PAI-2, and a series of prostaglandins, of which the most important in this regard may be thromboxane A2 (TxA2), TxA2 a powerful platelet aggregating agent and vasoconstrictor. TxA2 is rapidly metabolized to TxB2, which can be quantitated.

Some years ago, we found that premenopausal women produced 69% less TxB2 than men when their blood was subjected to low LPS-stimulation (Balteskard *et al.*, 1993). In postmenopausal women a linear rise in the concentration of TxB2 was observed over the years following menopause (Balteskard *et al.*, 1989). No such increase was observed in men with increasing age. Support for an effect of estrogen on platelet activation was observed where HRT over a one year period resulted in a significant reduction in TxB2 (35%). Interestingly this treatment brought women's TxB2 production back to that of premenopausal level (Aune *et al.*, 1995). Further support for an inhibitory effect on platelets by estrogen replacement therapy was provided in a study where a significant decrease in adrenalin-induced platelet aggregation, as well as ATP release, was observed after 3 months of therapy (Bar *et al.*, 1993).

## CONCLUDING REMARKS

Both aging and gender appear to be associated with important hemostatic effects, with evidence pointing to an influence of endogenous and exogenous sex hormones.

Although the levels of coagulation factors do not favour a preferable tendency of less clotting in premenopausal women as compared with men of the same age, this is probably counterbalanced by a more effective fibrinolytic system than that in men. Indeed, the higher potential of fibrinolytic activity in women may be an important factor in explaining the reduced risk of cardiovascular disease in premenopausal women. Whether this may be a secondary effect of less reactive blood cells in women, as compared with men as discussed in this chapter, remains to be established. Nevertheless, we believe that the striking difference in reactivity of blood cells in premenopausal women or women using HRT as compared to men of the same age, represents a major protection against the development of cardiovascular disease.

Our results suggest that HRT reestablishes several of the beneficial effects lost by the reduced hormone production of postmenopausal women.

## REFERENCES


Ahmad, S.S., Rewala, R., Chaung, N.F., Stafford, D.W. and Walsh, P.N. (1995) The role of the second growth-factor domain of human factor IXa in binding to platelets and in factor X activation. *Biochem. J.*, **310**, 427–431.

Ariens, R.A., Coppola, R., Potenza, I. and Manucci, P.M. (1995) The increase with age of the components of the tissue factor coagulation pathway is gender-dependent. *Blood Coagul. Fibrinolysis*, **6**, 433–437.

Aune, B., Øian, P. and Østerud, B. (1993) Enhanced sensitivity of extrinsic coagulation system during ovarian stimulation for *in-vitro* fertilization. *Hum. Reprod.*, **8**, 1349–1352.

Aune, B., Øian, P., Omsjø, I. and Østerud, B. (1995) Hormone replacement therapy reduces the reactivity of monocytes and platelets in whole blood-A beneficial effect on atherogenesis and thrombus formation? *Am. J. Obstet. Gynecol.*, **173**, 1816–1820.

Baeuerle, P.A. and Henkel, T. (1994) Function and activation of NFkB in the immune system. *Annu. Rev. Immunol.*, **12**, 11141–11178.

Balteskard, L., Brox, J.H. and Østerud, B. (1993) Thromboxane production in the blood of women increases after menopause whereas tumor necrosis factor is reduced in women compared with men. *Atherosclerosis*, **102**, 91–98.

Bar, T., Tepper, R., Fuchs, J., Pardo, V., Goldberger, S. and Ovadia, J. (1993) The effect of estrogen replacement therapy on platelet aggregation and adenosine triphosphate release in postmenopausal women. *Obstet. Gynecol.*, **81**, 261–264.

Barath, P., Fischbein, M.C., Cao, J., Berenson, J., Helfant, R.H. and Forrester, J.S. (1990) Detection and localization of tumor necrosis factor in human atheroma. *Am. J. Cardiol.*, **65**, 297–302.

Bertina, R.M., Koeleman, B.P.C., Koster, T., Rosendaal, F.R., Dirven, R.J., DeRonde, H., van der Velden, P.A. and Reitsma, P.M. (1994) Mutation in blood coagulation factor V associated with resistance to activated protein C. *Nature*, **369**, 64–67.

Bloemenkamp, K.W., Rosendaal, F.R., Helmerhorst, F.M., Buller, H.R. and Vandenbrouke, J.P. (1995) Enhancement by factor V Leiden mutation of risk of deep-vein thrombosis associated with oral contraceptives containing a third generation progestagen. *Lancet*, **346**, 1593–1596.

Broze, G.J. (1995) Tissue factor pathway inhibitor and the revised theory of coagulation. *Ann. rev. Med.*, **46**, 103–112.

Cardipan, R.C., Wang, B.Y., Buitrago, R., Tsao, P.S. and Coake, J.P. (1996) Regression or progression dependency on vascular nitric oxide. *Arterioscler. Thromb. Vasc. Biol.*, **16**, 44–50.

Clemetson, K.J. (1995) Platelet activation: signal transduction via membrane receptors. *Thromb. Haemost.*, **74**, 111–116.

Collen, D. and Lijnen, H.R. (1995) Molecular basis of fibrinolysis, as relevant for thrombolytic therapy. *Thromb. Haemost.*, **74**, 167–171.

Dahlback, B. (1995) The protein C anticoagulant system: Inherited defects as basis for venous thrombosis. *Thromb. Res.*, **77**, 1–43.

Dahlback, B. and Hildebrand, B. (1994) Inherited resistance to activated protein C is corrected by anticoagulant cofactor activity found to be a property of factor V. *Proc. Natl. Acad. Sci. USA*, **81**, 1396–1400.

Drake, T.A., Morrissey, J.H. and Edgington, T.S. (1989). Selective cellular expression of tissue factor in human tissues: Implications for disorders of haemostasis and thrombosis. *Am. J. Pathol.*, **134**, 1087–1097.

Folsom, A.R., Wu, K.K., Conlan, M.G., Finch, A., Davis, C.E., Marcucci, G., Sorlie, P.D. and Szklo, M. (1992) Distributions of hemostatic variables in blacks and whites: population reference values from the Atherosclerotic Risk in Communities (ARIC) study. *Ethn. Dis.*, **2**, 35–46.

Folsom, A.R., Wu, K.K., Davis, C.E., Conlan, M.G., Sorlie, P.D. and Szklo, M. (1991) Population correlates of plasma fibrinogen and factor VII, putative cardiovascular risk factors. *Atherosclerosis*, **91**, 191–205.

Francis, R.B.Jr., Kawanishi, D., Baruch, T., Mahrer, P., Rahimotoola, S. and Feinstein, D.I. (1988) Impaired fibrinolysis in coronary artery disease. *Am. Heart J.*, **115**, 776–780.

Frenette, P.S., Johnson, R.C., Hynes, R.O. and Wagner, D.D. (1995) Platelets roll on stimulated endothelium *in vivo*: an interaction mediated by endothelial P-selectin. *Proc. Natl. Acad. Sci. USA*, **92**, 7450–7454.

Gebara, O.C., Mittlerman, M.A., Sutherland, P., Lipinska, I., Matheney, T., Xu, P. *et al.* (1995) Association between increased estrogen status and increased fibrinolytic potential in the Framingham Offspring study. *Circulation*, **91**, 1952–1958.

Gilabert, J., Estelles, A., Cano, A., Espana, F., Barrachina, R., Grancha, S. *et al.* (1995) The effect of estrogen replacement therapy with or without progestogen on the fibrinolytic system and coagulation inhibitors in postmenopausal status. *Am. J. Obstet. Gynecol.*, **173**, 1849–1854.

Green, D., Ruth, K.J., Folsom, A.R. and Liu, K. (1994) Hemostatic factors in the coronary Artery Risk Development in young adults (CARDIA) study. *Arterioscler. Thromb.*, **14**, 686–693.

Halvorsen, H., Olsen, J.O. and Østerud, B. (1993) Granulocytes enhance LPS-induced tissue factor activity in monocytes via an interaction with platelets. *J. Leuk. Biol.*, **54**, 275–282.

Hamsten, A., Wiman, B., de Faire, U. and Blomback, M. (1985) Increased plasma levels of a rapid inhibitor of tissue plasminogen activator in young survivors of myocardial infarction. *N. Eng. J. Med.*, **313**, 1557–1563.

Hansen, J.B., Wilsgård, L., Olsen, J.O. and Østerud, B. (1989) The effect of lipopolysaccharides (LPS) on generation of thromboxane A2 and thromboplastin activity in whole blood of males, females and females on oral contraceptives. *Thromb. Haemost.*, **66**, 493–496.

Hansen, J.B., Wilsgård, L., Olsen, J.O. and Østerud, B. (1990) Formation and persistence of procoagulant activities in circulation after strenuous physical exercise. *Thromb. Haemost.*, **64**, 385–389.

Henkel, T., Machleidt, T, Alkalay, I., Kronke, M., Ben-Neriah, Y. and Baeuerle, P.A. (1993) Rapid proteolysis of I kappa B-alpha is neccessary for activation of transcription factor NF-kappaB. *Nature*, **365**, 182–185.

Henkens, C.M., Bom, V.J., Van der Schaaf, W., Pelsma, P.M., Sibinga, C.T., de Kamp, P.J. and van der Meer, J. (1995) Plasma levels of protein S, protein C, and factor X: effects of sex, hormonal state and age. *Thromb. Haemost.*, **74**, 1271–1275.

Hoffman, M., Monroe, D.M. and Roberts, H.R. (1996) Cellular interactions in hemostasis. *Haemostasis*, **26** (Suppl. 1), 12–16.

Hoffman, M., Monroe, D.M., Oliver, J.A. and Roberts, H.R. (1995) Factors IXa and Xa play distinct roles in tissue factor-dependent initiation of coagulation. *Blood*, 86, 1794–1801.

Holvoet, J. and Collen, D. (1994) Oxidized lipoproteins in atherosclerosis and thrombosis. *FASEB J.*, **8**, 1279–1284.

Ikeda, H., Nakayama, H., Oda, T., Kuwano, K., Muraishi, A., Sugi, K. *et al.* (1994) Soluble form of P-selectin in patients with acute myocardial infarction. *Coronary Artery Dis.*, **5**, 515–518.

Inauen, W., Stocker, G., Haeberli, A. and Straub, P.W. (1991) Effect of low and high dose contraceptives on blood coagulation and thrombogenesis induced by vascular subendothelium exposed to flowing human blood. *Contraception*, **43**, 435–446.

Jick, H., Jick, S.S., Gurewich, V., Myers, M.W. and Vasilakis, C. (1995) Risk of idiopathic cardiovascular death and non fatal venous thromboembolism in women using oral contraceptives with different progestogen components. *Lancet*, **346**, 1589–1593.

Juhan-Vague, I., Vague, P., Alessi, M.C., Bodier, C., Valadier, J., Aillaud, M.F. and Atlan, C. (1987) Relationship between plasma insulin triglyceride, body mass index, and plasminogen activator inhibitor-1. *Diabete Metab.*, **13**, 331–336.

Knopp, R.H., Zhu, X. and Bonet, B. (1994) Effects of estrogen on lipoprotein metabolism and cardiovascular disease in women. *Atherosclerosis*, **110**, Suppl. S83-91.

Kroon, U.B., Silverstolpe, G. and Tengborn, L. (1994) The effects of transdermal estradiol and oral conjugated estrogens on haemostasis variables. *Thromb. Haemost.*, **71**, 420–423.

Lane, D.A., Olds, R.R. and Thein, S.L. (1992) Antithrombin and its deficiency states. *Blood Coagul. Fibrinolysis*, **3**, 315–341.

Lawrence, J.B., Leifer, D.W., Moura, G.L., Southern, P., Emery, J.D., Bodenheimer, S.L. and Kramer, W.S. (1995) Sex differences in platelet adherence to subendothelium: relationship to platelet function tests and hematologic variables. Am. *J. Med. Sci.*, **309**, 201–207.

Lopes-Virella, M.F. (1993) Interations between bacterial lipopolysaccharides and serum lipoproteins and their possible role in coronary heart disease. *Eur. Heart J.*, **14**, Suppl. K, 118–124.

Malm, J., Laurell, M. and Dahlback, B. (1988) Changes in the plasma levels of vitamin K-dependent proteins C and S and of C4b-binding protein during pregnancy and oral contraception. *Brit. J. Haematol.*, **68**, 437–443.

Martin, J.F., Bath, P.M.W. and Burr, M.L. (1991) Influence of platelet size on outcome after myocardial infarction. *Lancet*, **338**, 1409–1411.

Meade, T.W., Dyer, S., Howarth, D.J., Imeson, J.D. and Stirling, Y. (1990) Antithrombin III and procoagulant activity:sex differences and effects of the menopause. *Br. J. Haematol.*, **74**, 77–81.

Meade, T.W., Mellows, S., Brozovic, M., Miller, G.J., Chakrabarti, R.R., North, W.R. *et al.* (1986) Haemostatic function and ischaemic heart disease: principal results of the Nortwick Park heart study. *Lancet*, **2**(8506), 533–537.

Mehta, J., Mehta, P., Lawson, D. and Saldeen, T. (1987) Plasma tissue plasminogen activator inhibitor levels in coronary artery disease: correlation with age and serum triglyceride concentrations. *J. Am. Coll. Cardiol.*, **9**, 263–268.

Meilahn, E.N., Kuller, L.H., Matthews, K.A. and Kiss, J.E. (1992) Hemostatic factors according to menopausal status and use for hormone replacement therapy. *Ann. Epidemiol.*, **2**, 445–455.

Miletich, J.P., Jackson, C.M. and Majerus, P.W. (1978) Properties of the factor Xa binding site on human platelets. *J. Biol. Chem.*, **253**, 6908–6916.

Morrissey, J.H., Macik, B.G., Neuenschwender, P.F. and Comp, P.C. (1993) Quantitation of activated factor VII levels in plasma using a tissue factor mutant selectively deficient in promoting factor VII activation. *Blood*, **81**, 734–744.

Myers, S.I., Russel, D.H., Parks, L. and Reed, M.K. (1991) Triphasic response of prostacyclin production in rabbit thoracic aorta in early atherosclerosis. *Prostaglandins Leukot. Essent. Fatty Acids*, **44**, 31–36.

Nabulsi, A.A., Folsom, A.R., White, A., Patsch, W., Heiss, G., Wu, K.K. and Szklo, M. (1993) Association of hormone replacement therapy with various cardiovascular risk factors in post-menopausal women. The Atherosclerosis Risk in Communities Study Investigators. *N. Eng. J. Med.*, **328**, 1069–1075.

Nielsen, L.B. (1996) Transfer of low density lipoprotein into the arterial wall and risk of atherosclerosis. *Atherosclerosis*, **123**, 1–15.

Niemetz, J. and Marcus, A.J. (1974) The stimulatory effect of platelets and platelet membranes on the procoagulant effect of leukocytes. *J. Clin. Invest.*, **54**, 1437–1443.

Nilsson, N. and Carlsten, H. (1994) Estrogen induces suppression of natural killer cell cytotoxicity and augmentation of polyclonal B cell activation. *Cell Immunol.*, **158**, 131–139.

Øian, P., Omsjø, I., Maltau, J.M. and Østerud, B. (1985) Reduced thromboplastin activity in blood monocytes and reduced sensitivity to stimuli *in vitro* of blood monocytes from pregnant women. *Brit. J. Haematol.*, **59**, 133–137.

Omsjø, I., Øian, P. and Østerud, B. (1986) Thromboplastin activity in blood monocytes from oral contraceptive users. *Gynecol. Obstet. Invest.*, **22**, 194–197.

Omsjø, I., Øian, P., Maltau, J.M. and Østerud, B. (1989) Effects of two triphasic oral contraceptives containing ethinyl estradiol plus levonorgestrol or gestodene on blood coagulation and fibrinolysis. *Acta Obstet. Gynecol. Scand.*, **68**, 27–30.

Østerud, B. (1995) The high responder phenomenon: Enhancement of LPS-induced tissue factor activity in monocytes by platelets and granulocytes. *Platelets*, **6**, 119–125.

Østerud, B. and Brox, J.H. (1983) The clotting time of whole blood in plastic tubes; the influence of exercise, prostcyclin and acteylsalicylic acid. *Thromb. Res.*, **29**, 425–435.

Østerud, B. and Rapaport, S.I. (1977) Activation of factor IX by the reaction product of tissue factor and factor VII: Additional pathway for initiating blood coagulation. *Proc. Natl. Acad. Sci. USA*, **74**, 5260–5264.

Østerud, B., Olsen, J.O. and Wislgård, L. (1990) The role of arachidonic acid release and lipoxygenase pathway in lipopolysaccharide-induced thromboplastin actovity in monocytes. *Blood Coagul. Fibrinolysis*, **1**, 41–46.

Pieters, J., Willems, G., Hemker, H.C. and Lindhout, T. (1988) Inhibition of factor IXa and factor Xa by antithrombin III/heparin during factor X activation. *J. Biol. Chem.*, **263**, 15313–15318.

Rao, L.V.R. and Rapaport, S.I. (1988) Activation of factor VII bound to tissue factor: A key early step in the tissue factor pathway of blood coagulation. *Proc. Natl. Acad. Sci. USA*, **85**, 6687–6691.

Rapaport, S.I. (1995) The tissue factor pathway: how it has become a "prima ballerina". *Thromb. Haemost.*, **74**, 7–17.

Rapaport, S.I. and Rao, L.V.M. (1992) Initiation and regulation of tissue factor-dependent blood coagulation. *Arterioscler. Thromb.*, **12**, 1111–1121.

Register, T.C., Bora, T.A. and Adams, M.R. (1995) Estrogen inhibits activation of arterial NF-kB transcription factor in early diet-induced atherosclerosis. *Circulation*, **92**, Suppl. 1, I-628.

Ross, R. (1993) Rous-Wipple Award Lecture. Atherosclerosis: a defense mechanism gone awry. *Am. J. Pathol.*, **143**, 987–1002.

Salomaa, V., Rasi, V., Pekkanen, J., Vahtera, E., Jauhiainen, M., Vartiainen, E., Myllyla, G. and Ehnholm, C. (1994) Haemostatic factors and prevalent coronary heart disease, the FINRISK haemostasis study. *Eur. Heart J.*, **15**, 1293–1299.

Scarabin, P.Y., Plu-Bureau, G., Bara, L., Bonthon-Kopp, C., Guize, L. and Samama, M.M. (1993) Haemostatic variables and menopausal status: influence of hormone replacement therapy. *Thromb. Haemost.*, **70**, 584–587.

Sissener Engstad, C., Lia, K., Rekdal, Ø., Olsen, J.O. and Østerud, B. (1995) A novel biological effect on platelet factor 4 (PF4): enhancement of LPS-induced tissue factor activity in monocytes. *J. Leuk. Biol.*, **58**, 575–581.

Smith, E.B. (1994) Lipids and plasma fibrinogen: early and late composition of the atherosclerotic plaque. *Cardiologica*, **39**, (12 Suppl. 1), 169–172.

Smyth, D.W., Martin, J.F., Michaelis, L., Bucknau, C.A. and Jevitt, D.E. (1993) Influence of platelets size before coronary angioplasty on subsequent restenosis. *Eur. J. Clin. Invest.*, **23**, 361–367.

Solberg, S., Østerud, B., Larsen, T. and Sørlie, D. (1990) Lack of ability to synthesize tissue factor by endothelial cell in intact human saphenous veins. *Blood Coagul. Fibrinolysis*, **1**, 595–600.

Spitzer, W.O., Lewis, M.A., Heinemann, L.A.J., Thorgood, M. and MacRae, K.D. (1996) Third generation oral contraceptives and risk of various thromboembolic disorders: an international case-control study. *Br. Med. J.*, **312**, 83–88.

Sporrong, T., Mattsson, L.A., Samsioe, G., Stigendal, L. and Hellgren, M. (1990) Haemostatic changes during oestradiol-progestogen treatment of postmenopausal women. *Br. J. Obstet. Gynecol.*, **97**, 939–944.

Stemme, S., Faber, B., Holm, J., Wicklund, O., Witztum, J.L. and Hansson, K.G. (1990) Detection and localization of tumor necrosis factor in human atheroma. *Am. J. Cardiol.*, **65**, 297–302.

Szymanski, L.M., Pate, R.R. and Durstine, J.L. (1994) Effects of maximal exercise and venous occlusion on fibrinolytic activity in physically active and inactive men. *J. Appl. Physiol.*, **77**, 2305–2310.

Tanner, F.C., Baulanger, C.N. and Luscher, T.F. (1993) Endothelium-derived nitric oxide, endothelin, and platelet vessel wall interaction alterations in hypercholesterolemia and atherosclerosis. *Semin. Thromb. Haemost.*, **19**, 167–175.

Tipping, P.G. and Hancock, W.W. (1993) Production of tumor necrosis factor and interleukin-l by macrophages from human atheromatous plaque. *Am. J. Pathol.*, **142**, 1721–1728.

Verstraete, M. (1995) The fibrinolytic system: from Petri dishes to genetic engineering. *Thromb. Haemost.*, **74**, 25–35.

Weiss, H. (1995) Flow related platelet deposition on subendothelium. *Thromb. Haemost.*, **74**, 117–122.

Weitz, J.I., Huang, A.J., Landman, L., Nicholson, S. and Silverstein, S.C. (1987) Elastase-mediated fibrinogenolysis by chemoattractant stimulated neutrophils occurs in the presence of physiologic concentrations of antiproteases. *J. Exp. Med.*, **166**, 1836–1850.

Wessler, S. (1992) Estrogen associated thromboembolism. *Ann. Epidemiol.*, **2**, 439–443.

Weyhrich, A.S., Elstad, M.R., McEver, R.P., McIntyre, T.M., Moore, K.L., Morrissey, J.H., Prescott, S.M. and Zimmerman, G.A. (1996) Activated platelets signal chemokine synthesis by human monocytes. *J. Clin. Invest.*, **97**, 1525–1534.

Wilcox, J.N., Smith, K.M., Schwartz, S.M. and Gordon, D. (1989) Localization of tissue factor in the normal vessel wall and in the atherosclerotic plaque. *Proc. Natl. Acad. Sci. USA*, **86**, 2839–2843.

Wiman, B. (1995) Plasminogen activator inhibitor 1 (PAI-1) in plasma: its role in thrombotic disease. *Thromb. Haemost.*, **74**, 71–76.

Winkler, U.H., Oberhoff, C., Bier, U., Schindler, A.E. and Gillain, D. (1995) Hemostatic effects of two oral contraceptives containing low doses of ethinyl estradiol and either gestodene or norgestimate: an open, randomized, parallel group study. *Int. J. Fert. Menopausal Stud.*, **40**, 260–268.

World Helath Organization Collaborative Study of Cardiovascular Disease and Steroid Hormone Contraception. (1995) Venous thromboembolic disease and combined oral contraceptives: results of an interaction multicenter case-control study. *Lancet*, **346**, 1575–1582.

Yla-Herttuala, S., Palinski, W., Rosenfeld, M.E., Parthasarathy, S., Carew, T.E., Butler, S., Witztum, J.L. and Steinberg, D. (1989) Evidence of the presence of oxidatively modified low density lipoprotein in atherosclerotic lesions of rabbit and man. *J. Clin. Invest.*, **84**, 1086–1095.

the beneficial effects of estrogen on the vascular system may be dependent on and independent of the classical estrogen receptor.

A number of the mechanisms which result in vascular relaxation will be reviewed, some of these mechanisms may also be involved in the long-term vascular protection which is conferred by estrogen.

## CARDIOVASCULAR CIRCULATION

### Acute Effects of Estrogen on Blood Flow

In the estrogen-deplete state, local and systemic administration of 17$\beta$-estradiol has effects on both reproductive and non-reproductive vascular beds. In a study of non-pregnant, ovariectomized ewes, 17$\beta$-estradiol was infused into a uterine artery (Magness and Rosenfeld, 1989). Thirty minutes after the uterine artery infusion there was an increase in uterine blood flow, a decrease in uterine vascular resistance with no change in cardiac output, mean arterial blood pressure or systemic vascular resistance. However, 30 minutes after a subsequent systemic infusion of 17$\beta$-estradiol there was an increase in cardiac output, a decrease in systemic vascular resistance with no change in mean arterial pressure. These changes were independent of any change in uterine blood flow or vascular resistance, demonstrating that the effects of 17$\beta$-estradiol on the systemic circulation are independent of changes in the reproductive circulation.

### Chronic Effects of Estrogen on Blood Flow

Aortic peak flow velocity gradually decreases with time since the onset of the menopause (Pines *et al.*, 1992), and there is a significant increase in aortic flow velocity and acceleration after 10 weeks of hormone replacement therapy (HRT) (Pines *et al.*, 1991). Treatment with 17$\beta$-estradiol (22 weeks) in normotensive postmenopausal women results in a decrease in pulsatility index of the carotid artery, which is thought to represent impedance to blood flow (Ganger *et al.*, 1991). Physiological doses of estrogen were used in this study and the resulting effect took several weeks to become apparent. The fall in pulsatility index with estrogen treatment was correlated to the duration since the menopause.

These peripheral vascular effects may involve one or a number of mechanisms, including ion channel modulation and/or by effects on the endothelium. The mechanism(s) may differ depending on the dose of estrogen used and the duration of use (Volterrani *et al.*, 1995; Ganger *et al.*, 1991).

## MECHANISMS INVOLVED IN THE VASCULAR EFFECTS OF ESTROGEN

Estrogen receptor-dependent mechanisms are involved in the response of the reproductive vasculature to gonadal hormones. In the cardiovascular system the role of the classic sex-steroid receptor is unclear, but the acute responses of coronary arteries to estrogen demonstrated in recent studies (Jiang *et al.*, 1991; 1992b; Cheng and Gruetter, 1992; Mügge *et al.*, 1993; Thomas *et al.*, 1995; Vargas *et al.*, 1995; Chester *et al.*, 1995; Han *et al.*, 1995) suggest that non-receptor as well as receptor mechanisms are involved. It seems unlikely that genomic mechanisms would be initiated within minutes, as in the responses in these studies. Chronic exposure to estrogen however may effect blood flow response via the classic estrogen receptor. The presence of estrogen receptors has been demonstrated in normal coronary arteries from women (Losordo *et al.*, 1994). There is variable expression in atherosclerotic coronary arteries from premenopausal women (Losordo *et al.*, 1994), while others have found no evidence for estrogen receptors in normal coronary arteries, using a different antibody technique (Collins *et al.*, 1995b). Other properties of estrogen have been demonstrated in recent studies which may also account for some of the beneficial effects of estrogen on the cardiovascular system, including calcium antagonism and angiotensin converting enzyme inhibition. This account will deal with direct vascular smooth muscle effects of estrogen and some of the mechanisms involved in these responses.

## Endothelium-Independent Direct Vascular Smooth Muscle Relaxation

17$\beta$-Estradiol has direct smooth muscle, endothelium-independent relaxing properties in coronary and cerebral blood vessels. In isolated rabbit coronary arteries Jiang *et al.* (1991) demonstrated direct vascular smooth muscle relaxation by 17$\beta$-estradiol. Inhibitors of EDRF did not affect 17$\beta$-estradiol-induced relaxation in endothelium-intact coronary arteries. Methylene blue, an inhibitor of EDRF-induced increase of cyclic GMP, also had no effect on relaxation by 17$\beta$-estradiol. These results confirmed that the *in vitro* acute relaxation of rabbit coronary arteries by 17$\beta$-estradiol is independent of EDRF (Jiang *et al.*, 1991). The concentrations of estrogen required to relax the vessels ($10^{-6}$ and $10^{-5}$ mol per liter) were greater than those found physiologically ($\approx 10^{-9}$ mol per liter).

Important differences in the relaxation response of isolated human coronary arteries to estrogen have been published (Mügge *et al.*, 1993; Chester *et al.*, 1995). Atherosclerosis-free male and female human epicardial arteries were removed from patients undergoing heart, or combined heart and lung transplantation (Chester *et al.*, 1995). The relaxing response to 17$\beta$-estradiol ($10^{-10}$–$10^{-5}$ mol per liter) was investigated and the effect of endothelium, $N^G$monomethyl-L-arginine (a nitric oxide synthase inhibitor) and indomethacin (a prostaglandin synthetase inhibitor) on the response of 17$\beta$-estradiol was assessed. 17$\beta$-Estradiol induced significant relaxation in coronary arteries pre-contracted with the thromboxane $A_2$ analog (U46619). Relaxation was significantly *greater* in coronary arteries from female compared to male patients. No significant differences were observed between arteries with or without endothelium, nor after nitric oxide synthase or cyclooxygenase inhibition. These results indicate that 17$\beta$-estradiol induces human coronary artery relaxation by an endothelium-independent mechanism *in vitro*. The gender of the patients significantly affecting the sensitivity of the coronary arterial rings to estrogen.

**Evidence for Cell Membrane Interactions of Estrogen**

A number of vascular responses in a variety of different blood vessels have been measured and the effect of estrogen on contractile agents has been investigated.

Treatment of rat aortic rings with 17$\beta$-estradiol (0.37–37 μmol per liter) for 15–180 minutes, and subsequent removal of estrogen, attenuated the vasoconstrictor responses to phenylephrine and potassium chloride in a time-dependent manner. The maximum inhibitory effect took 120 minutes to develop and was endothelium-independent, not affected by cylcooxygenase inhibitors such as indomethacin or the endothelium-derived relaxing factor inhibitor, L-NMMA. The effect was stereo-specific, in that the 17$\alpha$-isomer was less potent than the 17$\beta$-isomer of estradiol. The inhibitory effect of estradiol was, however, blocked completely by the pre-treatment with protein synthesis-inhibitors, cyclohexamide and puramycin, but not by actinomycin-D. There was an increase in ribosomal expression of the rough endoplasmic reticulum level after 120 minutes, indicating increased protein synthesis after exposure to 17$\beta$-estradiol may be related to an inhibitory effect on vascular smooth muscle at the protein translational level. It was speculated that one of the proteins induced by estradiol may be glucose-6-phosphate dehydrogenase. An effect on this enzyme has been reported in uterine smooth muscle by another group (Cummings and Barker, 1986). Glucose-6-phosphate dehydrogenase regulates cellular redox status and in turn modulates calcium transport across the plasma membrane. This enzyme may therefore relate to the mechanism by which estradiol inhibits smooth muscle contraction *in vitro* after prolonged exposure (Thomas *et al.*, 1995).

Short term pre-treatment of rat thoracic aorta with 17$\beta$-estradiol attenuates the contractile response to both phenylephrine and angiotensin-II. Application of $5 \times 10^{-6}$ mol per liter estradiol for 30 minutes compared to ethanol control significantly decreased the contractile responses to these constrictor agents. A membrane effect of estrogen was suggested. In contrast to a vasodilator effect, low doses of 17$\beta$-estradiol but not its stereo-isomer, 17$\alpha$-estradiol, or testosterone significantly *potentiated* the vasoconstrictor responses of rat isolated mesenteric preparations elicited by noradrenaline, potassium and the prostaglandin endoperoxide analogue, U-46619. The mechanism for this effect in this particular preparation is not known, but it does suggest differences between species and vascular beds (Vargas *et al.*, 1995).

These studies suggest that response to estrogen may not involve gene transcription and may be mediated by specific membrane receptors. The precise nature of membrane receptors and their relation to intracellular nuclear receptors remains to be elucidated. 17$\beta$-estradiol also increased, but did not significantly change vasoconstrictor responses of rat mesenteric preparations to electric field stimulation. Electric field stimulation is known to elicit vasoconstriction by releasing endogenous noradrenaline from nerve terminals. 17$\beta$-Estradiol inhibits potassium- and calcium-induced release of noradrenaline from nerve terminals of rat hypothalamus (Hyatt and Tyce, 1984). This has also been demonstrated in the portal vein (Bengtsson, 1978). Estradiol may increase the sensitivity of the vascular smooth muscle to noradrenaline in the mesenteric vascular bed but at the same time inhibits the release of endogenous pools of noradrenaline. This could explain why the electric field stimulation data did not achieve significance in the present study.

## Calcium Antagonism

Calcium antagonism as a mechanism of vascular smooth muscle relaxation by estrogen was proposed for uterine artery relaxation. Stice *et al.* (1987a) reported that estradiol decreased calcium entry into uterine vascular smooth muscle cells. Another study suggested calcium antagonistic properties of 17$\beta$-estradiol in isolated uterine arteries (Stice *et al.*, 1987b). Potential-sensitive calcium channels are activated by depolarisation of the plasma membrane when the extracellular potassium concentration is increased. Bay K8644 is a dihydro pyridine derivative which opens potential-sensitive calcium channels (Brown, Kunze and Yatani, 1984). Jiang *et al.* in 1991 found that 17$\beta$-estradiol-induced relaxation of coronary arterial rings contracted by both high extracellular potassium and Bay K8644. This evidence suggested that 17$\beta$-estradiol may have an inhibitory effect on potential-sensitive calcium channel activation. It was also shown that 17$\beta$-estradiol induced relaxation of contraction evoked by $PGF_{2\alpha}$, an agonist of receptor-operated calcium channels, indicating 17$\beta$-estradiol has similar relaxing effects on contraction induced both by activation of receptor-operated, and potential-operated calcium channels. Furthermore, incubation with 17$\beta$-estradiol shifted calcium concentration-dependent contraction curves to the right in high potassium-depolarisation medium in rabbit coronary arteries, and rat aorta. The curves shifted by incubation with different 17$\beta$-estradiol concentrations were parallel, although the maximal contractions were also affected. These findings suggested that 17$\beta$-estradiol may have a calcium antagonistic property in these arterial preparations.

Confirmation of a calcium antagonistic property of estrogen was provided by experiments on isolated guinea-pig cardiac myocytes (Jiang *et al.*, 1992a). The effect of 17$\beta$-estradiol on cardiac cell contraction, inward calcium current and intracellular free calcium was investigated using a photodiode array, voltage-clamp, and Fura-2 florescence techniques respectively. 17$\beta$-Estradiol was shown to have a negative inotropic effect on single ventricular myocytes by inhibiting inward calcium current and so reducing intracellular free calcium. This calcium antagonistic property could be one of the mechanisms of 17$\beta$-estradiol-induced endothelium-independent relaxation in isolated rabbit and human coronary arterial preparations.

It has been hypothesised that some of the cardiovascular benefit of estrogen replacement therapy may be due to a calcium antagonist effect of estrogen (Collins *et al.*, 1993). In a long-term study of estrogen in atherosclerotic, ovariectomized cynomolgus monkeys, a decrease in coronary atheroma was reported compared to the non-treated animals (Adams *et al.*, 1990). Long-term calcium antagonist treatment is known to decrease the progression of atheroma when given to patients with established coronary heart disease (Lichtlen *et al.*, 1990; Waters *et al.*, 1990), therefore estrogen may be cardioprotective via a beneficial effect on atheroma progression *in vivo* due to its calcium antagonistic properties.

### *Calcium antagonism in vascular smooth muscle*

Estrogen can inhibit the contraction of epicardial coronary arteries by inhibiting calcium influx without changing calcium sensitivity of contractile elements (Han *et al.*, 1995).

Cytosolic calcium concentration and contraction were measured simultaneously in fura-2 loaded porcine coronary arterial strips stimulated by the contractile agents thromboxane-$A_2$ analogue (U-46619), and high potassium depolarisation in the presence and absence of 17β-estradiol. 17β-Estradiol at concentrations ranging from 30 nmol per liter to 30 μmol per liter inhibited the sustained elevation of intracellular calcium and the sustained contraction induced by U46619. High concentrations (30 μmol per liter) of 17β-estradiol only partially inhibited the U46619-induced sustained contraction despite complete inhibition of the sustained increase in calcium. This suggests that 17β-estradiol inhibits the contraction of coronary vascular smooth muscle by inhibiting calcium influx without changing calcium sensitivity of contractile elements. This was further supported by the fact that 17β-estradiol (10 μmol per liter) inhibited the increase in calcium-induced contraction, and the contraction induced by the cumulative addition of calcium in strips pre-treated with 90 mmol per liter of potassium. However 17β-estradiol did not change the slope of the intracellular calcium tension curve and had no effect on levels of cyclic AMP and cyclic GMP. This report confirms earlier work in isolated cardiac myocytes demonstrating that 17β-estradiol has calcium channel-blocker action (Jiang *et al.*, 1992a). These data would further reinforce the possibility that a long-term calcium-antagonist mechanism of estrogen may play a role in its anti-atherosclerotic effect (Collins *et al.*, 1993).

Further evidence that 17β-estradiol has effects on calcium channels in vascular smooth muscle cells is provided by work on an isolated vascular smooth muscle cell line (A7r5) (Zhang *et al.*, 1994). Whole patch-clamp techniques were used to study effects of 17β-estradiol on voltage-dependent calcium channels in these cells. 17β-Estradiol, at a concentration of 10 μmol per liter, significantly reduced L-type barium currents and T-type calcium currents within 1–2 minutes. This was in contrast to 17α-estradiol, which caused significantly less reduction in both types of current. The inhibitory effect of 17β-estradiol on voltage-dependent calcium current may therefore contribute to previously demonstrated 17β-estradiol attenuation for vascular smooth muscle cells. Most of the acute effects in vascular smooth muscle cells occur at micromolar concentrations which are greater than those measured in plasma from normal females. However it is possible that tissues accumulate lipid soluble steroids and local concentrations may indeed reach this level.

17β-Estradiol has been shown to inhibit calcium influx through voltage-dependent calcium channels in Sprague–Dawley rat tail arteries. There was a correlation between the modulation of vascular smooth muscle cell intracellular calcium which led to an attenuation of noradrenaline and arginine-vasopressin constrictor response by estrogen (Shan *et al.*, 1994). As well as effects on calcium channels, estrogen affects large conductance chloride channels by a direct membrane effect (Hardy and Valverde, 1994). These channels, in cultured fibroblasts, were inhibited by exposure to extracellular, but not intracellular anti-estrogen, and this effect could be prevented by extracellular 17β-estradiol but not intracellular 17β-estradiol or extracellular 17α-estradiol. This demonstrates another regulatory role for ovarian steroids, affecting plasma ion channels via membrane binding sites distinct from the classical estrogen receptor and subsequent activation of intracellular second-messenger pathway(s).

*Effect on potassium channels*

In pig coronary arteries 17$\beta$-estradiol induces relaxation by an endothelium-independent mechanism involving calcium influx. Using patch clamp techniques it was confirmed that estrogen stimulated the large-conductance calcium- and voltage-activated potassium channel ($BK_{Ca}$) (White, Darkow and Lang, 1995). 17$\alpha$-Estradiol had no effect. Blocking $BK_{Ca}$ with a specific inhibitor iberiotoxin reversed the relaxing effect of estrogen, as did blockade of the cyclic-GMP-dependent protein kinase. Therefore another potential relaxing mechanism for estrogen in vascular smooth muscle exists. The mechanism of relaxation involves the opening of $BK_{Ca}$ by a cyclic GMP-dependent phosphorylation.

**Inhibition of Constrictor Factors**

*Endothelin-1*

Estrogen inhibits the constrictor responses to endothelin-1 in rabbit coronary arteries (Jiang *et al.*, 1992b). Plasma endothelin-1 levels increase following acetylcholine infusion in pigs with coronary atheroma (Lerman *et al.*, 1993), and although this has not been confirmed in humans, it may suggest that the reversal of the constrictor effect of acetylcholine in human atherosclerotic coronary arteries by estrogen (Reis *et al.*, 1994; Gilligan, Quyyumi and Cannon, 1994; Collins *et al.*, 1995a) may be partially explained by estrogen-induced inhibition of endothelin-1-induced contraction.

**Renin–Angiotensin System**

Estrogen inhibits angiotensin II-induced constrictor effects both *in vitro* and *in vivo*, suggesting an inhibitory effect on the renin–angiotensin system. Angiotensin II-induced concentration-response curves were compared in aortic rings with or without endothelium from ovariectomized rats pretreated with 17$\beta$-estradiol (1 mg per kilogram per day) or placebo for 14 days. Angiotensin II-induced contraction was significantly depressed in rings from the 17$\beta$-estradiol treated rats, which was independent of the vascular endothelium (Cheng and Gruetter, 1992). *In vivo* studies in ovariectomized ewes show a similar attenuation of the pressor responses to infused angiotensin II by chronic (2 weeks) treatment with 17$\beta$-estradiol (Magness, Parker and Rosenfeld, 1993). An interesting further observation in this study was that the attenuation was still demonstrable up to 2 weeks after the cessation of estradiol treatment which suggests residual effects of prolonged estrogen administration. The mechanism by which estrogen induces relaxation in the presence of these agents may involve calcium antagonism.

It has recently been suggested that estrogen is an angiotensin converting enzyme (ACE) inhibitor. By measuring ACE activity and the angiotensin peptides (angiotensin I, angiotensin II and angiotensin (1–7)) in cynomolgus monkeys, it was shown that 30 months of Premarin treatment decreased ACE activity whereas a combination of Premarin plus medroxyprogesterone acetate increased ACE activity (Brosnihan *et al.*, 1995). A study in postmenopausal women confirms an effect of hormone therapy on ACE activity in humans (Proudler *et al.*, 1995). After 6 months of estrogen/progestogen therapy, serum ACE activity was significantly reduced compared to non-treated controls. Gonadal steroids may therefore confer a beneficial effect on coronary heart disease risk by their effect on ACE.

## Possible Direct Vascular Smooth Muscle Relaxation by Estrogen *In Vivo*

### *Coronary circulation*

Supraphysiologic concentrations (>0.1 μmol per liter) of estrogen causes dilation of conductance and resistance coronary arteries in dogs when administered acutely into the coronary circulation (Sudhir *et al.*, 1995). By removing the endothelium and using inhibitors of adenosine triphosphate-sensitive potassium and calcium channels, it was shown that this effect is endothelium-independent and is partially mediated by effects on adenosine triphosphate-sensitive potassium and/or calcium channels. Experiments using a classic intracellular estrogen receptor antagonist suggest that this receptor is not involved in the response. More likely is an acute effect on other receptors on the plasma membrane, similar to those demonstrated for aldosterone (Wehling, Kasmayr and Theisen, 1991; Christ *et al.*, 1993).

### *Peripheral circulation*

In the estrogen-deplete state, local and systemic administration of 17β-estradiol has effects on both reproductive and non-reproductive vascular beds (Magness and Rosenfeld, 1989). The effect of estrogen administration on systemic hemodynamics has been studied in postmenopausal women. Forty minutes after the administration of sublingual 17β-estradiol to postmenopausal volunteers there was an increase in forearm blood flow and a reduction in forearm vascular resistance compared to placebo, with no difference in mean arterial blood pressure (Volterrani *et al.*, 1995). Relatively high plasma levels of 17β-estradiol (≈3000 pmol per liter) were achieved in these women (midcycle level ≈1800 pmol per liter), however the plasma levels are approximately 50% of those found in pregnancy. Similar changes in blood flow have been demonstrated in the blood supply to the leg (Riedel *et al.*, 1995). Using a double-blind design, postmenopausal women (mean age 59 years) received estradiol (1 mg) or placebo. Sixty to 80 minutes after administration, axial diameter and blood flow rates of the common femoral artery were measured. Estradiol induced a significant vasodilation of the femoral artery compared to placebo. Blood flow increased significantly by approximately 30%.

Both the coronary and peripheral effects on blood flow are likely to involve direct vascular smooth muscle relaxation. In the dog coronary removal of the coronary endothelium and inhibition of nitric oxide synthase did not affect the relaxing response. Also the rapid temporal effect of the relaxing response both in the dog coronary and human peripheral vasculature would argue against a mechanism involving the classical intracellular estrogen receptor.

Current research suggests that hormone therapy, either estrogen alone or in combination with progestogen, is cardioprotective. The vascular effects of hormone therapy is an area of great interest and potentially many mechanisms are involved.

However, the mechanisms apparent *in vitro* may not necessarily be relevant *in vivo*, and effects in animals may not be involved or reproduced in humans. Acute and chronic exposure to hormones may involve different mechanisms. The data argue against a single mechanism accounting for the vascular effects of hormone therapy, more likely is an interaction of mechanisms. The recently discovered direct effects of estrogen-induced vascular relaxation add support to the observation that estrogen engenders vascular health.

## REFERENCES


Adams, M.R., Kaplan, J.R., Manuck, S.B., Koritnik, D.R., Parks, J.S., Wolfe, M.S. *et al.* (1990) Inhibition of coronary artery atherosclerosis by 17-beta estradiol in ovariectomized monkeys. Lack of an effect of added progesterone. *Arteriosclerosis*, **10**, 1051–1057.

Bengtsson, B. (1978) Estrogen induced inhibition of 3H-noradrenaline release in the uterus and portal vein of the rat. *Acta Physiol. Scand.*, **104**, 287–298.

Brosnihan, K.B., Weedle, D., Anthony, M.S. and Ferrario, C.M. (1995) Does chronic estrogen act as a converting enzyme inhibitor in postmenopausal cynomolgous monkeys? *American Journal of Hypertension*, **8**, 107A.

Brown, A.M., Kunze, D.L. and Yatani, A. (1984) The agonist effect of dihydropyridines on $Ca^{++}$ channels. *Nature*, **311**, 570–572.

Cheng, D.Y. and Gruetter, C.A. (1992) Chronic estrogen alters contractile responsiveness to angiotensin II and norepinephrine in female rat aorta. *European Journal of Pharmacology*, **215**, 171–176.

Chester, A.H., Jiang, C., Borland, J.A., Yacoub, M.H. and Collins, P. (1995) Estrogen relaxes human epicardial coronary arteries through non-endothelium-dependent mechanisms. *Coronary Artery Disease*, **6**, 417–422.

Christ, M., Eisen, C., Aktas, J., Theisen, K. and Wehling, M. (1993) The inositol-1,4,5-trisphosphate system is involved in rapid nongenomic effects of aldosterone in human mononuclear leukocytes. *Journal of Clinical Endocrinology and Metabolism*, **77**, 1452–1457.

Collins, P., Rosano, G.M.C., Jiang, C., Lindsay, D., Sarrel, P.M. and Poole-Wilson, P.A. (1993) Hypothesis: Cardiovascular protection by oestrogen—a calcium antagonist effect? *Lancet*, **341**, 1264–1265.

Collins, P., Shay, J., Jiang, C. and Moss, J. (1994) Nitric oxide accounts for dose-dependent estrogen-mediated coronary relaxation following acute estrogen withdrawal. *Circulation*, **90**, 1964–1968.

Collins, P., Rosano, G.M.C., Sarrel, P.M., Ulrich, L., Adamopoulos, S., Beale, C.M. *et al.* (1995a) Estradiol-17$\beta$ attenuates acetylcholine-induced coronary arterial constriction in women but not men with coronary heart disease. *Circulation*, **92**, 24–30.

Collins, P., Sheppard, M., Beale, C.M. and Dowsett, M. (1995b) The classical estrogen receptor is not found in human coronary arteries. *Circulation*, **92**, I-37.

Cummings, A.M. and Barker, K.L. (1986) Isolation of a precursor and a nascent chain form of glucose-6-phosphate dehydrogenase from rat uterus and regulation of precursor processing by estradiol. *Biochemica et Biophysica Acta*, **880**, 226–241.

Ganger, K.F., Vyas, S., Whitehead, M., Crook, D., Meire, H. and Campbell, S. (1991) Pulsatility index in internal carotid artery in relation to transdermal oestradiol and time since menopause. *Lancet*, **338**, 839–842.

Gilligan, D.M., Quyyumi, A.A. and Cannon, R.O., III (1994) Effects of physiological levels of estrogen on coronary vasomotor function in postmenopausal women. *Circulation*, **89**, 2545–2551.

Gruchow, H.W., Anderson, A.J., Barboriak, J.J. and Sobocinski, K.A. (1988) Postmenopausal use of estrogen and occlusion of coronary arteries. *American Heart Journal*, **115**, 954–963.

Han, S.Z., Karaki, H., Ouchi, Y., Akishita, M. and Orimo, H. (1995) 17$\beta$-Estradiol inhibits $Ca^{2+}$ influx and $Ca^{2+}$ release induced by thromboxane A2 in porcine coronary artery. *Circulation*, **91**, 2619–2626.

Hardy, S.P. and Valverde, M.A. (1994) Novel plasma membrane action of estrogen and antiestrogens revealed by their regulation of a large conductance chloride channel. *FASEB J.*, **8**, 760–765.

Hyatt, M.C. and Tyce, G.M. (1984) Effects of estradiol on the basal and evoked efflux of norepinephrine and 5-hydroxytryptamine from slices of rat hypothalamus. *Life Sciences*, **35**, 2269–2274.

Jiang, C., Sarrel, P.M., Lindsay, D.C., Poole-Wilson, P.A. and Collins, P. (1991) Endothelium-independent relaxation of rabbit coronary artery by 17$\beta$-oestradiol *in vitro*. *British Journal of Pharmacology*, **104**, 1033–1037.

Jiang, C., Poole-Wilson, P.A., Sarrel, P.M., Mochizuki, S., Collins, P. and MacLeod, K.T. (1992a) Effect of 17$\beta$-oestradiol on contraction, $Ca^{2+}$ current and intracellular free $Ca^{2+}$ in guinea-pig isolated cardiac myocytes. *British Journal of Pharmacology*, **106**, 739–745.

Jiang, C., Sarrel, P.M., Poole-Wilson, P.A. and Collins, P. (1992b) Acute effect of 17$\beta$-estradiol on rabbit coronary artery contractile responses to endothelin-1. *American Journal of Physiology*, **263**, H271–H275.

Lerman, A., Webster, M.W.I., Chesebro, J.H., Edwards, W.D., Wei, C.-M., Fuster, V. *et al.* (1993) Circulating and tissue endothelin immunoreactivity in hypercholesterolemic pigs. *Circulation*, **88**, 2923–2928.

Lichtlen, P.R., Hugenholtz, P.G., Rafflenbeul, W., Hecker, H., Jost, S. and Deckers, J.W. (1990) Retardation of angiographic progression of coronary artery disease by nifedipine. *Lancet*, **335**, 1109–1113.

Losordo, D.W., Kearney, M., Kim, E.A., Jekanowski, J. and Isner, J.M. (1994) Variable expression of the estrogen receptor in normal and atherosclerotic coronary arteries of premenopausal women. *Circulation*, **89**, 1501–1510.

Magness, R.R., Parker, C.R., Jr. and Rosenfeld, C.R. (1993) Systemic and uterine responses to chronic infusion of estradiol-17 beta. *American Journal of Physiology*, **265**, E690–E698.

Magness, R.R. and Rosenfeld, C.R. (1989) Local and systemic estradiol-17 beta: effects on uterine and systemic vasodilation. *American Journal of Physiology*, **256**, E536–E542.

Mendelsohn, M.E. and Karas, R.H. (1994) Estrogen and the blood vessel wall. *Current Opinion in Cardiology*, **9**, 619–626.

Mikkola, T., Turunen, P., Avela, K., Orpana, A., Viinikka, L. and Ylikorkala, O. (1995) 17 beta-estradiol stimulates prostacyclin, but not endothelin-1, production in human vascular endothelial cells. *Journal of Clinical Endocrinology and Metabolism*, **80**, 1832–1836.

Mügge, A., Riedel, M., Barton, M., Kuhn, M. and Lichten, P.R. (1993) Endothelium independent relaxation of human coronary arteries by 17beta-oestradiol *in vitro*. *Cardiovascular Research*, **27**, 1939–1942.

Pines, A., Fisman, E.Z., Levo, Y., Averbuch, M., Lidor, A., Drory, Y. *et al.* (1991) The effects of hormone replacement therapy in normal postmenopausal women: measurements of Doppler-derived parameters of aortic flow. *American Journal of Obstetrics and Gynecology*, **164**, 806–812.

Pines, A., Fisman, E.Z., Drory, Y., Levo, Y., Shemesh, J., Ben-Ari, E. *et al.* (1992) Menopause-induced changes in doppler-derived parameters of aortic flow in healthy women. *American Journal of Cardiology*, **69**, 1104–1106.

Proudler, A.J., Hasib Ahmed, A.I., Crook, D., Fogelman, I., Rymer, J.M. and Stevenson, J.C. (1995) Hormone replacement therapy and serum angiotensin-converting-enzyme activity in postmenopausal women. *Lancet*, **346**, 89–90.

Reis, S.E., Gloth, S.T., Blumenthal, R.S., Resar, J.R., Zacur, H.A., Gerstenblith, G. *et al.* (1994) Ethinyl estradiol acutely attenuates abnormal coronary vasomotor responses to acetylcholine in post-menopausal women. *Circulation*, **89**, 52–60.

Riedel, M., Oeltermann, A., Mugge, A., Creutzig, A., Rafflenbeul, W. and Lichtlen, P. (1995) Vascular responses to 17 beta-oestradiol in postmenopausal women. *European Journal of Clinical Investigation*, **25**, 44–47.

Robson, S.C., Hunter, S., Boys, R.J. and Dunlop, W. (1989) Serial study of factors influencing changes in cardiac output during human pregnancy. *American Journal of Physiology*, **256**, H1060–H1065.

Shan, J., Resnick, L.M., Liu, Q.Y., Wu, X.C., Barbagallo, M. and Pang, P.K. (1994) Vascular effects of 17 beta-estradiol in male Sprague-Dawley rats. *American Journal of Physiology*, **266**, H967–H973.

Shay, J., Badrov, N., Attele, A., Levinson, M. and Moss, J. (1993) Estrogen antagonises endothelin-1 vasoconstriction in rabbit basilar artery. *Anesthesia and Analgesia*, A561.

Stampfer, M.J. and Colditz, G.A. (1991) Estrogen replacement therapy and coronary heart disease: a quantitative assessment of the epidemiologic evidence. *Preventive Medicine*, **20**, 47–63.

Stice, S.L., Ford, S.P., Rosazza, J.P. and Van-Orden, D.E. (1987a) Role of 4-hydroxylated estradiol in reducing $Ca^{2+}$ uptake of uterine arterial smooth muscle cells through potential-sensitive channels. *Biology of Reproduction*, **36**, 361–368.

Stice, S.L., Ford, S.P., Rosazza, J.P. and Van-Orden, D.E. (1987b) Interaction of 4-hydroxylated estradiol and potential-sensitive $Ca^{2+}$ channels in altering uterine blood flow during the estrous cycle and early pregnancy in gilts. *Biology of Reproduction*, **36**, 369–375.

Sudhir, K., Chou, T.M., Mullen, W.L., Hausmann, D., Collins, P., Yock, P.G. *et al.* (1995) Mechanisms of estrogen-induced vasodilation: *in vivo* studies in canine coronary conductance and resistance arteries. *Journal of the American College of Cardiology*, **26**(3), 807–814.

Sullivan, J.M., Vander Zwaag, R., Lemp, G.F., Hughes, J.P., Maddock, V., Kroetz, F.W. *et al.* (1988) Postmenopausal estrogen use and coronary atherosclerosis. *Annals of Internal Medicine*, **108**, 358–363.

Thomas, G., Ito, K., Zikic, E., Bhatti, T., Han, C. and Ramwell, P.W. (1995) Specific inhibition of the contraction of the rat aorta by estradiol 17 beta. *Journal of Pharmacology and Experimental Therapeutics*, **273**, 1544–1550.

Vargas, R., Delaney, M., Farhat, M.Y., Wolfe, R., Rego, A. and Ramwell, P.W. (1995) Effect of estradiol 17 beta on pressor responses of rat mesenteric bed to norepinephrine, K+, and U-46619. *Journal of Cardiovascular Pharmacology*, **25**, 200–206.

Volterrani, M., Rosano, G.M.C., Coats, A., Beale, C. and Collins, P. (1995) Estrogen acutely increases peripheral blood flow in postmenopausal women. *American Journal of Medicine*, **99**, 119–122.

Wagner, J.D., Clarkson, T.B., St. Clair, R.W., Schwenke, D.C., Shively, C.A. and Adams, M.R. (1991) Estrogen and progesterone replacement therapy reduces low density lipoprotein accumulation in the coronary arteries of surgically postmenopausal cynomolgus monkeys. *Journal of Clinical Investigation*, **88**, 1995–2002.

Waters, D., Lesperance, J., Francetich, M., Causey, D., Theroux, P., Chiang, Y.-K. *et al*. (1990) A controlled clinical trial to assess the effect of a calcium channel blocker on the progression of coronary atherosclerosis. *Circulation*, **82**, 1940–1953.

Wehling, M., Kasmayr, J. and Theisen, K. (1991) Rapid effects of mineralocorticoids on sodium-proton exchanger: genomic or nongenomic pathway? *American Journal of Physiology*, **260**, E719–E726.

White, R.E., Darkow, D.J. and Lang, J.L. (1995) Estrogen relaxes coronary arteries by opening BKCa channels through a cGMP-dependent mechanism. *Circulation Research*, **77**, 936–942.

Zhang, F., Ram, J.L., Standley, P.R. and Sowers, J.R. (1994) 17β-Estradiol attenuates voltage-dependent Ca2+ currents in A7r5 vascular smooth muscle cell line. *American Journal of Physiology*, **266**, C975–C980.

# 12 Endothelium-Dependent Vasoconstrictors

**Dustan A. Barber and Virginia M. Miller** *

*Medical Science Building 4-54, Mayo Clinic and Foundation,*
*200 First Street SW, Rochester, MN 55905, USA.*
** Corresponding author, Tel.: 001 507 284 2290, Fax: 001 507 284 5075,*
*E-mail. miller.virginia@mayo.edu.*

Endothelium-derived contractile factors influence the development and progression of cardiovascular disease by regulating vascular tone and growth. In addition to modulating actions of endothelium-derived relaxing factors like nitric oxide, estrogen may regulate endothelium-derived contractile factors. This chapter summarizes effects of estrogen on production, release and actions of metabolites of arachidonic acid by cyclooxygenase (thromboxane and prostacyclin) and the endothelins.

KEYWORDS: Eicosanoids, endothelin-1, estrogen, nitric oxide, prostacyclin, thromboxane.

## INTRODUCTION

Prior to menopause women have a lower incidence and severity of coronary artery disease compared to men of the same age (Barrett-Conner and Bush, 1991). After menopause this difference diminishes unless estrogen replacement therapy is initiated (Eaker *et al*., 1993; Stampfer and Colditz, 1991). Such observations suggest that estrogen provides protection against the development of cardiovascular disease [for reviews see, (Wenger, 1990; Barrett-Conner and Bush, 1991; Eaker *et al*., 1993; Stampfer and Colditz, 1991; Riedel *et al*., 1993)]. Estrogenic regulation of multiple components of the vessel wall has been proposed as a major mechanism to account for the "cardioprotective" effects of estrogen. Indeed, both the vascular endothelium (Kim-Schulze *et al*., 1996; Venkov *et al*., 1996) and smooth muscle (Horwitz and Horwitz, 1982; Orimo *et al*., 1993; Karas *et al*., 1994; Harder and Coulson, 1979; McGill and Sheridan, 1981; Losordo *et al*., 1994; Kaplan *et al*., 1992) express functional estrogen receptors which have the potential to regulate over 30 genes associated with the development of cardiovascular disease (Mendelsohn and Karas, 1994). The vascular endothelium acts as a dynamic regulatory interface between the blood and media and therefore represents a target through which estrogen might regulate vascular remodeling. Exactly how estrogen interacts with the vascular endothelium is unclear, but attention to date has focused on estrogenic regulation of those endothelium-derived relaxing factors which are generally believed to be salutary in nature such as nitric oxide and prostacyclin. Alternatively, estrogen may regulate endothelium-derived contractile factors, a number of which are mitogenic and/or thrombogenic (Figures 12.1–12.2). Thus, any estrogen-mediated regulation,

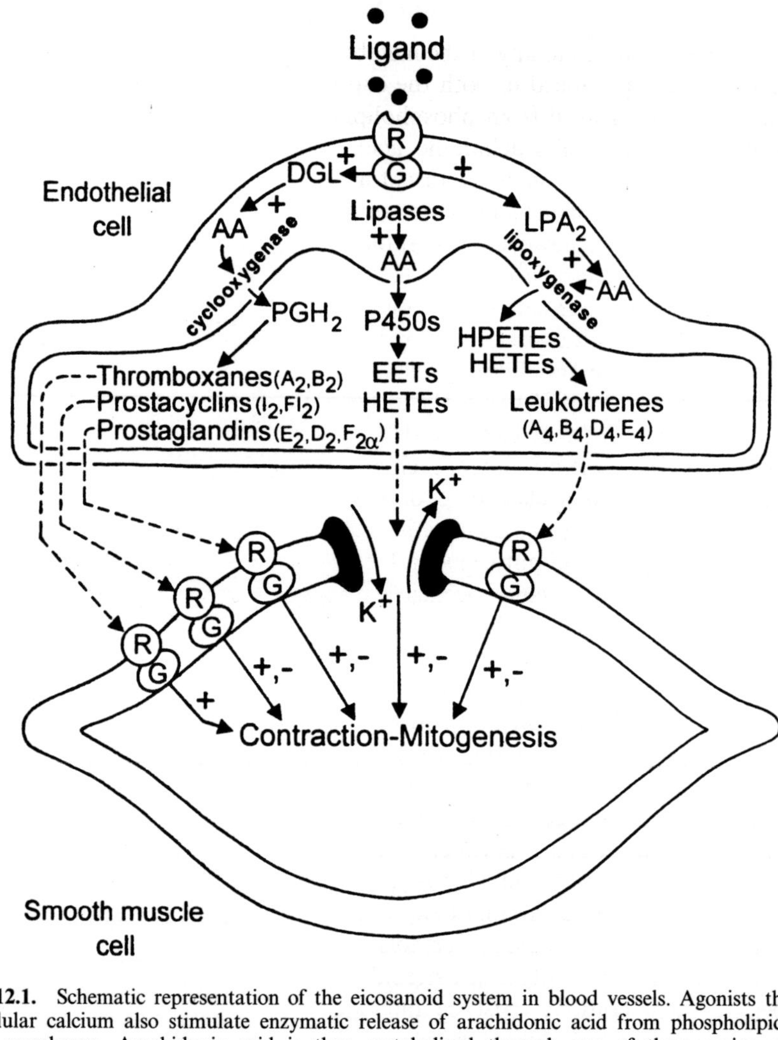


**Figure 12.1.** Schematic representation of the eicosanoid system in blood vessels. Agonists that affect intracellular calcium also stimulate enzymatic release of arachidonic acid from phospholipids of the plasma membrane. Arachidonic acid is then metabolized through one of three major pathways: (1) cyclooxygenase gives rise to the thromboxanes, prostacyclins and prostaglandins, (2) lipoxygenase produces several eicosatetraenoic acids and then the leukotrienes, and (3) cytochrome P450s in the cytosol produce a series of eicosatetraenoic acids. The eicosanoids act upon specific receptors and/or ion channels in the vascular smooth muscle to regulate smooth muscle tone and growth. The production, release and actions of the eicosanoids may be under regulation by estrogen (see text). AA = arachidonic acid, R = receptor, G = guanine nucleotide-binding regulatory protein, DGL = diacylglycerol lipase, $LPA_2$ = lipase $A_2$, HPETEs = hydroperoxyeicosatetraenoic acids, HETEs = hydroxyeicosatetraenoic acids, and EETs = epoxyeicosatrienoic acids.

be it direct or indirect, of endothelium-derived contractile factors may account for a portion of estrogen's cardioprotective actions. Estrogen's effects on the production and action of various endothelium-derived contractile factors will be the focus of this chapter.

## EICOSANOIDS

The term eicosanoid refers to any of the metabolites of arachidonic acid (Figure 12.1). Arachidonic acid is produced in both the endothelium and vascular smooth muscle when enzymatically cleaved from phospholipids of the plasma membrane. There is constitutive production of arachidonic acid, however, production is usually secondary to a signal event such as receptor-mediated guanine nucleotide-binding regulatory protein (G-protein) stimulation or ion channel activation. Arachidonic acid *per se* is not vasoactive, however, it is rapidly metabolized by either lipoxygenases to form hydroxyeicosatetraenoic acids (HETEs) and hydroperoxyeicosatetraenoic (HPETEs) and then to the leukotrienes or by cyclooxygenase to form prostaglandins and thromboxanes. Additionally, cytochrome P450 enzymes within the cytosol of the endothelium metabolize arachidonic acid to epoxyeicosatrienoic acids (EETs) and HETEs (Harder *et al.*, 1995). Nearly all eicosanoids are vasoactive with diverse mechanisms of action. Most eicosanoids act through distinct G-protein receptors residing in smooth muscle but others such as the EETs act through ion channels and may therefore represent a class of endothelium-derived hyperpolarizing factor(s) (Campbell *et al.*, 1996). Some contractile eicosanoids such as thromboxane $A_2$ are also potent smooth muscle mitogens and may therefore directly participate in the pathogenesis of several vascular occlusive diseases and atherosclerosis (Sachinidis *et al.*, 1995). The major endothelium-derived contractile eicosanoids are thromboxane $A_2$, prostacyclin $F_{1\alpha}$, prostaglandin $F_{2\alpha}$, and 20-HETE (Herman, Ed. *et al.*, 1982). However, prostacyclin, prostaglandin $E_2$, and prostaglandin $H_2$ may also increase vascular tone depending upon concentration, species, and vascular bed (Hadhazy *et al.*, 1984; Miller and Vanhoutte, 1990).

Studies examining the effect of estrogen on vascular thromboxane $A_2$ and prostacyclin are summarized in Table 12.1. Inconsistencies in the results may be attributable to the various experimental conditions such as gender and degree of sexual maturity, (Horwitz and Horwitz, 1982; Seillan *et al.*, 1983; Vargas *et al.*, 1993) concentration and duration of exposure to estrogen, experimental and culture conditions (Mikkola *et al.*, 1995; Berthois *et al.*, 1986), *in vivo* versus *in vitro*, and the absence or presence of other sex steroid hormones. Studies which fail to identify all experimental conditions must be interpreted in a limited fashion. Moreover, studies using cells from umbilical cords are inherently limited since such fetal tissue is not representative of other vascular tissues in many aspects (Fox and Khong, 1990; Kawano and Mori, 1990; Walters, 1992) and the fetal sex (gender of the umbilical cord cells) is usually either unknown or unreported by the investigator.

The mechanisms by which estrogen may affect eicosanoid production are unclear. Some studies suggest that rather than changing arachidonic acid synthesis or its cellular distribution, estrogen may affect basal eicosanoid levels by stimulating

enzymes such as smooth muscle cyclooxygenase (Chang *et al.*, 1980b) or enzymes which inactivate various eicosanoids (Chang and Tai, 1985). What conditions affect estrogen's actions on basal eicosanoid production remain to be defined and represents an important area for future research.

Most studies cited in Table 12.1 examine the effects of estrogen on basal rather than agonist-stimulated eicosanoid production. Nearly all cellular events that increase intracellular calcium also increase arachidonic acid production (Harder *et al.*, 1995). A critical issue then is whether estrogen affects agonist-stimulated eicosanoid production. Prostacyclin production after stimulation of cultured endothelial cells with either the calcium ionophore A23187, arachidonic acid, or thrombin is unaffected by estrogen (Corvazier *et al.*, 1984; Muck *et al.*, 1994; Muck *et al.*, 1993a). In homogenates of human umbilical vein or artery, prostacyclin production by stimulation with arachidonic acid is unaffected by estrogen (Muck *et al.*, 1994). However, estrogen-treatment of cultured smooth muscle cells has a stimulatory effect on prostacyclin production following the addition of either arachidonic acid or the immediate precursor of thromboxane $A_2$, prostaglandin $H_2$ (Chang *et al.*, 1980b). Such studies on agonist-stimulated eicosanoid production are important but analogous studies involving the direct measurement of non-prostacyclin eicosanoids after estrogen-treatment in stimulated vascular tissue are absent from the literature. In platelets from rabbits treated with estrogen, thrombin-stimulated conversion of arachidonic acid to thromboxane $B_2$ is unaffected by estrogen despite the fact that both uptake and incorporation of arachidonic acid are increased by estrogen-treatment (Skjaerlund *et al.*, 1983). Whether or not analogous effects of estrogen occur within the endothelium or vascular smooth muscle is unknown.

Rather than directly measuring individual eicosanoids, an alternative approach to addressing effects of estrogen on eicosanoid production is to evaluate indirectly responses of blood vessels to agents known to stimulate eicosanoid production. For example, in isolated rabbit and rat aorta, contractions to exogenously administered arachidonic acid are abolished if either the endothelium is removed or if indomethacin is added to the organ bath. These data suggest that the endothelium metabolizes arachidonic acid via cyclooygenase to a contractile eicosanoid. If rabbits are treated with estrogen for two weeks prior to testing, contractions to arachidonic acid are more than doubled and remain endothelium-dependent and indomethacin sensitive (Figure 12.2). These data suggest that estrogen increases contractile eicosanoids from a cyclooxygenase metabolite of arachidonic acid in the endothelium. Alternatively, estrogen may increase the sensitivity of the smooth muscle to contractile eicosanoids. For example, in the same study (Miller and Vanhoutte, 1990) estrogen-treatment increased contractions of aorta without endothelium to prostacyclin (Figure 12.3). However, this effect of estrogen did not encompass changes in sensitivity to all eicosanoids since contractions to either a thromboxane $A_2$ mimetic (U46619), $PGF_{2\alpha}$, or $PGE_2$ where unaffected by the estrogen-treatment.

Another example of indirect assessment of estrogen's effects on eicosanoids comes from perfusion studies demonstrating greater pressor responses of lung and mesenteric beds to the thromboxane $A_2$ mimetic U46619 in female rats compared to their male counterparts; ovariectomy of the females eliminates this gender difference (Farhat and Ramwell, 1992; Vargas *et al.*, 1995). Moreover, estrogen replacement,

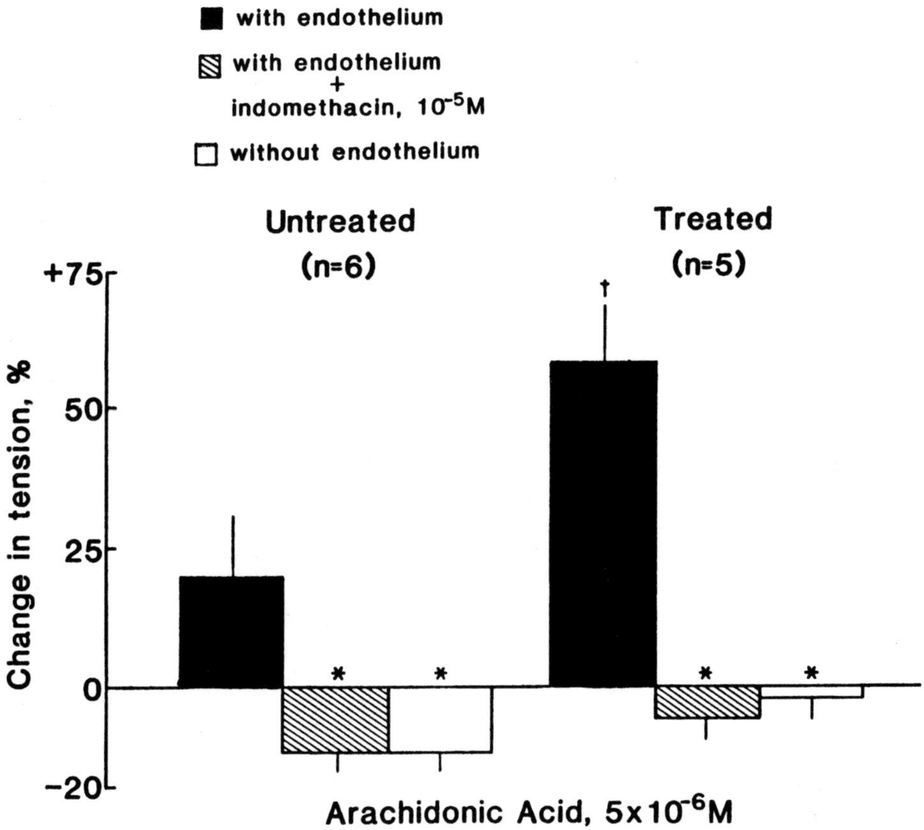


**Figure 12.2.** Responses to arachidonic acid in rings of aorta from ovariectomized rabbits either untreated or treated with 17$\beta$-estradiol by subcutaneous time release tablet (100 mg) for 14 days. Data are expressed as percent change in tension from contraction to norepinephrine and shown as mean $\pm$ SEM. * = significant difference from contractions in rings with endothelium; † = significant differences from contractions in untreated rings with endothelium, $p<0.05$. (Reprinted with permission from *American Journal of Physiology*, 258: R1502–R1506, 1990).

either acutely in the perfusion media or chronically in the ovariectomized rats, restores the gender difference. Interestingly, this gender difference to U46619-mediated contractions was indomethacin-sensitive suggesting involvement of other eicosanoids secondary to stimulation by U46619. On the contrary, gender differences in contractions to U46619 and/or $PGF_{2\alpha}$ are greater in isolated canine coronary and renal arteries and rat aorta in male compared to female animals (Karanian and Ramwell, 1996; Cunard *et al.*, 1987; Maddox *et al.*, 1987). Consistent with these results, in the canine arteries, anti-estrogens increase contractile responses to U46619 (Maddox *et al.*, 1987). However, in aorta from female rat, estrogen-treatment increases contractions to $PGF_{2\alpha}$ (Maddox *et al.*, 1987).

Drugs and physical stimuli which elicit endothelium-dependent relaxations also stimulate the production of arachidonic acid and subsequent activation of the eicosanoid pathways. Estrogenic regulation of the endothelium-derived contractile

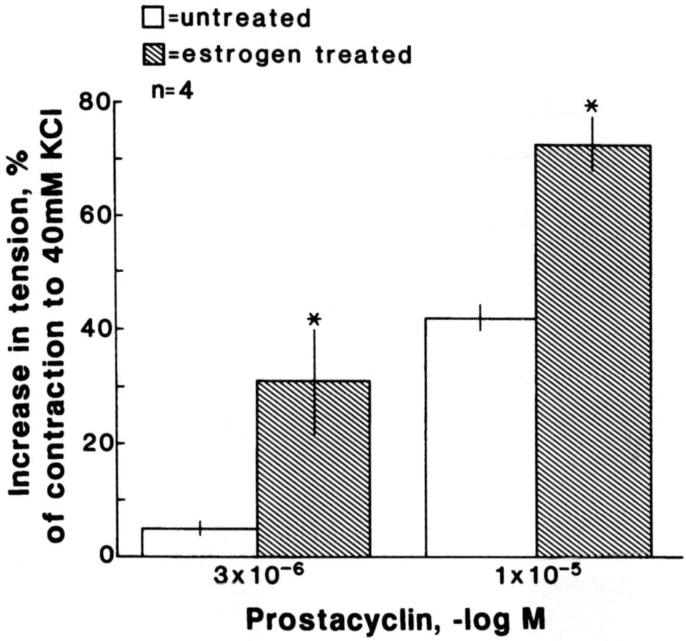


**Figure 12.3.** Contractions to prostacyclin in rings of aorta from ovariectomized rabbits either untreated or treated with 17β-estradiol by subcutaneous time release tablet (100 mg) for 14 days. Indomethacin ($10^{-5}$ M), propranolol ($5\times10^{-6}$ M), and phentolamine ($10^{-6}$ M) were present during the experiment. Data are expressed as percent contraction to 40 mole/L KCl and shown as mean ± SEM. * = significant difference between untreated and estrogen-treated groups, $p<0.05$. (Reprinted with permission from *American Journal of Physiology*, 258: R1502–R1506, 1990.)

eicosanoids, therefore, has the potential to influence endothelium-dependent relaxations. For example, in femoral arteries from ovariectomized rabbits treated with estrogen for four days, endothelium-dependent relaxations to adenosine diphosphate (ADP) are reduced compared to untreated ovariectomized animals (Gisclard *et al.*, 1988) (Figure 12.4). This difference is indomethacin sensitive suggesting that chronic estrogen replacement increases the production of an endothelium-derived contractile eicosanoid(s) which opposes ADP-mediated endothelium-dependent relaxations.

Functional antagonism of endothelium-dependent relaxations by contractile eicosanoids may also contribute to gender differences in vascular responses. Endothelium-dependent relaxations are greater in isolated porcine coronary arteries from sexually mature female compared to male pigs. This gender difference is reduced when the arterial rings are treated with indomethacin (Figure 12.5) suggesting that production of cyclooxygenase-dependent contractile eicosanoids antagonize endothelium-dependent relaxations. Radioimmunoassay measurements of prostanoids produced by the coronary arteries suggest that the reduced endothelium-dependent relaxations may be related, in part, to higher basal and agonist-stimulated release of thromboxane $B_2$ and 6-keto-prostaglandin $F_{1\alpha}$ in coronary arteries of male compared to female pigs (unpublished observation).

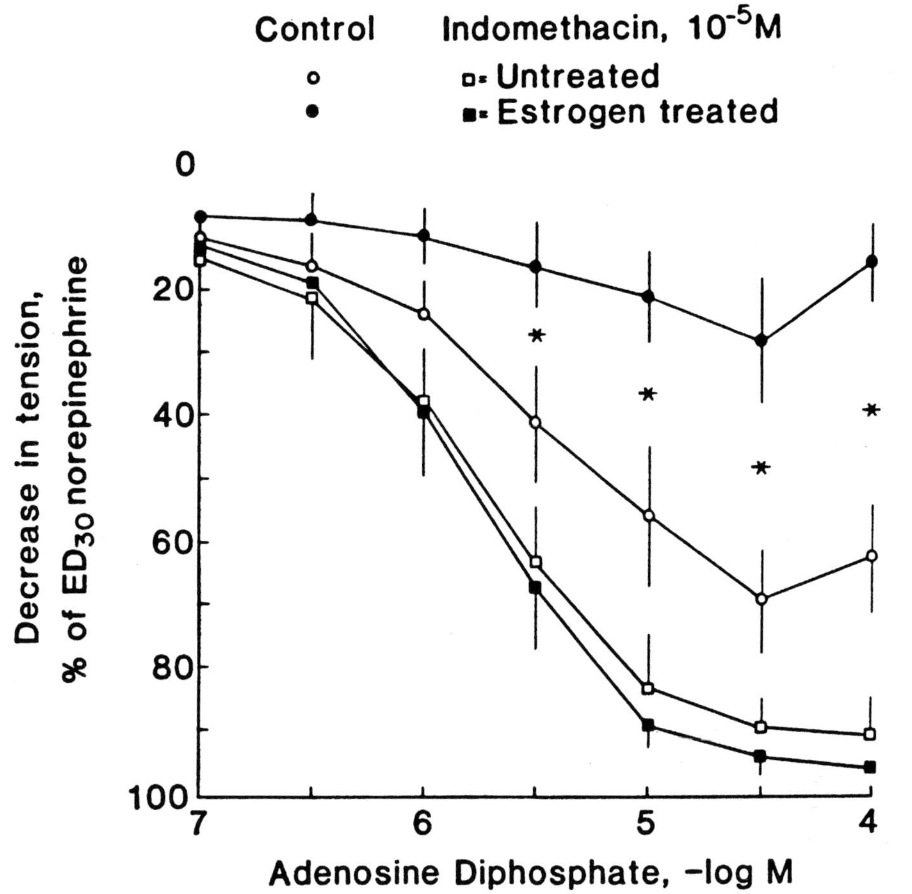


**Figure 12.4.** Effect of 17β-estradiol (100 μg i.m. for four days) treatment of ovariectomized rabbits on the concentration response curves to adenosine diphosphate in rings of femoral arteries with endothelium contracted with norepinephrine ($ED_{30}$), in the absence (○) or presence (□) of indomethacin ($10^{-5}$ M). The responses of estrogen-treated rings are represented by filled symbols; those of untreated rings by open symbols. Data are expressed as a percentage of the response to norepinephrine and shown as mean ± SEM. * denotes that the responses of the estrogen-treated group are statistically different from those obtained in the untreated group in the absence of indomethacin. Indomethacin significantly enhanced the relaxant effect of adenosine diphosphate in both untreated and treated groups, $p < 0.05$. (Reprinted with permission from *The Journal of Pharmacology and Experimental Therapeutics*, 244: 19–22, 1988.)

## ENDOTHELINS

In 1988, Yanagisawa *et al.* isolated and determined the structure of a small endothelium-derived peptide named endothelin (Yanagisawa *et al.*, 1988). Subsequent studies have identified three endothelin isopeptides (endothelin-1, -2, and -3), each derived from different genes (Inoue *et al.*, 1989; Masaki *et al.*, 1990). Endothelin-1 is produced by endothelial cells and is the most potent vasoconstrictor

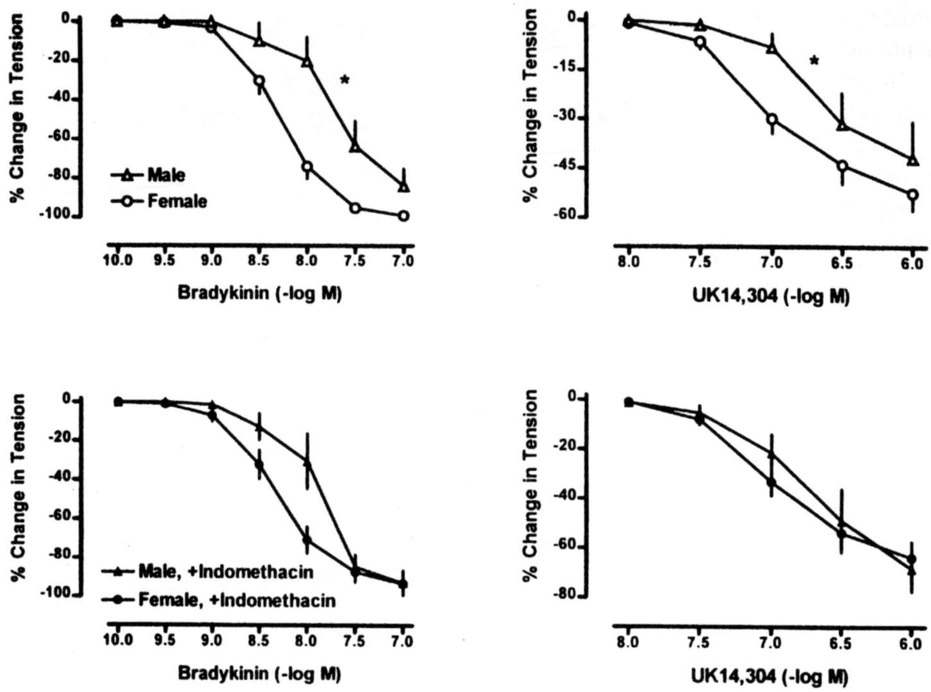


**Figure 12.5.** Effect of gender and indomethacin ($10^{-5}$ M) on endothelium-dependent relaxations in porcine coronary arteries with endothelium from sexually mature pigs. Concentration-response curves to the endothelium-dependent agonists bradykinin, and the $\alpha_2$ adrenergic agonist UK14,304 are shown as percent change in tension from contraction to prostaglandin $F_{2\alpha}$ ($2\times10^{-6}$ M) and are expressed as mean$\pm$SEM. In untreated rings from female pigs, relaxations were significantly greater and/or more sensitive to the endothelium-dependent agonists compared to untreated rings of male pigs. Treatment of rings with indomethacin ($10^{-5}$ M) to eliminate cyclooxygenase-dependent products attenuated the gender differences to the endothelium-dependent agonists. Male ($n=8$) are represented as triangles and female ($n=22$) as circles. Filled symbols indicate indomethacin treated. Asterisk denotes a significant difference between rings of male and female animals in the effective concentration producing half-maximal relaxation ($EC_{50}$) for both UK14,304 and bradykinin and in the maximal relaxation for bradykinin (unpaired *t*-test, $p<0.05$).

known with a potency ten times greater than that of angiotensin II (Yanagisawa *et al.*, 1988). There are at least two distinct endothelin receptor subtypes: endothelin-A and endothelin-B. The endothelin-A receptor resides in vascular smooth muscle and its activation produces contraction via increases in intracellular calcium secondary to the activation of a number of second messenger systems such as phospholipase C, inositol triphosphate and protein kinase C (Griendling *et al.*, 1989; Kasuya *et al.*, 1989; Lee *et al.*, 1989; Muldoon *et al.*, 1989). The endothelin-B receptor is found on both the endothelium and vascular smooth muscle. Activation of endothelin-B receptors in the endothelium stimulates the release of nitric oxide and prostacyclin which causes relaxation of the underlying smooth muscle. In contrast, activation of endothelin-B receptors on the smooth muscle elicits a contraction. Although no molecular data are yet available, there are pharmacological data to suggest the existence of multiple endothelin-B receptor subtypes (Gray *et al.*, 1994). The endothelin system is illustrated schematically in Figure 12.6.

In addition to its effects on vascular tone, endothelin is a smooth muscle mitogen that acts synergistically with other mitogens (Dubin *et al.*, 1989; Hirata *et al.*, 1989; Eguchi *et al.*, 1995). Endothelins may participate in the development and/or progression of cardiovascular disease since these endogenous coronary spasmogens/mitogens (Klemm *et al.*, 1995; Igarashi *et al.*, 1989) are found in atherosclerotic vessels in elevated levels (Lerman *et al.*, 1991; Bacon *et al.*, 1995; Timm *et al.*, 1995).

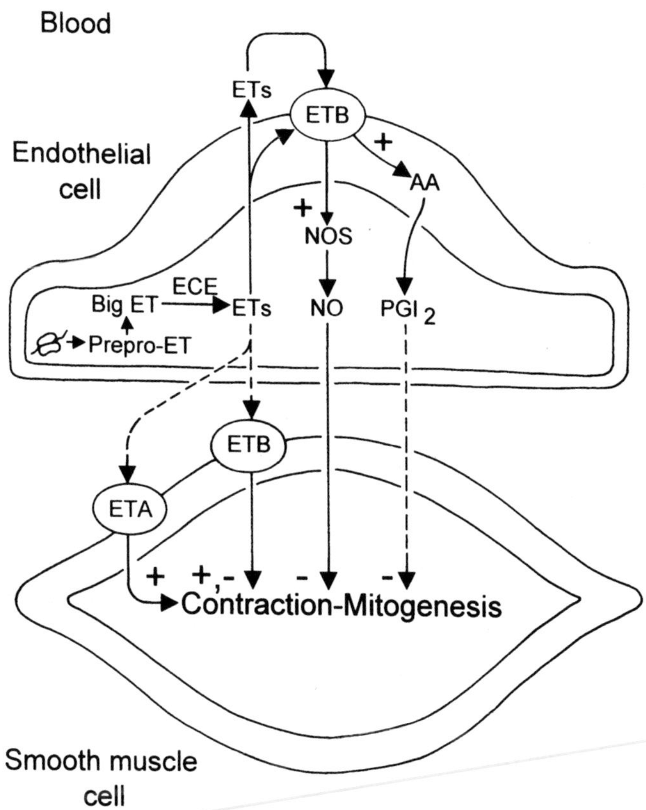


**Figure 12.6.** Schematic representation of the endothelin peptide system in blood vessels. Endothelin synthesis begins with transcription of mRNA for Prepro-Endothelin within the endothelium. The newly synthesized peptide is converted to Big Endothelin which is further cleaved by endothelin converting enzyme (ECE) to form endothelin-1 which may or may not be converted to endothelin-2 or endothelin-3. The endothelins can either be released into the blood and/or act in an autocrine or paracrine fashion on specific endothelin receptors residing in both the endothelium and vascular smooth muscle. Activation of endothelin-B receptors in the endothelium causes nitric oxide (NO) and prostacyclin ($PGI_2$) release which in turn diffuse to the smooth muscle where they inhibit contraction and cellular growth. In contrast, activation of endothelin-A receptors on the vascular smooth muscle causes contraction and stimulates mitogenesis. Activation of the endothelin-B receptor on the smooth muscle elicits a contraction but has variable effects on cellular growth and proliferation. Estrogen may regulate multiple components of the endothelin system (see text). NOS = nitric oxide synthase, AA = arachidonic acid, ET = endothelin, ETA = endothelin-A receptor subtype, ETB = endothelin-B receptor subtype.

Moreover, estrogen may either directly or indirectly affect the production or actions of the endothelins. Plasma endothelin levels are lower in women compared to men and that during pregnancy when estrogen is elevated, plasma endothelin levels are reduced below pre-pregnancy values (Polderman *et al.*, 1993). Further, plasma endothelin levels are decreased in male-to-female transsexuals receiving estradiol and cyproterone (testosterone antagonist) and increased in female-to-male transsexuals receiving testosterone (Polderman *et al.*, 1993). Ylikorkala *et al.*, 1995, examined plasma endothelin levels in healthy post-menopausal women before and during a one year hormone replacement therapy. Compared to pre-treatment levels, plasma endothelin decreased significantly at both six and 12 months post-therapy regardless of whether the hormone replacement consisted of either estrogen alone or estrogen plus progestin (Ylikorkala *et al.*, 1995).

Other studies have suggested that estrogen may have direct effects on endothelin contractions of smooth muscle. Acute administration of estradiol to isolated rabbit coronary arteries reduces contractions to endothelin-1 (Jiang *et al.*, 1992). However, this effect was observed using concentrations of estradiol in the micromolar range. At these concentrations, effects of estrogen are not specific to endothelin (Han *et al.*, 1995; Kondo *et al.*, 1980) since estradiol causes direct relaxation of smooth muscle via non-genomic mechanisms involving inhibition of voltage-gated calcium channels (White *et al.*, 1995; Han *et al.*, 1995; Salas *et al.*, 1994). In isolated coronary arteries from sexually mature pigs of either sex, preliminary data from our laboratory did not show an effect of a one hour incubation with physiological concentrations of estradiol ($10^{-9}$–$10^{-7}$ M) on endothelin-mediated contractions (unpublished observations). And despite the fact that contractile responses to endothelin in uterine arteries are blunted in pregnancy, this attenuated response is not reproduced by estrogen-treatment (Yang and Clark, 1992).

Estrogen may, however, affect endothelin-mediated contractions via indirect mechanisms. For example, prostacyclin production resulting from the addition of endothelin-1 to endothelial cultures from human umbilical cord veins is increased in the presence of estrogen (Muck *et al.*, 1994). Other indirect effects of estrogen may relate to estrogen affecting ion channels which in turn regulate responses to the endothelins and other endothelium-derived contractile factors. For example, estrogen directly activates $K^+$ channels (Harder and Coulson, 1979) and affects the expression of mRNA for the delayed rectifier $K^+$ channel in heart (Drici *et al.*, 1996). Moreover, in freshly dissociated porcine coronary smooth muscle stimulated with endothelin-1 (Prakash *et al.*, 1996) or U46619 (Han *et al.*, 1995), estradiol inhibits $Ca^{++}$ influx and increases $Ca^{++}$ efflux by mechanisms dependent upon the presence of the estrogen receptor (Sieck *et al.*, 1997). The extent to which estrogen affects $Ca^{++}$ channels *in vivo* remains to be determined but any effect of estrogen on systems with broad physiological functions such as ion channels is consistent with the wide variety of non-genomic actions attributed to estrogen.

There is also evidence for gender differences in vascular responses to the endothelins. Maximal contractions to endothelin-1 are greater in coronary arteries without endothelium from sexually mature female compared to male animals (Figure 12.7, Top Panel) (Barber *et al.*, 1996). These differences are not due to non-specific gender differences of the smooth muscle since contractions to depolarization with

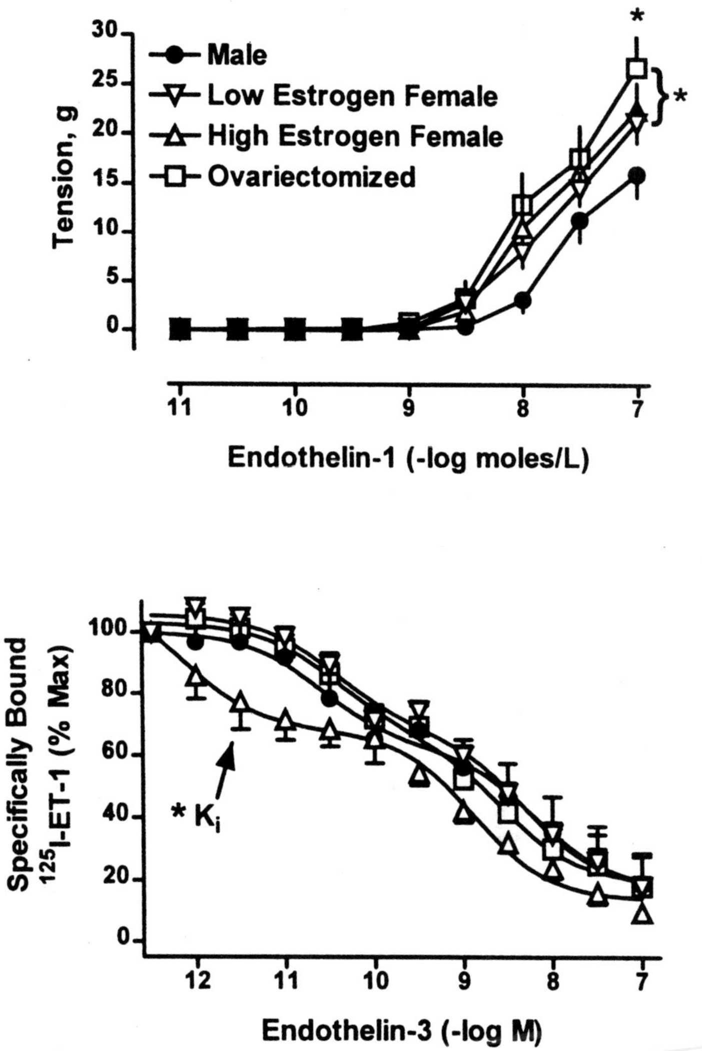


**Figure 12.7.** *Upper Panel*: Cumulative concentration-response curves for endothelin-1 in isolated coronary arteries without endothelium from sexually mature male and female pigs. Female pigs were grouped according to their endogenous plasma 17$\beta$-estradiol level at the time of the experiment. Compared to rings from male pigs, maximal contractions to endothelin-1 were significantly greater in coronary arteries from female pigs irrespective of their plasma 17$\beta$-estradiol levels. Asterisks denotes a statistically significant difference in maximal tension compared to arteries from male pigs; male $n=8$, low estrogen female $n=11$, high estrogen female $n=6$ and ovariectomized $n=5$; $p<0.05$, ANOVA, Bonferroni post hoc. Half-bracket } indicates all female groups combined. All points represent the means $\pm$ SEM. *Lower Panel*: Competitive inhibition of $^{125}$I-endothelin-1 binding by unlabeled endothelin-3 in membranes prepared from coronary arteries without endothelium from pigs as described above. Nonlinear regression analysis for $^{125}$I-endothelin-1 inhibition by endothelin-3 was significant for a two-site binding model. In membranes from the high estrogen females, the $K_i$ for the high affinity binding site (presumably the endothelin-B receptor) for endothelin-3 was significantly lower compared to all other groups (male $n=11$, low estrogen female $n=16$, high estrogen female $n=5$, ovariectomized $n=5$, $p<0.05$). Receptor numbers ($B_{max}$) were not different among any groups. (Reprinted with permission from *American Journal of Physiology*, 271: H1999–H2006; 1996.)

potassium chloride did not differ by sex. Moreover, the gender differences to endothelin-1 do not directly correlate with the absolute plasma levels of individual sex steroid hormones since neither ovariectomy nor the estrogen status of the female pigs correlated with the gender difference under these experimental conditions (Figure 12.7, Top Panel). These differences may therefore relate more to gender-associated ratios of sex steroid hormones (i.e. ratio of estrogen to testosterone). One possible mechanism for these differences to endothelin relates to calcium sensitivity of the contractile proteins. In $\beta$-escin permeabilized coronary arterial strips, force-calcium relationships are the same between sexes in the absence and presence of endothelin-1 (Miller *et al.*, 1996). These data suggest that gender differences in force generation to endothelin-1 are not due to intrinsic differences in calcium sensitivity of the contractile proteins.

Estrogen may also affect actions of the endothelins through regulation of expression of receptor subtypes. In particular, the mitogenic functions of the endothelins may be mediated primarily through activation of endothelin-B receptors (Eguchi *et al.*, 1994; Morbidelli *et al.*, 1995). In membranes prepared from porcine coronary arteries, a selective and significant increase in the binding affinity of the endothelin-B receptor (or another as of yet unidentified high affinity binding site) occurs during the high estrogen phase of the normal ovulatory cycle (Figure 12.7, Lower Panel) (Barber *et al.*, 1996). This change in binding affinity does not necessarily correlate with contractile responses of the vessels to endothelins (Figure 12.7, Top Panel). The significance of this change in endothelin-B binding affinity that occurs in conjunction with endogenous estrogen fluctuations is unclear, but is speculated to be involved in growth regulatory pathways. If this speculation is confirmed, it would be consistent with the hypothesis that endogenous estrogen regulates vascular growth and hence the development of coronary artery disease via interactions with mitogenic pathways controlled by the endothelins.

## OTHER ENDOTHELIUM-DERIVED CONTRACTING FACTORS

Oxygen-derived free radicals such as superoxide anion are putative endothelium-derived contractile factors (Katusic and Vanhoutte, 1989; Cosentino *et al.*, 1994). 17$\beta$-Estradiol is a molecule with moderate antioxidant properties and therefore has the potential to reduce endothelium-derived free radicals. Indeed, ethinylestradiol decreases superoxide production in a dose- and time-dependent manner in cultured endothelial cells (Arnal *et al.*, 1996). Since superoxide inactivates nitric oxide as a vasodilator (Rubanyi and Vanhoutte, 1986) and inhibits nitric oxide synthase, (Rengasamy and Johns, 1993) reports of estrogen-induced increases in endothelium-derived nitric oxide (Hishikawa *et al.*, 1995; Hayashi *et al.*, 1995; Conrad *et al.*, 1993; Goetz *et al.*, 1994; Weiner *et al.*, 1994) may be due in part to estrogen's lowering of endothelium-derived superoxide.

The renin angiotensin system (RAS) within the vessel wall functions independently of the circulating RAS (Falkenhahn *et al.*, 1994). Whether or not sex steroid hormones regulate production of the endothelium-derived contractile/ mitogenic factor angiotensin II or any component of the vascular RAS has not been extensively investigated. In both humans and a variety of animals, pressor responses to infused angiotensin II are decreased during pregnancy and/or following treatment with estrogen (Gant *et al.*, 1976; Hettiaratchi and Pickford, 1968; Paller, 1987; Baylis and Collins, 1986; Inao, 1985; Chesley and Tepper, 1967). Isolated aortic rings from ovariectomized rats treated with estrogen for 14 days have decreased contractile responses to angiotensin II (Cheng and Gruetter, 1992). This blunted response remains when the endothelium is removed suggesting estrogen-induced changes in the smooth muscle (Cheng and Gruetter, 1992). Interestingly, the blunted response to angiotensin II in these models of elevated estrogen may not require prolonged exposure to estrogen since acute infusion of estrogen in non-pregnant ewes reduces subsequent pressor responses to infused angiotensin II (Rosenfeld and Jackson, 1984). Collectively, these studies suggest that contractile responses to exogenous angiotensin II are reduced by estrogen and that this effect may be acute and endothelium-independent. The mechanism of this effect is unclear, but one possibility is down-regulation of angiotensin receptors and/or converting enzymes since estrogen down-regulates both the activity of angiotensin-converting enzyme (ACE) and angiotensin receptors in the pituitary and arcuate nucleus (Seltzer *et al.*, 1993; 1992). Moreover, in post-menopausal cynomolgous monkeys treated with estrogen for 30 months, plasma ACE activity is decreased and angiotensin I levels are increased (Brosnihan *et al.*, 1995). Whether estrogen has similar effects in the vascular RAS remains to be determined.

Estrogen may also interact with RAS indirectly through nitric oxide. For example, in transgenic hypertensive rats expressing the *REN*-2 gene, pressor responses of systemic injections of the nitric oxide synthase inhibitor, $N^G$-monomethyl-L-arginine are greater compared to non-transgenic rats. This difference is abrogated if the transgenic rats are ovariectomized and returns with estrogen replacement suggesting that estrogen counter-modulates the renin-mediated hypertension by enhancing endothelium-derived nitric oxide (Brosnihan *et al.*, 1994).

All of the above studies describe effects of estrogen on the circulating endocrine RAS and do little in terms of defining the mechanisms by which estrogen regulates the local vascular RAS.

## CONCLUSIONS

In summary, estrogen affects production, release and action of endothelium-derived contractile factors such as various eicosanoids and the endothelins. Estrogen may also affect endothelium-derived free radicals and the vascular RAS although the evidence to date is less definitive. The regulation of endothelium-derived contractile factors by estrogen is sensitive to a variety of experimental conditions which are complicated by gender, species, and anatomical origin of the blood vessels. Moreover, effects of estrogen are not limited to single isolated pathways, but rather, entail regulation at multiple sites among a cascade of pathways involved in the regulation of the vessel wall. These effects are further complicated by the fact that the sites of interaction for estrogen are often systems which have broad cellular effects such as ion channels. Collectively, these complexities call for more extensive research with more stringent control over the many factors influencing the effects of estrogen. Such studies should define and clarify those criteria required for estrogen to affect the endothelium which in turn will aid in defining the role of estrogen in the development and progression of cardiovascular disease.

## ACKNOWLEDGEMENTS

Grant support provided by NIH HL51736 and HL07111. The authors thank Mr. Robert Lorenz for drawing the schematics.

## REFERENCES

Arnal, J.F., Clamens, S., Pechet, C., Negre-Salvayre, A., Allera, C., Girolami, J.P., Salvayre, R. and Bayard, F. (1996) Ethinylestradiol does not enhance the expression of nitric oxide synthase in bovine endothelial cells but increases the release of bioactive nitric oxide by inhibiting superoxide anion production. *Proc. Natl. Acad. Sci.* **93**, 4108–4113.

Bacon, C.R., Cary, N.R.B. and Davenport, A.P. (1995) Distribution of endothelin receptors in atherosclerotic human coronary arteries. *J. Cardiovasc. Pharmacol.* **26**, S439–S441.

Barber, D.A., Sieck, G.C., Fitzpatrick, L.A. and Miller, V.M. (1996) Endothelin receptors are modulated in association with endogenous fluctuations in estrogen. *Am. J. Physiol.* **271**, H1999–H2006.

Barrett-Conner, E. and Bush, T.L. (1991) Estrogen and coronary heart disease in women. *JAMA* **265**, 1861–1867.

Baylis, C. and Collins, R.C. (1986) Angiotensin II inhibition on blood pressure and renal hemodynamics in pregnant rats. *Am. J. Physiol.* **250**, F308–F314.

Berge, L.N., Hansen, J., Svensson, B., Lyngmo, V. and Nordoy, A. (1990) Female sex hormones and platelet/endothelial cell interactions. *Haemostasis* **20**, 313–320.

Berthois, Y., Katzenellenbogen, J.A. and Katzenellenbogen, B.S. (1986) Phenol red in tissue culture media is a weak estrogen: implications concerning the study of estrogen-responsive cells in culture. *Proc. Natl. Acad. Sci.* **83**, 2496–2500.

Brosnihan, K.B., Moriguchi, A., Nakamoto, H., Dean, R.H., Ganten, D. and Ferrario, C.M. (1994) Estrogen augments the contribution of nitric xxide to blood pressure regulation in transgenic hypertensive rats expressing the mouse *ren*-2 Gene. *Am. J. Hypertens.* **7**, 576–582.

Brosnihan, K.B., Weddle, D., Anthony, M.S. and Ferrario, C.M. (1995) Does chronic estrogen act as a converting enzyme inhibitor in postmenopausal cynomolgous monkeys? *Am. J. Hypertens.* **8**, 107A(Abstract).

Campbell, W.B., Gebremedhin, D., Pratt, P.F. and Harder, D.R. (1996) Identification of epoxyeicosatrienoic acids as endothelium-derived hyperpolarizing factors. *Circ. Res.* **78**, 415–423.

Chang, W., Nakao, J., Orimo, H. and Murota, S. (1980a) Stimulation of prostacyclin biosynthetic activity by estradiol in rat aortic smooth muscle cells in culture. *Biochim. Biophys. Acta* **619**, 107–118.

Chang, W., Nakao, J., Orimo, H. and Murota, S. (1980b) Stimulation of prostaglandin cyclooxygenase and prostacyclin synthetase activities by estradiol in rat aortic smooth muscle cells. *Biochim. Biophys. Acta* **620**, 472–482.

Chang, W., Nakao, J., Neichi, T., Orimo, H. and Murota, S. (1981) Effects if estradiol on the metabolism of arachidonic acid by aortas and platelets in rats. *Biochim. Biophys. Acta* **664**, 291–297.

Chang, W. and Tai, H. (1985) Induction of a decrease in renal NAD+-dependent 15-hydroxyprostaglandin dehydrogenase activity by estradiol in rats. *Biochem. Pharmacol.* **34**, 2073–2076.

Cheng, D.Y. and Gruetter, C.A. (1992) Chronic estrogen alters contractile responsiveness to angiotensin II and norepinephrine in female rat aorta. *Eur. J. Pharmacol.* **215**, 171–176.

Chesley, L.C. and Tepper, I.H. (1967) Effects of progesterone and estrogen on the sensitivity to angiotensin II. *J. Clin. Endocr.* **27**, 576–581.

Conrad, K.P., Joffe, G.M., Kruszyna, H., Kruszyna, R., Rochelle, L.G., Smith, R.P., Chavez, J.E. and Mosher, M.D. (1993) Identification of increased nitric oxide biosynthesis during pregnancy in rats. *FASEB J.* **7**, 566–571.

Corvazier, E., Dupuy, E., Dosne, A.M. and Maclouf, J. (1984) Minimal effect of estrogens on endothelial cell growth and production of prostacyclin. *Thromb. Res.* **34**, 303–310.

Cosentino, F., Sill, J.C. and Katusic, Z.S. (1994) Role of superoxide anions in the mediation of endothelium-dependent contractions. *Hypertension* **23**, 229–235.

Cunard, C., Maddox, Y., Falcon, J., Ridinger, M. and Ramwell, P.W. (1987) Eicosanoid expression of vascular sexual dimorphism. In: Gryglewski, R.J. and Stock, G. (Eds.) *Prostacyclin and Its Stable Analogue Iloprost*, pp. 115–121. Berlin Heidelberg: Springer-Verlag].

David, M., Griesmacher, A. and Muller, M.M. (1989) 17 alpha-ethinylestradiol decreases production and release of prostacyclin in cultured human umbilical vein endothelial cells. *Prostaglandins* **38**, 431–438.

Drici, M.D., Burklow, T.R., Haridasse, B., Glazer, R.I. and Woosley, R.L. (1996) Sex hormones prolong the QT interval and downregulate potassium channel expression in the rabbit heart. *Circulation* **94**, 1471–1474.

Dubin, D., Pratt, R.E., Cooke, J.P. and Dzau, V.J. (1989) Endothelin, a potent vasoconstrictor, is a vascular smooth muscle mitogen. *J. Vasc. Med. Biol.* **1**, 150–154.

Eaker, E.D., Chesebro, J.H., Sacks, F.M., Wenger, N.K., Whisnant, J.P. and Winston, M. (1993) AHA medical/scientific statement: cardiovascular disease in women. Special report. *Circulation* **88**, 1999–2009.

Eguchi, S., Hirata, Y., Imai, T., Kanno, K. and Marumo, F. (1994) Phenotypic changes of endothelin receptor sybtype in cultured rat vascular smooth muscle cells. *Endocrinology* **134**, 222–228.

Eguchi, S., Hirata, Y., Imai, T. and Marumo, F. (1995) Endothelin-1 as an autocrine growth factor for endothelial cells. *J. Cardiovasc. Pharmacol.* **26**, S279–S283.

Elam, M.B., Lipscomb, G.E., Chesney, C.M., Terragno, D.A. and Terragno, N.A. (1980) Effect of synthetic estrogen on platelet aggregation and vascular release of $PGI_2$-like material in the rabbit. *Prostaglandins* **20**, 1039–1051.

Falkenhahn, M., Gohlke, P., Paul, M., Stoll, M. and Unger, T. (1994) The renin-angiotensin system in the heart and vascular wall: new therapeutic aspects. *J. Cardiovasc. Pharmacol.* **24**, S6–S13.

Farhat, M.Y. and Ramwell, P.W. (1992) Estradiol potentiates the vasopressor response of the isolated perfused rat lung to the thromboxane mimic U-46619. *J. Pharmacol. Exp. Ther.* **261**, 686–691.

Fogelberg, M., Vesterqvist, O., Diczfalusy, U. and Henriksson, P. (1990) Experimental atherosclerosis: effects of oestrogen and atherosclerosis on thromboxane and prostacyclin formation. *Eur. J. Clin. Invest.* **20**, 105–110.

Fox, S.B. and Khong, T.Y. (1990) Lack of innervation of human umbilical cord. An immunohistological and histochemical study. *Placenta* **11**, 59–62.

Gant, N., Chand, S., Cunningham, S.G. and MacDonald, P.C. (1976) Control of vascular reactivity to angiotensin II in human pregnancy. In: Lindheimer, M.D., Katz, A.I. and Zuspan, F.P. (Eds.) *Hypertension in Pregnancy*, pp. 377–389. New York: Wiley.

Gisclard, V., Miller, V.M. and Vanhoutte, P.M. (1988) Effect of 17B-estradiol on endothelium-dependent responses in the rabbit. *J. Pharmacol. Exp. Ther.* **244**, 19–22.

Goetz, R.M., Morano, I., Calovini, T., Studer, R. and Holtz, J. (1994) Increased expression of endothelial constitutive nitric oxide synthase in rat aorta during pregnancy. *Biochem. Biophys. Res. Commun.* **205**, 905–910.

Gray, G.A., Löffler, B. and Clozel, M. (1994) Characterization of endothelin receptors mediating contraction of rabbit saphenous vein. *Am. J. Physiol.* **266**, H959–H966.

Griendling, K.K., Tsuda, T. and Alexander, R.W. (1989) Endothelin stimulates diacylglycerol accumulation and activates protein kinase C in cultured vascular smooth muscle cells. *J. Biol. Chem.* **264**, 8237–8240.

Hadhazy, P., Malmoolgyi, B. and Magyar, K. (1984) Differential contractile responsiveness of isolated rabbit arteries from different vascular beds to cyclooxygenase inhibitors and PGI2. *Eur. J. Pharmacol.* **98**, 323–330.

Han, S., Karaki, H., Ouchi, Y., Akishita, M. and Orimo, H. (1995) 17b-Estradiol inhibits $Ca^{2+}$ Influx and $Ca^{2+}$ release induced by thromboxane $A_2$ in porcine coronary artery. *Circulation* **91**, 2619–2626.

Harder, D.R., Campbell, W.B. and Roman, R.J. (1995) Role of cytochrome P-450 enzymes and metabolites of arachidonic acid in the control of vascular tone. *J. Vasc. Res.* **32**, 79–92.

Harder, D.R. and Coulson, P.B. (1979) Estrogen receptors and effects of estrogen on membrane electrical properties of coronary vascular smooth muscle. *J. Cell. Physiol.* **100**, 375–382.

Hayashi, T., Yamada, K., Esaki, T., Kuzuya, M., Satake, S., Ishikawa, T., Hidaka, H. and Iguchi, A. (1995) Estrogen increases endothelial nitric oxide by a receptor-mediated system. *Biochem. Biophys. Res. Commun.* **214**, 847–855.

Herman, A.G., Vanhoutte, P.M., Denolin, H., and Goossens, A., Eds. (1982) *Cardiovascular Pharmacology of the Prostaglandins*, New York: Raven Press.

Hettiaratchi, E.S.G. and Pickford, M. (1968) The effect of oestrogen and progesterone on the pressor action of angiotensin in the rat. *J. Physiol.* **196**, 447–451.

Hirata, Y., Takagi, Y., Fukuda, Y. and Marumo, F. (1989) Endothelin is a potent mitogen for rat vascular smooth muscle cells. *Arteriosclerosis* **78**, 225–228.

Hishikawa, K., Nakaki, T., Marumo, T., Suzuki, H., Kato, R. and Saruta, T. (1995) Up-regulation of nitric oxide synthase by estradiol in human aortic endothelial cells. *FEBS Lett.* **360**, 291–293.

Horwitz, K.B. and Horwitz, L.D. (1982) Canine vascular tissues are targets for androgens, estrogens, progestins, and glucocorticoids. *J. Clin. Invest.* **69**, 750–758.

Igarashi, Y., Aizawa, Y., Tamura, M., Ebe, K., Yamaguchi, T. and Shibata, A. (1989) Vasoconstrictor effect of endothelin on the canine coronary artery: Is a novel endogenous peptide involved in regulating myocardial blood flow and coronary spasm? *Am. Heart J.* **118**, 674–678.

Inao, H. (1985) Effects of sex steroid hormones on vascular reactivity to angiotensin II in castrated rat. *Acta Obst. Gynaec. Jpn.* **37**, 2806–2812.

Inoue, A., Yanagisawa, M., Kimura, S., Kasuya, Y., Miyauchi, T., Goto, K. and Masaki, T. (1989) The human endothelin family: Three structurally and pharmacologically distinct isopeptides predicted by three separate genes. *Proc. Natl. Acad. Sci.* **86**, 2863–2867.

Jiang, C., Sarrel, P.M., Poole-Wilson, P.A. and Collins, P. (1992) Acute effect of 17b-estradiol on rabbit coronary artery contractile responses to endothelin-1. *Am. J. Physiol.* **263**, H271–H275.

Kaplan, J.D., Trulock, E.P., Cooper, J.D. and Schuster, D.P. (1992) Pulmonary vascular permeability after lung transplantation. A positron emission tomographic study. *Am. Rev. Respir. Dis.* **145**, 954–957.

Karanian, J.W. and Ramwell, P.W. (1996) Effect of gender and sex steroids on the contractile response of canine coronary and renal blood vessels. *J. Cardiovasc. Pharmacol.* **27**, 312–319.

Karas, R.H., Patterson, B.L. and Mendelsohn, M.E. (1994) Human vascular smooth muscle cells contain functional estrogen receptor. *Circulation* **89**, 1943–1950.

Karpati, L., Chow, F.P.R., Woollard, M.L., Hutton, R.A. and Dandona, P. (1980) Prostacyclin-like activity in the female rat thoracic aorta and the inferior vena cava after ethinyloestradiol and norethisterone. *Clin. Science* **59**, 369–372.

Kasuya, Y., Takuwa, Y., Yanagisawa, M., Kimura, S., Goto, S. and Masaki, T. (1989) Endothelin-1 induces vasoconstriction through two functionally distinct pathways in porcine coronary artery: contribution of phosphoinositide turnover. *Biochem. Biophys. Res. Commun.* **161**, 1049–1055.

Katusic, Z.S. and Vanhoutte, P.M. (1989) Superoxide anion is an endothelium-derived contracting factor. *Am. J. Physiol.* **257**, H33–H37.

Kawano, M. and Mori, N. (1990) Prostacyclin producing activity of human umbilical blood vessels in adrenergic innervated and non-innervated portions. *Prostaglad. Leukotrien. Essen. Fatty Acids* **39**, 239–245.

Kim-Schulze, S., McGowan, K.A., Hubchak, S.C., Cid, M.C., Martin, M.B., Kleinman, H.K., Greene, G.L. and Schnaper, H.W. (1996) Expression of an estrogen receptor by human coronary artery and umbilical vein endothelial cells. *Circulation* **94**, 1402–1407.

Klemm, P., Warner, T.D., Corder, R. and Vane, J.R. (1995) Endothelin-1 mediates coronary vasoconstriction caused by exogenous and endogenous cytokines. *J. Cardiovasc. Pharmacol.* **26**, S419–S421.

Kondo, K., Okuno, T., Eguchi, T., Yasui, T., Suzuki, H., Nagahama, S. and Saruta, T. (1980) Vascular action of high dose estrogen in rats. *Endocrinol. Japan* **27**, 307–313.

Lee, T.S., Chao, T., Hu, K. and King, G.L. (1989) Endothelin stimulates a sustained 1,2-Diacylglycerol increase and protein kinase C activation in bovine aortic smooth muscle cells. *Biochem. Biophys. Res. Commun.* **162**, 381–386.

Lerman, A., Edwards, B.S., Hallett, J.W., Heublein, D.M., Sandberg, S.M. and Burnett, J.C., Jr. (1991) Circulating and tissue endothelin immunoreactivity in advanced atherosclerosis. *N. Engl. J. Med.* **325**, 997–1001.

Losordo, D.W., Kearney, M., Kim, E.A., Jekanowski, J. and Isner, J.M. (1994) Variable expression of the estrogen receptor in normal and atherosclerotic coronary arteries of premenopausal women. *Circulation* **89**, 1501–1510.

Maddox, Y.T., Falcon, J.G., Ridinger, M., Cunard, C.M. and Ramwell, P.W. (1987) Endothelium-dependent gender differences in the response of the rat aorta. *J. Pharmacol. Exp. Ther.* **240**, 392–395.

Makila, U., Wahlberg, L., Viinikka, L. and Ylikorkala, O. (1982) Regulation of prostacyclin and thromboxane production by human umbilical vessels: The Effect of Estradiol and Progesterone in a Superfusion Model. *Prostagland. Leukotrienes Med.* **8**, 115–124.

Masaki, T., Yanagisawa, M., Inoue, A., Kimura, S. and Goto, K. (1990) Study on the endothelin gene and its expression. *Jap. Circ. J.* **54**, 1221–1225.

McGill, H.C. and Sheridan, P.J. (1981) Nuclear uptake of sex steroid hormones in the cardiovascular system of the baboon. *Circ. Res.* **48**, 238–244.

Mendelsohn, M.E. and Karas, R.H. (1994) Estrogen and the blood vessel wall. *Curr. Opinion Cardiol.* **9**(5), 619–626.

Mikkola, T., Turunen, P., Avela, K., Orpana, A., Viinikka, L. and Ylikorkala, O. (1995) 17b-estradiol stimulates prostacyclin, but not endothlein-1, production in human vascular endothelial cells . *J. Clin. Endocrinol. Metab.* **80**, 1832–1836.

Miller, V.M., Barber, D.A., Fenton, A.M., Wang, X. and Sieck, G.C. (1996) Gender differences in response to endothelin-1 in coronary arteries: transcription, receptors and calcium regulation. *Clin. Exp. Pharmacol. Physiol.* **23**, 256–259.

Miller, V.M. and Vanhoutte, P.M. (1990) 17B-estradiol augments endothelium-dependent contractions to arachidonic acid in rabbit aorta. *Am. J. Physiol.* **258**, R1502–R1507.

Morbidelli, L., Orlando, C., Maggi, C.A., Ledda, F. and Ziche, M. (1995) Proliferation and migration of endothelial cells is promoted by endothelins via activation of $ET_B$ receptors. *Am. J. Physiol.* **269**, H686–H695.

Muck, A.O., Seeger, H., Korte, K., Dartsch, P.C. and Lippert, T.H. (1993a) Natural and synthetic estrogens and prostacyclin production in human endothelial cells from umbilical cord and leg veins. *Prostaglandins* **45**, 517–525.

Muck, A.O., Seeger, H., Korte, K. and Lippert, T.H. (1993b) The effect of 17b-estradiol and endothelin-1 on prostacyclin and thromboxane production in human endothelial cell cultures. *Clin. Exp. Obst. Gyn.* **14**, 203–206.

Muck, A.O., Seeger, H., Korte, K. and Lippert, T.H. (1994) Cardiovascular protection by post-menopausal estrogen replacement therapy: possible mechanisms of thc estrogen action. *Clin. Exp. Obst. Gyn.* **39**, 143–149.

Muldoon, L.L., Rodland, K.D., Forsythe, M.L. and Magun, B.E. (1989) Stimulation of phosphatidylinositol hydrolysis, diacylglycerol release, and gene expression in response to endothelin, a potent new agonist for fibroblasts and smooth muscle cells. *J. Biol. Chem.* **264**, 8529–8536.

Neuer, A., Seeger, H., Muck, A.O. and Lippert, T.H. (1994) The effect of zearalenone and 17 beta-estradiol on prostacyclin and thromboxane production in cell cultures from human umbilical cord veins. *Prostaglad. Leukotrien. Essen. Fatty Acids* **51**, 241–244.

Orimo, A., Inoue, S., Ikegami, A., Hosoi, T., Akishita, M., Ouchi, Y., Muramatsu, M. and Orimo, H. (1993) Vascular smooth muscle cells as target for estrogen. *Biochem. Biophys. Res. Commun.* **195**(2), 730–736.

Paller, M.S. (1987) Decreased pressor responsiveness in pregnancy: studies in experimental animals. *Am. J. Kidney Diseases* **9**, 308–311.

Polderman, K.H., Stehouwer, C.D.A., van Kamp, G.J., Dekker, G.A., Verheugt, F.W.A. and Gooren, L.J.G. (1993) Influence of sex hormones on plasma endothelin levels. *Ann. Intern. Med.* **118**, 429–432.

Pomerantz, K., Maddox, Y., Maggi, F., Ramey, E. and Ramwell, P. (1980) Sex and hormonal modification of 6-keto-$PGF_{1a}$ release by rat aorta. *Life Sciences* **27**, 1233–1236.

Redmond, E.M., Cherian, M.N. and Wetzel, R.C. (1994) 17b-estradiol inhibits flow- and acute hypoxia-induced prostacyclin release from perfused endocardial endothelial cells. *Circulation* **90**, 2519–2524.

Rengasamy, A. and Johns, R.A. (1993) Inhibition of nitric oxide synthase by a superoxide generating system. *J. Pharmacol. Exp. Ther.* **267**, 1024–1027.

Riedel, M., Rafflenbeul, W. and Lichtlen, P. (1993) Ovarian sex steroids and atherosclerosis. *Clin. Investig.* **71**, 406–412.

Roncaglioni, M.C., di Minno, G., Reyers, I., De Gaetano, G. and Donati, M.B. (1979) Increased prostacyclin-like activity in vascular tissues from rats on long-term treatment with an oestrogen-progestagen combination. *Thromb. Res.* **14**, 793–797.

Rosenfeld, C.R. and Jackson, G.M. (1984) Estrogen-induced refractoriness to the pressor effects of infused angiotensin II. *Am. J. Obstet. Gynecol.* **148**, 429–435.

Rubanyi, C.M. and Vanhoutte, P.M. (1986) Superoxide anions and hyperoxia inactivate endothelium-derived relaxing factor(s). *Am. J. Physiol.* **250**, H822–H827.

Sachinidis, A., Flesch, M., Ko, Y., Schror, K., Bohm, M., Dusing, R. and Vetter, H. (1995) Thromboxane $A_2$ and vascular smooth muscle cell proliferation. *Hypertension* **26**, 771–780.

Salas, E., Lopez, M.G., Villarroya, M., Sanchez-Garcia, P., DePascual, R., Dixon, W.R. and Garcia, A.G. (1994) Endothelium-independent relaxation by 17-alpha-estradiol of pig coronary arteries. *Eur. J. Pharmacol.* **258**, 47–55.

Seillan, C., Ody, C., Russo-Marie, F. and Duval, D. (1983) Differential effects of sex steroids on prostaglandin secretion by male and female cultured piglet endothelial cells. *Prostaglandins* **26**(1), 3–12.

Seltzer, A., Pinto, J.E.B., Viglione, P.N., Correa, F.M.A., Libertun, C., Tsutsumi, K., Steele, M.K. and Saavedra, J.M. (1992) Estrogens regulate angiotensin-converting enzyme and angiotensin receptors in female rat anterior pituitary. *Neuroendocrinology* **55**, 460–467.

Seltzer, A., Tsutsumi, K., Shigematsu, K. and Saavedra, J.M. (1993) Reproductive hormones modulate angiotensin II $AT_1$ receptors in the dorsomedial arcuate nucleus of the female rat. *Endocrinology* **133**, 939–941.

Sieck, G.C., Prakash, Y.S., Kannan, M.S. and Miller, V.M. (1997) Mechanisms of estrogen influence on $[Ca^{2+}]_i$ regulation in porcine coronary artery smooth muscle. *J. Vas. Res.* **34** (Suppl. 1): 129.

Skjaerlund, J.M., Gallon, L. and Subbiah, M.T.R. (1983) Effect of ethinylestradiol on arachidonic acid incorporation, release, and conversion to prostaglandins in rabbit platelets. *Biochem. Med.* **30**, 357–362.

Stampfer, M.J. and Colditz, G.A. (1991) Estrogen replacement therapy and coronary heart disease: a quantitative assessment of the epidemiologic evidence. *Prev. Med.* **20**, 47–63.

Steinleiter, A., Stanczyk, F.Z., Levin, J.H., d'Ablaing, G., III., Vijod, M.A., Shahbazian, V.L. and Lobo, R.A. (1989) Decreased *in vitro* production of 6-keto-prostaglandin $F_{1a}$ by uterine arteries from postmenopausal women. *Am. J. Obstet. Gynecol.* **161**, 1677–1681.

Subbiah, M.T.R., Deitemeyer, D., Yunker, R. and Gallon, L. (1981) Effect of specific estrogens on prostaglandin synthesis in aorta and thrombocytes of female pigeons (41064). *Proc. Soc. Exp. Biol. Med.* **166**, 300–304.

Sylvester, J.T., Gordon, J.B., Malamet, R.L. and Wetzel, R.C. (1985) Prostaglandins and estradiol-induced attenuation of hypoxic pulmonary vasoconstriction. *Chest* **88**, 252S–254S.

Timm, M., Kaski, J.C. and Dashwood, M.R. (1995) Endothelin-Like immunoreactivity in atherosclerotic human coronary arteries. *J. Cardiovasc. Pharmacol.* **26**, S442–S444.

Vargas, R., Wroblewska, B., Rego, A., Hatch, J. and Ramwell, P.W. (1993) Oestradiol inhibits smooth muscle cell proliferation of pig coronary artery. *Br. J. Pharmacol.* **109**, 612–617.

Vargas, R., Delaney, M., Farhat, M.Y., Wolfe, R., Rego, A. and Ramwell, P.W. (1995) Effect of estradiol 17b on pressor responses of rat mesenteric bed to norepinephrine, K+, and U-46619. *J. Cardiovasc. Pharmacol.* **25**, 200–206.

Venkov, C.D., Rankin, A.B. and Vaughan, D.E. (1996) Identification of authentic estrogen receptor in cultured endothelial cells. a potential mechanism for steroid hormone regulation of endothelial function. *Circulation* **84**, 727–733.

Wakasugi, M., Noguchi, T., Kazama, Y.I., Kanemaru, Y. and Onaya, T. (1989) The effects of sex hormones on the synthesis of prostacyclin (PGI2) by vascular tissues. *Prostaglandins* **37**, 401–410.

Walters, W.A. (1992) Autacoids affecting vascular tone in the human fetal extracorporeal circulation. *Clin. Exp. Pharmacol. Physiol.* **19**, 3–8.

Weiner, C.P., Lizasoain, I., Baylis, S.A., Knowles, R.G., Charles, I.G. and Moncada, S. (1994) Induction of calcium-dependent nitric oxide synthases by sex hormones. *Proc. Natl. Acad. Sci.* **91**, 5212–5216.

Wenger, N.K. (1990) Editorial: Gender, coronary artery disease, and coronary bypass surgery. *Ann. Intern. Med.* **112**, 557–558.

White, R.E., Darkow, D.J. and Falvo Lang, J.L. (1995) Estrogen relaxes coronary arteries by opening $BK_{Ca}$ channels through a cGMP-dependent mechanism. *Circ. Res.* **77**, 936–942.

Witter, F.R. and DiBlasi, M.C. (1984) Effect of steroid hormones on arachidonic acid metabolites of endothelial cells. *Obstet. Gynecol.* **63**, 747–751.

Yanagisawa, M., Kurihara, H., Kimura, S., Tomobe, Y., Kobayashi, M., Mitsui, Y., Yuzaki, Y., Goto, K. and Masaki, T. (1988) A novel potent vasoconstrictor peptide produced by vascular endothelial cells. *Nature* **332**, 411–415.

Yang, D. and Clark, K.E. (1992) Effect of endothelin-1 on the uterine vasculature of the pregnant and estrogen-treated nonpregnant sheep. *Am. J. Obstet. Gynecol.* **167**, 1642–1650.

Ylikorkala, O., Orpana, A., Puolakka, J., Pyorala, T. and Viinikka, L. (1995) Postmenopausal hormonal replacement decreases plasma levels of endothelin-1. *J. Clin. Endocrinol. Metab.* **80**, 3384–3387.

# 13 Animal Models of Atherosclerosis ☆

**Michael R. Adams** *, **Thomas C. Register, Randolph L. Geary, Janice D. Wagner and J. Koudy Williams**

*Comparative Medicine Clinical Research Center, Bowman Gray School of Medicine of Wake Forest University, Winston-Salem, NC 27157-1040, USA.*
*☆ Supported in part by Grants P01 HL45666 and R01 HL38964 from the National Heart, Lung, and Blood Institute, National Institutes of Health, Bethesda, MD, USA.*
** Corresponding author, Tel.: 001 910 716 1580, Fax: 001 910 716 1515, E-mail: madams@cpm.bgsm.edu.*

Estrogen inhibits the progression of atherosclerosis in animal models. In a primate model, coronary artery atherosclerosis extent is reduced by 50% in ovariectomized monkeys treated with parenteral estradiol and by 70% in similar animals treated with oral conjugated equine estrogens. While progestins administered alone have no influence on atherosclerosis, and coadministration of progesterone to estradiol-treated monkeys has no effect on atherosclerosis, we have found that coadministration of medroxyprogesterone acetate to monkeys treated with conjugated equine estrogens antagonized estrogen's atheroinhibitory effect. This suggests that effects of progestin coadministered with estrogen may depend on type, dose, and route or pattern of administration, and that some forms of progestin cotherapy may antagonize favorable effects of estrogen. Further studies using this model indicate that estrogen acts, in part, by inhibiting the arterial uptake and degradation of plasma LDL. Evidence also is accumulating which indicates that estrogen inhibits the expression of proinflammatory and mitogenic cytokines implicated in atherogenesis. While there is no longer any doubt that estrogen inhibits atherosclerosis, we have found no effect of estrogen (oral conjugated equine estrogens) on the neointimal or structural responses of the artery to balloon catheter injury in monkeys with preexisting advanced atherosclerosis. While this seems to contradict the results of studies with nonatherosclerotic nonprimate animals, it is likely that this indicates that atherosclerotic and nonatherosclerotic vessels respond differently to mechanical injury and that estrogen is not effective in inhibiting arterial neointimal response to balloon-catheter injury in a model that more closely mimics the clinical situation of balloon angioplasty and restenosis.

KEYWORDS: Atherosclerosis, estrogens, nonhuman primates, progestins, restenosis.

## INTRODUCTION

In most Western societies, the incidence of coronary heart disease and coronary artery atherosclerosis (McGill and Stern, 1979) are much lower in premenopausal women than in men of similar age. The widespread belief that ovarian estrogen is responsible for this relative sparing of the coronary arteries is supported by compelling evidence that estrogen replacement therapy is associated with a marked reduction in angiographically-defined extent of coronary artery atherosclerosis (Adams *et al.*, 1987a) and risk of coronary heart disease in postmenopausal women (Bush, 1990). The mechanisms involved in estrogen's apparent protective effects remain unclear. Among multiple possibilities are inhibitory effects on atherosclerosis progression and coronary thrombosis, and beneficial effects on vasomotor function of coronary arteries. Also unclear are effects of combination hormone treatment, i.e., estrogen plus progestin or androgen, on atherosclerosis and coronary heart disease risk.

We summarize here the experimental evidence regarding effects of sex hormone deficiency and hormone replacement therapy on the initiation and progression of atherosclerosis and mechanisms by which sex hormones influence the pathogenesis of atherosclerosis.

## ANIMAL MODEL STUDIES

### Effects of Estrogen on Coronary Artery Atherosclerosis

*Overview*

This discussion emphasizes coronary arteries because: (1) The gender difference in atherosclerosis extent is confined to coronary arteries (McGill and Stern, 1979); (2) Experimental evidence indicates that effects of sex hormones are confined to coronary arteries and, perhaps, femoral arteries (McGill and Stern, 1979; Adams *et al.*, 1987b; Clarkson *et al.*, 1990); and (3) Effects on coronary arteries are of greatest relevance to coronary heart disease, the major clinical sequela of atherosclerosis in human beings. Effects on aortic atherosclerosis, for example, may be of limited relevance and, in fact, may result in misleading conclusions regarding coronary heart disease.

Studies done in the 1950s at the Michael Reese Research Institute provided some of the initial evidence that estrogen inhibits the progression of coronary artery atherosclerosis. Among the important findings from these studies was the resistance of hens to development of coronary artery atherosclerosis relative to roosters (Stamler, Pick and Katz, 1954). Furthermore, ligation of the hen oviduct, which results in a marked elevation in plasma cholesterol concentrations, had no effect on the relative resistance of hens to atherosclerosis, whereas ovariectomy resulted in a marked exacerbation of atherosclerosis (Stamler, Pick and Katz, 1954). In addition, exogenous estradiol in physiologic doses inhibited progression of atherosclerosis (Pick *et al.*, 1952a) and promoted its regression (Pick *et al.*, 1952b) in this species. Subsequent studies at our institution using White Carneau pigeons (Prichard, Clarkson and Lofland, 1966) also reported inhibitory effects of physiologic doses of estrogen on coronary artery atherosclerosis.

In 1977, McGill and colleagues studied the effects of exogenous estrogens on atherosclerosis extent in ovariectomized baboons, and reported that ovariectomy was not associated with increased extent or severity of diet-induced atherosclerosis. Nor were there any significant effects of treatment with either physiologic or pharmacologic estrogen replacement therapy. It is important to note that, unlike many other primate species, baboons are relatively resistant to diet-induced hyperlipidemia and atherosclerosis. Furthermore, there is no sex difference in the extent of diet-induced coronary artery atherosclerosis in baboons.

## The Cynomolgus Macaque Model

In our laboratory, we have used cynomolgus macaques (*Macaca fascicularis*) to study the effects of reproductive steroids on coronary artery atherosclerosis. This nonhuman primate species has been used in atherosclerosis research for approximately 30 years, principally because of its susceptibility to diet-induced atherosclerotic involvement of main branch coronary arteries. We chose it for our research because, in addition to its susceptibility to atherosclerosis, its reproductive physiology is very similar to that of human beings; the female has a 28-day menstrual cycle and circulating sex hormone patterns similar to those of women (Adams *et al.*, 1987a). Also, a natural menopause occurs in aged monkeys.

In an initial study of the relationship between endogenous sex steroids and atherosclerosis, we studied cynomolgus macaques fed a moderately atherogenic diet for 30 months containing 40% of calories as fat and 0.4 mg cholesterol per kcal. There were four experimental groups: males ($n=15$); intact, nonpregnant females ($n=23$); surgically postmenopausal (i.e., ovariectomized) females ($n=21$); and pregnant females ($n=27$). Total plasma cholesterol (TPC) and plasma high density lipoprotein (HDL) cholesterol concentrations and blood pressure were determined periodically. After 30 months, all animals were necropsied and extent of atherosclerosis (lesion cross-sectional area) was determined morphometrically.

As in a previous study (Hamm *et al.*, 1983) males had more extensive coronary artery atherosclerosis relative to intact nonpregnant females (Kaplan *et al.*, 1984). Males also had significantly lower plasma HDL cholesterol concentrations and higher systolic blood pressure. Ovariectomy (i.e., estrogen deficiency) resulted in a more atherogenic plasma lipoprotein pattern (decreased plasma HDL cholesterol and increased TPC concentrations), and a doubling in coronary artery atherosclerosis extent (Adams *et al.*, 1985). The hyperestrogenic state of pregnancy was associated with a marked reduction in extent of coronary artery atherosclerosis (Adams *et al.*, 1987c), but both TPC and HDL cholesterol concentrations were markedly decreased during pregnancy. These findings are summarized in Figure 13.1 (Adams *et al.*, 1994).

The results of this study provide indirect evidence regarding effects of endogenous sex hormones on atherosclerosis extent. Males and ovariectomized females did not differ in regard to coronary artery atherosclerosis extent, and both groups also had consistently low plasma estradiol concentrations (around 20 pg/ml). Atherosclerosis extent was reduced by approximately one-half in intact, nonpregnant females, which had much higher plasma estradiol concentrations (range of 60–300 pg/ml depending on time of menstrual cycle). Relative to these intact females, atherosclerosis extent was reduced by approximately 50 percent in pregnant females, a group with large, sustained elevations in plasma estradiol concentrations (300–1000 pg/ml).

## Experimental Effects of Hormone Therapy

### *Study 1*

Direct evidence for an inhibitory effect of endogenous estrogen on progression of coronary artery atherosclerosis is provided by the results of a subsequent study (Adams *et al.*, 1990). In this study, ovariectomized monkeys fed an atherogenic diet

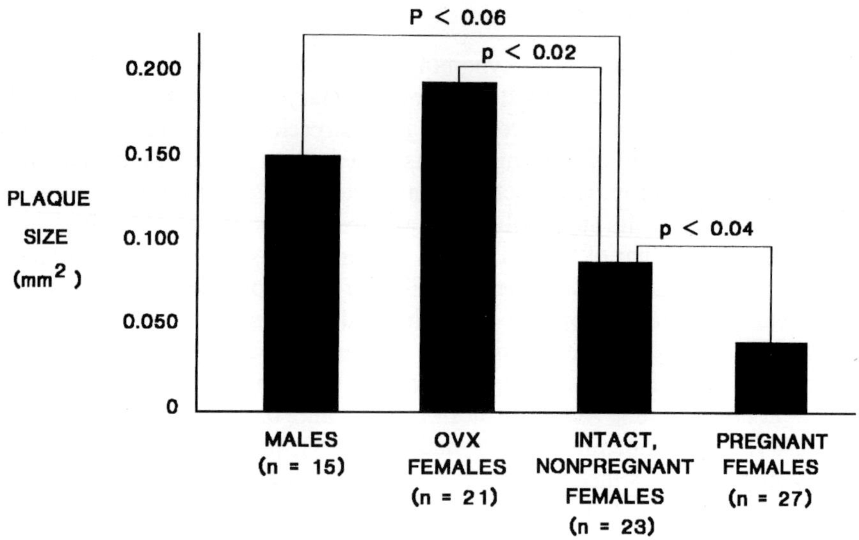


**Figure 13.1.** Coronary artery plaque extent in four groups of cynomolgus macaques with diet-induced atherosclerosis. Although ovariectomized monkeys (OVX) had a far more atherogenic profile (i.e., lower plasma concentrations of high density lipoprotein and higher plasma concentrations of low density lipoproteins), their plaque extent did not differ from that of males. Reprinted with permission from Adams *et al*., 1994.

were assigned randomly to one of three treatment groups: (1) no hormone replacement ($n=17$), (2) continually administered 17-$\beta$ estradiol plus cyclically administered progesterone ($n=20$), and (3) continuously administered 17-$\beta$ estradiol ($n=18$). Physiologic patterns of plasma estradiol and progesterone concentrations were maintained by administering the hormones in sustained-release subcutaneous Silastic® implants. The experiment lasted 30 months. At necropsy, coronary artery atherosclerosis was inhibited similarly (reduced by approximately half) in animals in both hormone replacement groups (Figure 13.2). Antiatherogenic effects of hormone replacement were independent of variations in TPC, lipoprotein cholesterol, and apolipoprotein A-I and B concentrations; average low density lipoprotein (LDL) particle size; and HDL subfractional heterogeneity. Similarly, effects of hormone replacement on atherosclerosis were not explained by other risk variables, i.e., blood pressure or carbohydrate tolerance.

Whereas effects of hormone replacement observed in this study are consistent with results of epidemiologic studies that associate estrogen use with reduced risk of coronary heart disease, the hormones administered and their mode of administration are quite different. For this reason, we conducted a second study of the effects of the compounds most commonly prescribed for prevention or treatment of post-menopausal symptoms.

### *Study 2*

In Study 2, ovariectomized monkeys were fed an atherogenic diet and assigned randomly to one of four treatment groups; (1) no treatment ($n=21$), (2) conjugated

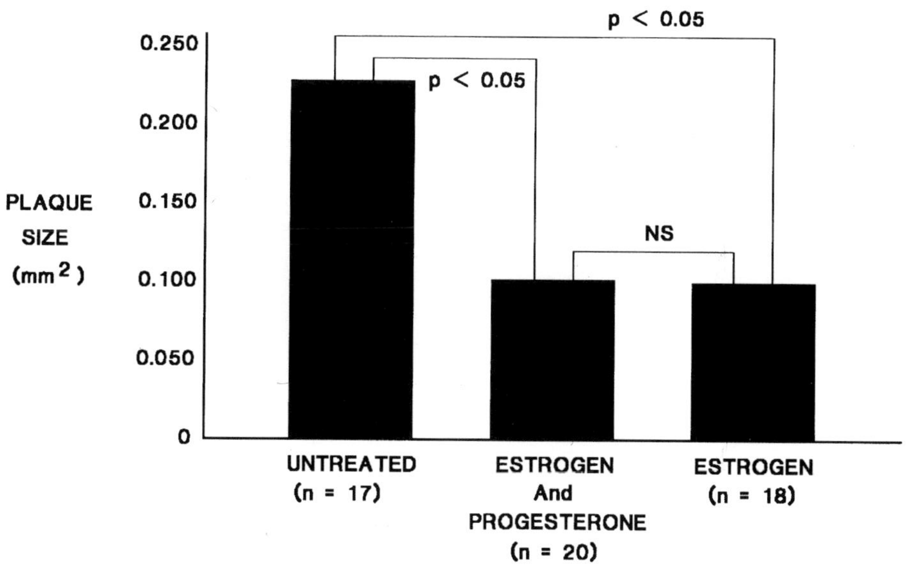


**Figure 13.2.** Effect of hormone replacement therapy on coronary artery plaque extent of surgically postmenopausal cynomolgus macaques fed a moderately atherogenic diet. Hormone treatment reduced plaque extent by half. NS = not significant. Reprinted with permission from Adams *et al.*, 1994.

equine estrogens (human equivalent of 0.625 mg/day) ($n=25$), (3) medroxyprogesterone acetate (MPA; human equivalent of 2.5 mg/day) ($n=19$), and (4) conjugated equine estrogens and MPA ($n=26$). Treatment continued for 30 months. Preliminary analyses of these data (Adams *et al.*, 1997) indicate that treatment with estrogen alone resulted in a 70% decrease ($p<0.01$) in atherosclerosis extent, whereas treatment with estrogen and MPA or MPA alone resulted in an atherosclerosis extent not different from that of untreated controls and was increased 170% and 290%, respectively, relative to animals treated with estrogen alone ($p<0.01$) (Figure 13.3). Preliminary analyses also indicate that these effects are largely independent of variation in plasma lipoproteins.

The results of these two studies (Figure 13.3) agree incompletely with the epidemiologic data. Although the results of Study 1 support a protective effect of estrogen monotherapy or combination estrogen/sequential MPA therapy against coronary heart disease, the effects seen in Study 2 (combined hormone replacement with estrogen and continuous MPA) are inconsistent. How might this be explained? Unlike the epidemiologic data for estrogen monotherapy and coronary heart disease risk, the data for combined hormone therapy with MPA are somewhat limited. Some (Psaty *et al.*, 1994; Stampfer M. personal communication), although not all (Nachtigall *et al.*, 1979; Rosenberg *et al.*, 1979), indicate a decrease in risk similar to that seen with estrogen alone. However, these data are limited by the fact that the predominant prescribing pattern represented in the data for combined hormone replacement was continuous estrogen plus sequential MPA administration (i.e., 7–14 days per month). Recently, prescriptions for combined continuous replacement therapy, similar to that used in our second monkey study, have increased. Observational data describing how

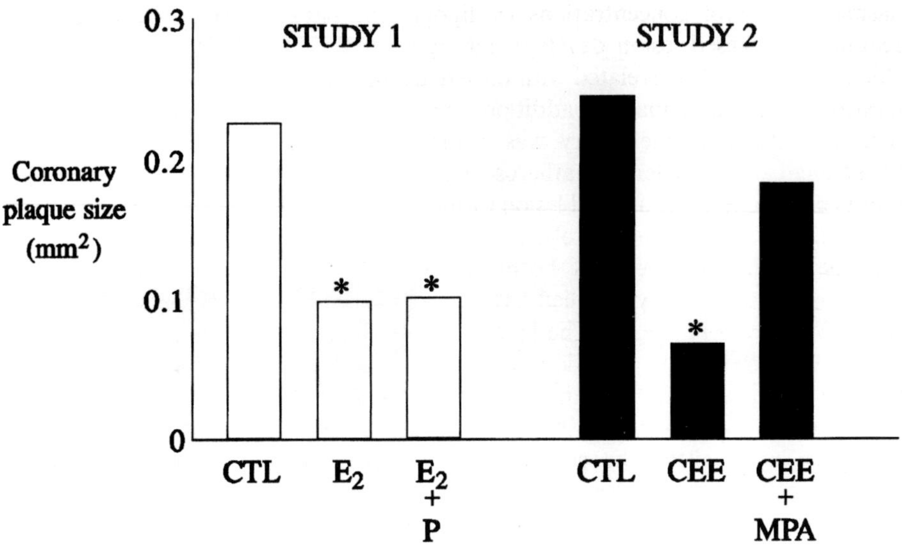


**Figure 13.3.** Comparison of effects of two types of estrogen monotherapy and combined hormone replacement therapy on atherosclerosis in ovariectomized cynomolgus monkeys. Study 1: CTL = untreated controls; $E_2$ = 17-$\beta$ estradiol administered continuously by subcutaneous implant; P = progesterone administered sequentially by subcutaneous implant. Study 2: CTL = untreated controls; CEE = conjugated equine estrogens (human equivalent of 0.625 mg/day); MPA = oral medroxyprogesterone acetate (human equivalent of 2.5 mg/day). * = $p < 0.05$ compared to untreated controls in each study. Adapted with permission from Adams *et al.*, 1990.

this form of therapy affects coronary risk are not yet available. The monkey data suggest that this form of therapy may result in increased risk relative to users of estrogen alone or estrogen and sequentially administered progestin.

Our findings also suggest that inhibitory influences of hormone replacement on atherogenesis and coronary heart disease must be mediated either through risk factors not measured in these studies (or not yet described), or through direct influences on cellular or biochemical events occurring in the arterial intima. Subsequent studies in our laboratory have addressed further the effects of sex hormones on pathophysiologic processes of the arterial wall.

## Effect of Sex Hormones on Arterial Pathophysiology

### *Sex hormones and arterial lipoprotein metabolism*

While effects of estrogen replacement on atherosclerosis are incompletely accounted for by plasma lipoprotein patterns, evidence exists that estrogens may directly influence arterial lipoprotein metabolism. Several groups have performed studies in rabbits (Kushwaha and Hazzard, 1981; Haarbo *et al.*, 1991; Hough and Zilversmit, 1986). Among them, Hough and Zilversmit (1986) investigated the effects of 17-$\beta$ estradiol cypionate treatment on net arterial cholesterol influx and hydrolysis of plasma cholesteryl ester in cholesterol-fed rabbits. Although no significant effects on plasma cholesterol concentrations or lipoprotein patterns were found, estrogen treatment decreased lesion development significantly. Net cholesteryl ester influx, which was positively correlated with the extent of atherosclerosis, also was decreased in estrogen-treated animals. In addition, the percentage of newly entering cholesteryl ester hydrolyzed by the artery was reduced significantly by estrogen treatment independent of the extent of atherosclerosis. These results indicate that estrogen treatment may decrease arterial lesion formation by directly inhibiting the accumulation of cholesterol and/or hydrolysis of cholesteryl ester in the arterial wall.

To determine the effects of hormone treatment on lipoprotein metabolism in cynomolgus macaques, we studied LDL coupled to radiolabeled tyramine cellobiose (TC) (Wagner *et al.*, 1991). The labeled TC-LDL, originally described by Pittman *et al.* (1983), allows quantification of the accumulation of products of LDL degradation and undegraded LDL in tissue or arterial samples. When lipoproteins coupled to iodinated TC are degraded by the cell, the labeled TC remains trapped in the lysosomes, representing the accumulation of products of LDL degradation and undegraded LDL over time (Carew *et al.*, 1984; Wagner *et al.*, 1992).

We studied ovariectomized monkeys fed a moderately atherogenic diet for 18 weeks to study the effects of hormone treatment on early atherogenesis. Estrogen and cyclic progesterone were administered via Silastic® implants to one group ($n=9$), resulting in physiologic hormone concentrations, whereas controls ($n=8$) had low to undetectable hormone levels. Radiolabeled LDL was injected 24 hours before necropsy, and whole body and arterial metabolism were measured. Hormone replacement therapy resulted in significantly reduced accumulation of LDL and products of LDL accumulation in coronary arteries (Wagner *et al.*, 1991) as well as other arterial sites (Wagner *et al.*, 1992). This occurred despite no significant difference in plasma lipid and lipoprotein concentrations. Furthermore, the reduction in LDL metabolism occurred before any changes in intimal thickening or other indices of endothelial injury, such as increased endothelial cell turnover or numbers of leukocytes adherent to the endothelium. Subsequent studies of contraceptive estrogens and progestins (Wagner *et al.*, 1993) and an estrogen/androgen combination used for hormone replacement therapy (Wagner *et al.*, 1997) are consistent with these findings. Furthermore, they implicate the estrogen component as responsible for the protective effects observed when combined treatment (estrogen/progestin, estrogen/androgen) is used.

Estrogen's effects on atherosclerosis may occur through a number of mechanisms. The study of Hough and Zilversmit (1986) and our studies in monkeys (Wagner *et al.*, 1991; 1992) suggest that estrogen may act intracellularly by inhibiting the hydrolysis of cholesterol and degradation of LDL, respectively.

## Sex Hormones and Pathobiology of the Arterial Intima

Sex hormones may also have direct effects on cellular or molecular processes in the arterial intima implicated in the initiation and progression of atherosclerosis. There is plentiful evidence that sex hormones are modulators of immune and inflammatory reactions (Ahmed, Penhale and Talal, 1985; Loy *et al.*, 1992; Polan *et al.*, 1989; Schuurs and Verheul, 1990), both of which have been implicated in atherosclerosis

injured artery (intimal area or any indices of structural response or cell proliferation) ($p$'s$>0.4$). In contrast to previous studies of nonatherosclerotic animals, the results show that in the circumstance of transmural injury to arteries of primates with preexisting atherosclerosis, estrogen does not appear to suppress intimal or arterial structural responses to injury.

## SUMMARY AND CONCLUSIONS

Although it has long been known that women have a lower incidence of coronary heart disease, until recently clinical and epidemiologic studies have not investigated the possible role of sex hormones in cardiovascular disease, particularly coronary artery atherosclerosis. We have reviewed the current experimental evidence concerning the roles of sex hormone deficiency and hormone replacement therapy in the initiation and progression of coronary artery atherosclerosis.

Early experimental models of estrogen-deficient animals with coronary artery atherosclerosis used chickens and pigeons. However, nonhuman primate models replaced these earlier models because of greater similarities to human beings. Our studies have focused on female cynomolgus macaques, especially surgically postmenopausal (i.e., estrogen-deficient) animals fed an atherogenic diet. We found that male and ovariectomized female cynomolgus macaques did not differ with respect to coronary artery atherosclerosis, while premenopausal females had half the coronary artery atherosclerosis of their ovariectomized counterparts. In two studies of hormone replacement therapy of ovariectomized cynomolgus macaques, we discovered that estrogen replacement therapy (either parenteral estradiol or oral conjugated equine estrogens) resulted in 50–70% reductions in the extent of coronary artery atherosclerosis compared to control animals. However, while continuous estradiol/sequential progesterone replacement, administered parenterally, caused a 50% reduction in atherosclerosis extent, continuous conjugated equine estrogens and MPA given orally resulted in atherosclerosis extent that did not differ from untreated controls. These results indicate that effects of progestin coadministered with estrogen on coronary arteries may depend on type, dose, route of administration or pattern (continuous versus sequential) of administration. Furthermore, some forms of progestin cotherapy may antagonize favorable effects of estrogen.

Further investigations have indicated that direct arterial actions may explain sex hormones' effects on atherosclerosis. Estrogens inhibit arterial uptake and degradation of plasma LDL, preserve endothelium-dependent vasomotor responses, and inhibit the expression of proinflammatory and mitogenic cytokines implicated in atherogenesis. Thus, estrogen may directly inhibit the arterial accumulation of cholesterol, inhibit inflammatory and proliferative processes of the arterial intima, and promote vasodilation, thereby retarding atherosclerosis progression and reducing the risk of clinically evident coronary heart disease by all of these means. Effects of progestins on these pathophysiologic processes of the artery wall are unclear.

Estrogen inhibits the intimal proliferative response to arterial injury in nonatherosclerotic rodents and rabbits and inhibits the *in vitro* proliferation of smooth muscle cells. In contrast, although long-term estrogen treatment inhibits progression of diet-induced atherosclerosis, it had no effect on short-term (28 day) neointimal or remodeling responses of the artery to balloon catheter injury in a nonhuman primate model (Geary *et al.*, 1996). This apparent contradiction is probably due to substantial differences in experimental design. Previous studies utilized nonatherosclerotic animals, which have limited relevance to the clinical situation where response of vessels with advanced atherosclerosis is involved. It is now clear that compounds effective in preventing myointimal response to arterial injury in nonatherosclerotic animals have typically been ineffective in preventing restenosis in clinical trials. Among possible explanations for this incongruence are species differences in smooth muscle migration or proliferation, or differences in responses of atherosclerotic and nonatherosclerotic arteries. Furthermore, the endpoints typically addressed in animal studies, i.e., intimal proliferation and mass, may be minor determinants of restenosis in human patients.

## ACKNOWLEDGMENTS

The authors gratefully acknowledge the editorial assistance of Karen Potvin Klein.

## REFERENCES

Adams, M.R., Clarkson, T.B., Kaplan, J.R. and Koritnik, D.R. (1985) Ovariectomy, social status, and coronary artery atherosclerosis in cynomolgus monkeys. *Arteriosclerosis*, **5**, 192–200.

Adams, M.R., Clarkson, T.B., Koritnik, D.R. and Nash, H.A. (1987b) Contraceptive steroids and coronary artery atherosclerosis in cynomolgus macaques. *Fertility and Sterility*, **47**, 1010–1018.

Adams, M.R., Kaplan, J.R., Clarkson, T.B. and Koritnik, D.R. (1987c) Pregnancy-associated inhibition of coronary artery atherosclerosis in monkeys: Evidence of a relationship with endogenous estrogen. *Arteriosclerosis*, **7**, 378–384.

Adams, M.R., Kaplan, J.R., Clarkson, T.B. and Koritnik, D.R. (1987a) Effects of psychosocial stress, menopause and pregnancy on coronary artery atherosclerosis. In *Coronary Heart Disease in Women*, edited by E.D. Eaker, B. Packard, N.K. Wenger, H.A. Tyroler and T.B. Clarkson, pp. 151–157. New York, Haymarket Doyma.

Adams, M.R., Kaplan, J.R., Manuck, S.B., Koritnik, D.R., Parks, J.S., Wolfe, M.S. *et al.* (1990) Inhibition of coronary artery atherosclerosis by 17-beta estradiol in ovariectomized monkeys. Lack of effect of an added progesterone. *Arteriosclerosis*, **10**, 1051–1057.

Adams, M.R., Register, T.C., Golden, D.L., Wagner, J.D. and Williams, J.K. (1997) Medroxyprogesterone acetate antagonizes inhibitory effects of conjugated equine estrogens on coronary artery atherosclerosis. *Arterioscler Thromb Vasc Biol*, **17**, 217–221.

Adams, M.R., Washburn, S.A., Wagner, J.D., Williams, J.K. and Clarkson, T.B. (1994) Arterial changes: Estrogen deficiency and effects of hormone replacement. In *Treatment of the Postmenopausal Woman: Basic and Clinical Aspects*, edited by R.A. Lobo, pp. 243–250. New York, Raven Press.

Ahmed, S.A., Penhale, W.J. and Talal, N. (1985) Sex hormones, immune responses, and autoimmune diseases: Mechanisms of sex hormone action. *American Journal of Pathology*, **121**, 531–551.

Bush, T.L. (1990) The epidemiology of cardiovascular disease in postmenopausal women. *Annals of the New York Academy of Sciences*, **592**, 263–271.

Carew, T.C., Pittman, R.C., Marchand, E.R. and Steinberg, D. (1984) Measurement *in vivo* of irreversible degradation of low density lipoprotein in the rabbit aorta. *Arteriosclerosis*, **4**, 214–224.

Caulin-Glaser, T., Watson, C.A., Pardi, R. and Bender, J.R. (1996) Effects of 17$\beta$-estradiol on cytokine-induced endothelial cell adhesion molecule expression. *Journal of Clinical Investigation*, **98**, 36–42.

Cid, M.C., Kleinman, H.K., Grant, D.S., Schnaper, H.W., Fauci, A.S. and Hoffman, G.S. (1994) Estradiol enhanced leukocyte binding to tumor necrosis factor (TNF)-stimulated endothelial cells via an increase in TNF-induced adhesion molecules E-selectin, intercellular adhesion molecule type 1, and vascular cell adhesion molecule type 1. *Journal of Clinical Investigation*, **93**, 17–25.

Clarkson, T.B., Shively, C.A., Morgan, T.M., Koritnik, D.R., Adams, M.R. and Kaplan, J.R. (1990) Oral contraceptives and coronary artery atherosclerosis of cynomolgus monkeys. *Obstetrics and Gynecology*, **75**, 217–222.

Clowes, A.W. and Karnovsky, M.J. (1977) Suppression by heparin of smooth muscle cell proliferation in injured arteries. *Nature*, **265**, 625–626.

Faxon, D.P. (1995) Effect of high dose angiotensin-converting enzyme inhibition on restenosis: Final results of the MERCATOR study, a multicenter, double-blind, placebo-controlled trial of cilazapril. *Journal of the American College of Cardiology*, **25**, 362–369.

Faxon, D.P., Spiro, T.E., Minor, S., Cote, G., Douglas, J., Gottlieb, R. *et al.* (1994) Low molecular weight heparin in prevention of restenosis after angioplasty: Results of enoxaparin restenosis (ERA) trial. *Circulation*, **90**, 908–914.

Fischer-Dzoga, K., Wissler, R. and Vesslinovitch, D. (1983) The effect of estradiol on the proliferation of rabbit aortic medial tissue cells induced by hyperlipemic serum. *Experimental and Molecular Pathology*, **39**, 355–362.

Foegh, M.L., Khirabadi, B.S., Nakanishi, T., Vargas, R. and Ramwell, P.W. (1987) Estradiol protects against experimental cardiac transplant atherosclerosis. *Transplant Proceedings*, **19**, 90–95.

Frazier-Jessen, M.R. and Kovacs, E.J. (1995) Estrogen modulation of JE/monocyte chemoattractant protein-1 expression in murine macrophages. *Journal of Immunology*, **154**, 1838–1845.

Geary, R.L., Williams, J.K., Golden, D.L., Brown, D.G., Benjamin, M.E. and Adams, M.R. (1996) Time course of cellular proliferation, intimal hyperplasia, and remodeling following angioplasty in monkeys with established atherosclerosis: A nonhuman primate model of restenosis. *Arteriosclerosis, Thrombosis, and Vascular Biology*, **16**, 34–43.

Geary, R.L., Williams, J.K., Golden, D.L., Brown, D.G., Benjamin, M.E. and Adams, M.R. (1995) Estrogen inhibits progression of atherosclerosis but not angioplasty-associated restenosis in a postmenopausal primate model (abstract). *Circulation*, **92**(Suppl. I), I-695.

Girasole, G., Jilka, R.L., Passeri, G., Boswell, S., Boder, G. and Williams, D.C. (1992) 17$\beta$-estradiol inhibits interleukin-6 production by bone marrow-derived stromal cells and osteoblasts *in vitro*: A potential mechanism for the antiosteoporotic effect of estrogens. *Journal of Clinical Investigation*, **89**, 883–891.

Haarbo, J., Leth-Espensen, P., Stender, S. and Christiansen, C. (1991) Estrogen monotherapy and combined estrogen-progestogen replacement therapy attenuate aortic accumulation of cholesterol in ovariectomized cholesterol-fed rabbits. *Journal of Clinical Investigation*, **87**, 1274–1279.

Hamm, T.E., Jr., Kaplan, J.R., Clarkson, T.B. and Bullock, B.C. (1983) Effects of gender and social behavior on the development of coronary artery atherosclerosis in cynomolgus monkeys. *Atherosclerosis*, **48**, 221–233.

Hough, J.L. and Zilversmit, D.B. (1986) Effect of 17-beta estradiol on aortic cholesterol content and metabolism in cholesterol-fed rabbits. *Arteriosclerosis*, **6**, 57–63.

Hu, S.-K., Mitcho, Y.L. and Rath, N.C. (1988) Effect of estradiol on interleukin-1 synthesis by macrophages. *International Journal of Immunopharmacology*, **10**, 247–252.

Jilka, R.L., Hangoc, G., Girasole, G., Passeri, G., Williams, D.C., Abrams, J.S. *et al.* (1992) Increased osteoclast development after estrogen loss: Mediation by interleukin-6. *Science*, **257**, 88–91.

Kaplan, J.R., Adams, M.R., Clarkson, T.B. and Koritnik, D.R. (1984) Psychosocial influences on female protection among cynomolgus macaques. *Atherosclerosis*, **53**, 283–295.

Kushwaha, R.S. and Hazzard, W.R. (1981) Exogenous estrogens attenuate dietary hypercholesterolemia and atherosclerosis in the rabbit. *Metabolism*, **30**, 359–366.

Lehman, K.G., Doria, R.J., Feuer, J.M., Hall, P.X. and Hoang, D.T. (1991) Paradoxical increase in restenosis rate with chronic heparin use: Final results of a randomized trial (abstract). *Journal of the American College of Cardiology*, **17**, 181.

Loy, R.A., Loukides, J.A. and Polan, M.L. (1992) Ovarian steroids modulate human monocyte tumor necrosis factor alpha messenger ribonucleic acid levels in cultured human peripheral monocytes. *Fertility and Sterility*, **58**, 733–739.

McGill, H.C., Jr., Axelrod, L.R., McMahan, C.A., Wigodsky, H.S. and Mott, G.E. (1977) Estrogens and experimental atherosclerosis in the baboon (*Papio cynocephalus*). *Circulation*, **56**, 657–662.

McGill, H.C., Jr. and Stern, M.P. (1979) Sex and atherosclerosis. *Atherosclerosis Reviews*, **4**, 157–242.

MERCATOR Study Group. (1992) Does the new angiotensin converting enzyme inhibitor cilazapril prevent restenosis after percutaneous transluminal angioplasty? Results of the MERCATOR Study: A multicenter, randomized, double-blind placebo-controlled trial. *Circulation*, **86**, 100–110.

Nachtigall, L.E., Nachtigall, R.H., Nachtigall, R.D. and Beckman, E.M. (1979) Estrogen replacement therapy II: A prospective study in the relationship to carcinoma and cardiovascular and metabolic problems. *Obstetrics and Gynecology*, **54**, 74–79.

Nakai, K., Itoh, C., Hotta, K., Itoh, T., Yoshizumi, M. and Hiramori, K. (1994) Estradiol-17$\beta$ regulates the induction of VCAM-1 mRNA expression by interleukin-1$\beta$ in human umbilical vein endothelial cells. *Life Sciences*, **54**, 221–227.

Pick, R., Stamler, J., Rodbard, S. and Katz, L.N. (1952a) The inhibition of coronary atherosclerosis by estrogens in cholesterol-fed chicks. *Circulation*, **6**, 276–280.

Pick, R., Stamler, J., Rodbard, S. and Katz, L.N. (1952b) Estrogen-induced regression of coronary atherosclerosis in cholesterol-fed chicks. *Circulation*, **6**, 858–861.

Pittman, R.C., Carew, T.E., Glass, C.K., Green, S.R., Taylor, C.A., Jr. and Attie, A.D. (1983) A radioiodinated, intracellularly trapped ligand for determining the sites of plasma protein degradation *in vivo*. *Biochemical Journal*, **212**, 791–800.

Polan, M.L., Daniele, A. and Kuo, A. (1989) Gonadal steroids modulate human monocyte interleukin (IL-1) activity. *Fertility and Sterility*, **49**, 964–968.

Polan, M.L., Loukides, J., Nelson, P., Carding, S., Diamond, M., Walsh, A. *et al.* (1989) Progesterone and estradiol modulate interleukin-1 beta messenger ribonucleic acid levels in cultured human peripheral monocytes. *Journal of Clinical Endocrinology and Metabolism*, **69**, 1200–1206.

Powell, J.S., Clozel, J.P., Muller, E.K.M., Kuhn, H., Hefti, F., Hosang, M. *et al.* (1989) Inhibitors of angiotensin-converting enzyme prevent myointimal proliferation after vascular injury. *Science*, **245**, 186–188.

Prichard, R.W., Clarkson, T.B. and Lofland, H.B. (1966) Estrogen in pigeon atherosclerosis. Estradiol valerate effects at several dose levels on cholesterol-fed male white Carneau pigeons. *Archives of Pathology*, **82**, 15–17.

Psaty, B.M., Heckbert, S.R., Atkins, D., Lemaitre, R., Koepsell, T.D., Wahl, P.W. *et al.* (1994) The risk of myocardial infarction associated with the combined use of estrogens and progestins in postmenopausal women. *Archives of Internal Medicine*, **154**, 1333–1339.

Register, T.C., Bora, T.A. and Adams, M.R. (1995) Estrogen inhibits activation of arterial NF-kB transcription factor in early diet-induced atherogenesis (abstract). *Circulation*, **92**(Suppl. I), I-628.

Rhee, C.Y., Spaet, T.H., Stemmerman, M.B., Lajam, F. and Shiang, H.H. (1977) Estrogen suppression of surgically induced vascular intimal hyperplasia in rabbits. *Journal of Laboratory and Clinical Medicine*, **90**, 77–84.

Rosenberg, L., Shapiro, S., Kaufman, D.W., Slone, D., Miettinen, O.S. and Stolley, P.D. (1979) Patterns and determinants of conjugated estrogen use. *American Journal of Epidemiology*, **109**, 676–686.

Schuurs, A.H.W.M. and Verheul, H.A.M. (1990) Effects of gender and sex steroids on the immune response. *Journal of Steroid Biochemistry*, **35**, 157–172.

Shanker, G., Sorci-Thomas, M. and Adams, M.R. (1994) Estrogen modulates the expression of tumor necrosis factor-alpha mRNA in phorbol ester- and lipopolysaccharide-stimulated human monocytic THP-1 cells. *Lymphokine and Cytokine Research*, **13**, 377–382.

Shanker, G., Sorci-Thomas, M. and Adams, M.R. (1995) Estrogen modulates the inducible expression of platelet-derived growth factor mRNA by monocyte/macrophages. *Life Sciences*, **56**, 499–507.

Shanker, G., Sorci-Thomas, M., Register, T.C. and Adams, M.R. (1994) The inducible expression of THP-1 cell interleukin-1 mRNA: Effects of estrogen on differential response to phorbol ester and lipopolysaccharide. *Lymphokine and Cytokine Research*, **13**, 1–7.

Stamler, J., Pick, R. and Katz, L.N. (1954) Inhibition of cholesterol-induced coronary atherogenesis in the egg-producing hen. *Circulation*, **10**, 251–254.

Stein, B. and Yang, M.X. (1995) Repression of the interleukin-6 promoter by estrogen receptor is mediated by NF-kB and C/EBP$\beta$. *Molecular and Cellular Biology*, **15**, 4971–4979.

Stock, J.L., Coderre, J.A., McDonald, B. and Rosenwasser, L.J. (1989) Effect of estrogen *in vitro* and *in vivo* on spontaneous interleukin-1 release by monocytes from postmenopausal women. *Journal of Clinical Endocrinology and Metabolism*, **68**, 364–368.

Sullivan, T.R., Karas, R.H., Aronovitz, M., Faller, G.T., Ziar, J.P., Smith, J.J. *et al.* (1995) Estrogen inhibits the response-to-injury in a mouse carotid artery model. *Journal of Clinical Investigation*, **96**, 2482–2488.

Vargas, R., Wroblewska, B., Rego, A., Hatch, J. and Ramwell, P.W. (1993) Oestradiol inhibits smooth cell proliferation of pig coronary artery. *British Journal of Pharmacology*, **109**, 612–617.

Wagner, J.D., Adams, M.R., Schwenke, D.C. and Clarkson, T.B. (1993) Oral contraceptive treatment decreases arterial LDL degradation in female cynomolgus monkeys. *Circulation Research*, **72**, 1300–1307.

Wagner, J.D., Clarkson, T.B., St. Clair, R.W., Schwenke, D.C., Shively, C.A. and Adams, M.R. (1991) Estrogen and progesterone replacement therapy reduces LDL accumulation in the coronary arteries of surgically postmenopausal cynomolgus monkeys. *Journal of Clinical Investigation*, **88**, 1995–2002.

Wagner, J.D., St. Clair, R.W., Schwenke, D.C., Shively, C.A., Adams, M.R. and Clarkson, T.B. (1992) Regional differences in arterial low density lipoprotein metabolism in surgically postmenopausal cynomolgus monkeys: Effects of estrogen and progesterone therapy. *Arteriosclerosis and Thrombosis*, **12**, 717–726.

Wagner, J.D., Zhang, L., Williams, J.K., Register, T.C., Ackerman, D.M., Wiita, B. *et al.* (1997) Esterified estrogens with and without methyltestosterone decrease arterial LDL metabolism in cynomolgus monkeys. *Arterioscler Thromb Vasc Biol*, **16**, 1473–1480.

Weigensberg, B.I., Lough, H., More, R.H., Katz, E., Pugash, E. and Peniston, C. (1984) Effects of estradiol on myointimal thickening from catheter injury and on organizing white mural nonocclusive thrombi. *Atherosclerosis*, **52**, 253–265.

# 14 Effects of Estrogen and Raloxifene, a Selective Estrogen Receptor Modulator, in Animal Models of Vascular Injury

**Raymond F. Kauffman** *, **William R. Bensch and James S. Bean**

*Lilly Research Laboratories, Eli Lilly and Company, Lilly Corporate Center, Indianapolis, IN 46285, USA.*
**Corresponding author, Tel.: 001 317 276 5754, Fax: 001 317 277 0892, E-mail: kauffman_raymond_f@lilly.com.*

Estrogen modulates arterial response to mechanical/balloon injury in a variety of species. In various animal models, estrogen inhibits intimal and/or medial thickening that occurs in response to endothelial denudation and medial damage. *In vitro* studies support potential mechanisms involving direct inhibitory effects of estrogen on smooth muscle cell migration and enhanced rates of reendothelialization. In addition, *in vivo* studies in rabbit and mouse have demonstrated estrogen-mediated reductions of medial smooth muscle cell proliferation following injury. In the rat model of carotid artery balloon injury, the pure estrogen antagonist ICI 182,780 produced specific inhibition of estrogen-mediated reduction of intimal thickening, providing *in vivo* evidence of a role for estrogen receptors in this response. The availability of a mouse model of vascular injury that responds to estrogen opens up new possibilities for probing mechanisms underlying *in vivo* vascular effects of estrogen through the use of transgenic and/or knock-out mice.

A novel class of estrogenic compounds, i.e., selective estrogen receptor modulators (SERMs), has recently been characterized as producing estrogen agonist effects upon bone and serum lipids together with estrogen antagonism in reproductive tissues. Vascular effects of SERMs are largely uncharacterized. Thus, effects of the SERM raloxifene HCl were characterized in the rat model of balloon injury to the carotid artery. Raloxifene HCl, like estrogen, produced significant, dose-dependent reductions in intimal thickening. Thus, the tissue selective estrogenic profile of raloxifene HCl includes an estrogen agonist-like effect upon the injured vessel wall. The mechanism responsible for this vascular effect of raloxifene is currently under investigation.

KEYWORDS: Estrogen, vascular injury, intimal thickening, raloxifene, SERM.

## INTRODUCTION

A large body of epidemiological data supports the ability of estrogen replacement therapy to protect postmenopausal women from cardiovascular disease (Barrett-Connor and Bush, 1991; Stampfer and Colditz, 1991). These findings are in good accord with the observations that prior to menopause women are relatively protected from estrogen as compared to age-matched men, while in the postmenopausal state (natural or surgical) women develop a risk for cardiovascular disease comparable to

gonadectomized rats to normal adult values by using subcutaneous implants produced substantial inhibition of injury-induced intimal thickening in both sexes. These findings support the ability of physiological levels of estrogen to produce vascular protection in settings of injury.

In a follow-up to the initial rat study, the Oparil group went on to demonstrate that the progestin, medroxyprogesterone acetate (MPA), attenuated the inhibitory effect of estrogen on intimal thickening in response to balloon injury in gonadectomized rats of both sexes (Levine *et al.*, 1996). In this study, MPA alone had no effect on intimal thickening, and the inhibitory effect of MPA on estrogen was not mediated by reduction in estradiol blood levels. These observations suggested that continuous progestin may antagonize effects of estrogen on vascular response to injury. Moreover, analogous to the findings in rats, continuous administration of MPA also antagonized the vascular protective effect of estrogen in a primate model of atherosclerosis (Adams and Golden, 1996). Thus, findings in the rat injury model may have broader implications with respect to protective effects of estrogen on the vessel wall in cardiovascular disease.

The effects of estrogen upon vascular response to injury were recently described in a mouse carotid injury model (Sullivan *et al.*, 1995). Because of limited size of the mouse carotid artery, injury and endothelial denudation in this model is produced by multiple passes of an angioplasty guide wire with rotation. Analogous to balloon injury models, the injured vessel responds with development of intimal and medial thickening accompanied by proliferation of intimal and medial smooth muscle cells. Daily subcutaneous injections of estradiol (100 μg; approx. 4 μg/kg) substantially inhibited thickening of both the medial and intimal compartments. In addition, estrogen treatment nearly abolished cellular proliferation in the medial wall, whereas it had no significant effect on proliferation in the intima. The importance of the mouse model of vascular injury for studying estrogen's effect on the vessel wall deserves further comment. The mouse model opens up the field of transgenics for studying vascular activity of estrogen in modified genetic settings. Such an approach was recently undertaken by the research group of Mendelsohn to examine the effect of estrogen on response to carotid injury in estrogen receptor knockout mice (data presented at this Symposium). Interestingly, their findings raised the possibility that vascular response to injury may not be entirely dependent on the presence of the classical estrogen receptor (ERα, according to nomenclature of Kuiper *et al.*, 1996). Future studies will undoubtedly explore other aspects of *in vivo* vascular mechanisms of estrogen in injury and atherosclerosis, including the potential role of a novel, transcriptionally active estrogen receptor (ERβ) recently described by Gustafsson (Kuiper *et al.*, 1996; see also chapter herein).

Prospective clinical trials of estrogen replacement therapy for prevention of restenosis following percutaneous coronary interventions (angioplasty and atherectomy) have not been reported to date. Nevertheless, postmenopausal women on estrogen replacement therapy were followed and compared with the corresponding cohort not taking estrogen as part of a subgroup analysis of the Coronary Angioplasty Versus Excisional Atherectomy Trial (CAVEAT) results (O'Brien *et al.*, 1996). Angiographic follow up at 6 months revealed that estrogen users had significantly less late loss in minimal lumen diameter than non-users. Interestingly, the angiographic benefit was largely restricted to those women undergoing atherectomy, with minimal effects on restenosis endpoints in the angioplasty group. In a retrospective study of clinical endpoints in postmenopausal women following coronary angioplasty, estrogen replacement therapy had no effect on revascularization rates; nevertheless, the cardiovascular event rate (death, nonfatal myocardial infarction or nonfatal stroke) was significantly lower at the 7 year follow-up period (O'Keefe *et al.*, 1997). These data point out the potential utility of estrogen in preventing chronic complications of coronary interventions in women; however, angiographic benefit may depend heavily on the particular type of intervention employed. The importance of these findings to the practice of cardiology in postmenopausal women must await confirmation of these findings in a prospective, randomized, placebo-controlled clinical trial.

## POSSIBLE MECHANISM(S) UNDERLYING ESTROGEN EFFECTS IN VASCULAR INJURY

The *in vivo* biology of estrogen as it relates to the cardiovascular system is potentially quite complex, likely involving direct vascular effects as well as a variety of indirect systems including hormones, vasoactive factors, coagulation factors, sympathetic and parasympathetic tone, etc. Although such indirect actions cannot and should not be ignored, insufficient information is currently available to assess the contribution of these actions in estrogen's effects on the vasculature. Future *in vivo* studies with appropriate pharmacological tools and in transgenic and/or knock-out mouse models will be instrumental in this regard.

Mechanistic studies of vascular estrogen effects have focused to date on the hypothesis that estrogen exerts direct effects on vascular cells that contribute to observed *in vivo* effects in settings of injury and atherosclerosis. Support for direct vascular effects of estrogen is derived from the demonstration of functional estrogen receptors in cultured vascular smooth muscle (Karas *et al.*, 1994; Bayard *et al.*, 1995) and endothelial cells (Kim-Schulze *et al.*, 1996), and in vascular tissues *ex vivo* (Knauthe *et al.*, 1996).

### Effects of Estrogen on Vascular Smooth Muscle Cells

Relevant to biological response to injury, 17$\beta$-estradiol produced potent inhibition of directed migration of female rat aortic smooth muscle cells *in vitro* (Kolodgie *et al.*, 1996). The role of estrogen receptors in this response was implicated by the ability of the pure estrogen antagonist ICI 164,384 to inhibit this action of 17$\beta$-estradiol. These findings were recently confirmed and extended to both male and female human vascular smooth muscle cells (Do *et al.*, 1996).

Estrogen has also been reported to inhibit smooth muscle cell proliferation; however, the potency and efficacy of 17$\beta$-estradiol vs. proliferation have generally been found to be low (Vargas *et al.*, 1993; Do *et al.*, 1996), in contrast to *in vitro* findings in migration studies. This observation seems at variance with the reported ability of estrogen to inhibit proliferation in *in vivo* models of vascular injury as described above (Foegh *et al.*, 1994; Sullivan *et al.*, 1995). It is however possible that cultured cells do not replicate precisely the status of vascular cells *in vivo*. Alternatively, indirect effects of estrogen may be responsible for the *in vivo* antiproliferative action, or 17β-estradiol may be metabolized *in vivo* to a more potent inhibitor of proliferation as suggested by studies of 2-methoxyestradiol upon cultured smooth muscle cells (Nishigaki *et al*., 1995). Further studies are needed to clarify which of these possible explanations is in fact correct.

## Effects of Estrogen on Vascular Endothelial Cells

Physiological concentrations of 17β-estradiol increase endothelial cell migration and proliferation *in vitro* (Morales *et al*., 1995). Such an effect, were it to occur *in vivo*, would be expected to produce a more rapid reendothelialization following injury-induced endothelial denudation. Because the endothelial cell layer produces substances that are known to limit smooth muscle proliferation (e.g., glycosaminoglycans, prostacyclin, nitric oxide, etc.), an enhanced rate of reendothelialization would be expected to limit the vascular response to injury. Such an inhibitory effect of reendothelialization on intimal thickening has been demonstrated by local delivery of vascular endothelial growth factor to injured arteries (Asahara *et al*., 1995).

Estrogen appears to modulate the "activation state" of endothelial cells (i.e., the balance between pro- vs. antiinflammatory, or pro- vs. anticoagulant properties of the endothelium). Thus, in addition to producing stimulatory effects upon endothelial cell migration and proliferation, estrogen also attenuates interleukin-1β-induced endothelial cell adhesion molecule expression (Nakai *et al*., 1994; Caulin-Glaser *et al*., 1996) and increases nitric oxide availability *in vitro*. Effects of estrogen on cellular adhesion molecules appear to be complex, because in other studies estrogen increased endothelial expression of leukocyte adhesion molecules in response to TNF-α (Cid *et al*., 1994). Concerning the effect on nitric oxide, estrogen appears to be able to increase endothelial nitric oxide synthase (Hishikawa *et al*., 1995; Hayashi *et al*., 1995), or inhibit degradation of nitric oxide by decreasing superoxide anion production (Arnal *et al*., 1996), although the contribution of these two effects seems to vary in particular *in vitro* systems. Antiestrogens inhibited both these effects of 17β-estradiol on nitric oxide pathways, suggesting a role for estrogen receptors in effects on NO availability. The precise role of modulating endothelial cell "activation state" in estrogen's effects upon response to injury is at present unclear. Additional insights into potential mechanism(s) of estrogen effects on the vessel wall in the setting of injury are available elsewhere in this monograph in chapters describing more recent studies of *in vitro* and *in vivo* actions of estrogen on vasculature.

## Role of Estrogen Receptors

The role of estrogen receptors in many of the above described effects of estrogen on vascular cells has been implicated through the ability of either partial (e.g., tamoxifen) or pure (e.g., ICI 182,780) antiestrogens to inhibit these responses. Although such findings are generally taken to indicate a role for the classical estrogen receptor, it must be kept in mind that a novel, transcriptionally active estrogen receptor (ER$\beta$) has recently been identified (Kuiper *et al.*, 1996). Thus, the possibility exists that other estrogen receptors may in fact contribute to the vascular biology of 17$\beta$-estradiol. Relevant to this discussion, the research group of Levin recently presented evidence that antiproliferative effects of 17$\beta$-estradiol upon vascular smooth muscle cells are mediated by plasma membrane receptors rather than the classical cytoplasmic estrogen receptor (Morey *et al.*, 1996). Thus, rapid, non-genomic actions may also contribute to the effects of estrogen in vascular injury (for a review, see Farhat *et al.*, 1996).

In addition to studying the role of estrogen receptors in *in vitro* systems, it is essential to address the role of estrogen receptors in *in vivo* models of vascular injury. Toward this end, we evaluated effects of the pure estrogen antagonist, ICI 182,780 (Wakeling *et al.*, 1991), on the actions of estrogen in the rat carotid balloon injury model. The doses of ICI 182,780 were chosen on the basis of previous studies of antagonism of estrogen responses in non-reproductive tissues in rats (Wade *et al.*, 1993). Following balloon injury, ovariectomized female rats were administered daily doses of 17$\alpha$-ethinylestradiol (0.1 μg/kg/d, p.o.) in the presence or absence of ICI 182,780 (0.3 or 4.0 μg/kg/d, s.c.) for two weeks, at which time animals were sacrificed and carotid intimal thickening was evaluated by quantitative histomorphometry. 17$\alpha$-ethinylestradiol reduced intimal thickening by approximately 50%, and ICI 182,780 produced significant, dose-dependent inhibition of estrogen's effect, with maximal inhibition of approximately 70% of the effect of estrogen at the dose of 4 μg/kg/d (Kauffman and Bean, unpublished data). The lack of effect of ICI 182,780 alone on intimal thickening was consistent with specific antagonism of estrogen by ICI 182,780 in this model.

Thus the effects of estrogen upon vascular response to injury appear to be mediated by estrogen receptors. Less clear, however, is the specific estrogen receptor(s) involved. Because ICI 182,780 is a structural analog of 17$\beta$-estradiol, it remains possible that this antiestrogen could be non-selective with respect to estrogen receptors other than the classical ER. Furthermore, the *in vivo* data do not clarify the location of receptors mediating estrogen effects in the intact animal. Therefore, the question of whether or not estrogen acts directly on vascular cells to modulate vascular response to injury remains unanswered. Clearly, additional *in vivo* studies are required, particularly in transgenic and knock-out mouse models, to provide answers to these important mechanistic questions.

## RELEVANCE OF ESTROGEN EFFECTS IN VASCULAR INJURY MODELS

On one level, the effects of estrogen in mechanical/balloon injury models may provide information concerning potential utility of estrogen in preventing chronic complications following percutaneous coronary interventions. The finding that estrogen use by postmenopausal women is associated with reduced angiographic restenosis endpoints following coronary atherectomy (O'Brien *et al.*, 1996) lends support to this interpretation of the injury model data, despite the well known limitations of preclinical models of restenosis.

Upon further consideration, however, studies of estrogen in vascular injury models may have broader implications relevant to protective effects in postmenopausal against cardiovascular disease. This hypothesis is based on the observation that many of the actions of vascular cells that are modulated by estrogen are generally considered to play a role in both vascular injury and atherosclerosis. Included in the list of such cellular actions are smooth muscle cell migration and proliferation, and endothelial cell migration, proliferation/reendothelialization, and nitric oxide synthesis/availability. Consequently, studies of estrogen in vascular injury models may provide insight into estrogen's ability to protect against cardiovascular disease and events in women. Because injury models are relatively well-characterized and generally of shorter duration than traditional atherosclerosis models, studies in the former may provide a facile model for developing a better understanding of vascular estrogen effects and mechanisms underlying these vascular actions.

## SELECTIVE ESTROGEN RECEPTOR MODULATORS IN VASCULAR INJURY

Although a large body of data supports protective effects of estrogen replacement therapy on bone and the cardiovascular system, these effects are accompanied by real and/or perceived liabilities, the most serious of which is cancer in breast and uterine reproductive tissues (reviewed in Kauffman and Bryant, 1995). Consequently, prescription of and compliance with estrogen replacement therapy has been poor (Hammond, 1994). Selective estrogen receptor modulators (SERMs) represent a novel class of estrogenic compound that produces one or more beneficial estrogenic effects on target tissues such as bone, liver (serum lipids), etc., while having minimal or no agonist effect on reproductive tissues such as breast and uterus (Kauffman and Bryant, 1995). Such a profile may offer advantages over hormonal replacement therapy with respect to safety and compliance in prevention and treatment of postmenopausal complications in women.

The benzothiophene raloxifene, originally discovered as an estrogen antagonist for treatment of breast cancer (Black *et al*., 1983), was subsequently shown to possess the *in vivo* profile of a SERM in ovariectomized rats (Black *et al*., 1994). Furthermore, cholesterol lowering by raloxifene HCl in ovariectomized rats was recently shown by pharmacological characterization to be mediated by estrogen receptors (Kauffman *et al*., 1997), thereby justifying classification of this compound as a SERM. Details on the molecular basis for estrogenic actions of raloxifene have recently begun to appear in the literature (Yang *et al*., 1996). Available clinical data, although limited, suggest that raloxifene HCl acts as a SERM in postmenopausal women with a similar profile of tissue selectivity as observed in preclinical studies (Draper *et al*., 1996).

Based on classification of raloxifene HCl as a SERM, it was of interest to determine whether this compound produced estrogen agonist-like effects on the vessel wall in the setting of vascular injury. Thus, we evaluated effects of raloxifene HCl on neointima formation in the balloon-injured rat carotid artery model. Raloxifene HCl (0.003–3 μg/kg) or vehicle were administered to male rats once daily by oral gavage from one day prior to, until two weeks after, balloon surgery. Subsequently arteries were perfusion-fixed and cross-sectional intimal areas were determined by quantitative histomorphometry. Raloxifene HCl produced dose-dependent reductions in intimal thickening with half maximal and maximal (60%) reductions effects occurring at 0.03 μg/kg/d and 3 μg/kg/d, respectively (Kauffman *et al.*, 1995). In separate studies, 17α-ethinylestradiol produced a comparable maximal reduction in intimal thickening in this model (approximately 50%). Therefore, raloxifene HCl was characterized as producing an estrogen agonist-like response in the balloon injured rat carotid artery. Reduction of intimal thickening was also observed with local adventitial delivery of raloxifene HCl to the injured carotid artery using osmotic minipumps (Kauffman *et al.*, 1995), suggesting a role for direct vascular effects of this compound. To the authors' knowledge, this represents the first demonstration that the tissue selective profile of a SERM includes an estrogen agonist-like effect upon the injured vessel wall. The results suggest that raloxifene produces beneficial effects upon the vessel wall that are mediated by estrogen receptors. However, further *in vitro* and *in vivo* studies are required to substantiate this point. The relevance of these preclinical findings to potential cardiovascular benefit of raloxifene in postmenopausal women is also under investigation.

In summary, estrogen modulates vascular response to mechanical/balloon injury in a variety of preclinical models. Studies in cultured vascular cells support a mechanism involving direct effects of estrogen on both endothelial and smooth muscle cells. Based on selective *in vivo* antagonism by the pure estrogen antagonist ICI 182,780 in the rat carotid balloon injury model, effects of estrogen upon vascular injury appear to be mediated by estrogen receptors. However, the precise identity and location of these receptors remains to be determined *in vivo*. The selective estrogen receptor modulator, raloxifene HCl, produced significant, dose-dependent reductions in intimal thickening in the rat model of balloon injury to the carotid artery. Thus, the tissue selective estrogenic profile of raloxifene HCl includes an estrogen agonist-like effect upon the injured vessel wall. These observations raise the possibility that raloxifene HCl, like estrogen, may produce beneficial vascular effects in addition to lowering serum lipids in postmenopausal women.

## REFERENCES

Adams, M.R. and Golden, D.L. (1995) Atheroprotective effects of estrogen replacement therapy are antagonized by medroxyprogesterone acetate in monkeys. *Circulation* **92** (suppl. I), I-8. Abstract.

Arnal, J.F., Clamens, S., Pechet, C., Negre-Salvayre, A., Allera, C., Girolami, J.-P., Salvayre, R. and Bayard, F. (1996) Ethinylestradiol does not enhance the expression of nitric oxide synthase in bovine endothelial cells but increases the release of bioactive nitric oxide by inhibiting superoxide anion production. *Proceedings of the National Academy of Science U.S.A.* **93**, 4108–4113.

Asahara, T., Bauters, C., Pastore, C., Kearney, M., Rossow, S., Bunting, S., Ferrara, N., Symes, J.F. and Isner, J.M. (1995) Local delivery of vascular endothelial growth factor accelerates reendothelialization and attenuates intimal hyperplasia in balloon-injured rat carotid artery. *Circulation* **91**, 2793–2801.

Barrett-Connor, E. and Bush, T.L. (1991) Estrogen and coronary heart disease in women. *Journal of the American Medical Association* **265**, 1861–1867.

Bayard, F., Clamens, S., Megetto, F., Blaes, N., Delsol, G. and Faye, J.-C. (1995) Estrogen synthesis, estrogen metabolism, and functional estrogen receptors in rat arterial smooth muscle cells in culture. *Endocrinology* **136**, 1523–1529.

Black, L.J., Jones, C.D. and Falcone, J.F. (1983) Antagonism of estrogen action with a new benzothiophene derived antiestrogen. *Life Sciences* **32**, 1031–1036.

Black, L.J., Sato, M., Rowley, E.R., Magee, D.E., Bekele, A., Williams, D.C., Cullinan, G.J., Bendele, R., Kauffman, R.F., Bensch, W.R., Frolik, C.A., Termine, J.D. and Bryant, H.U. (1994) Raloxifene (LY139481 HCl) prevents bone loss and reduces serum cholesterol without causing uterine hypertrophy in ovariectomized rats. *Journal of Clinical Investigation* **93**, 63–69.

Bochaton-Piallat, M.-L., Gabbiani, F., Redard, M., Desmouliere, A. and Gabbiani, G. (1995) Apoptosis participates in cellularity regulation during rat aortic intimal thickening. *American Journal of Pathology* **146**, 1059–1064.

Caulin-Glaser, T., Watson, C.A., Pardi, R. and Bender, J.R. (1996) Effects of 17$\beta$-estradiol on cytokine-induced endothelial cell adhesion molecule expression. *Journal of Clinical Investigation* **98**, 36–42.

Chen, S.-J., Li, H., Durand, J., Oparil, S. and Chen, Y.-F. (1996) Estrogen reduces myointimal proliferation after balloon injury of rat carotid artery. *Circulation* **93**, 577–584.

Cid, M.C., Kleinman, H.K., Grant, D.S., Schnaper, H.W., Fauci, A.S. and Hoffman, G.S. (1994) Estradiol enhances leukocyte binding to tumor necrosis factor (TNF)-stimulated endothelial cells via an increase in TNF-induced adhesion molecules E-selectin, intercellular adhesion molecule type I and vascular cell adhesion molecule type I. *Journal of Clinical Investigation* **93**, 17–25.

Clowes, A.W., Reidy, M.A. and Clowes, M.M. (1983) Kinetics of cellular proliferation after arterial injury: smooth muscle cell growth in the absence of endothelium. *Laboratory Investigation* **49**, 327–333.

Do, D.D., Espinosa. E., Liu, G.Z., Rabelink, T.J., Julmy, F., Yang, Z.H., Mahler, F. and Luscher, T.F. (1996) 17$\beta$-estradiol inhibits proliferation and migration of human vascular smooth muscle cells: similar effects in cells from postmenopausal females and in males. *Cardiovascular Research* **32**, 980–985.

Draper, M.W., Flowers, D.E., Huster, W.J., Neild, J.A., Harper, K.D. and Arnaud, C. (1996) A controlled clinical trial of raloxifene (LY139481) HCl: impact on bone turnover and serum lipid profile in healthy postmenopausal women. *Journal of Bone and Mineral Research* **11**, 835–842.

Espinosa, E., Oemar, B.S. and Luscher, T.F. (1996) 17$\beta$-estradiol and smooth muscle cell proliferation in aortic cells of male and female rats. *Biochemical and Biophysical Research Communications* **221**, 8–14.

Farhat, M.Y., Abi-Younes, S. and Ramwell, P.W. (1996) Non-genomic effects of estrogen and the vessel wall. *Biochemical Pharmacology* **51**, 571–576.

Foegh, M.L., Asotra, S., Howell, M.H. and Ramwell, P.W. (1994) Estradiol inhibition of arterial neointimal hyperplasia after balloon injury. *Journal of Vascular Surgery* **19**, 722–726.

Foegh, M.L., Zhao, Y., Farhat, M. and Ramwell, P.W. (1995) Oestradiol inhibition of vascular myointimal proliferation following immune, chemical and mechanical injury. In *Non-reproductive actions of sex steroids. Ciba Foundation Symposium 191*, pp. 139–149. Chichester: Wiley.

Guzman, L.A., Mick, M.J., Arnold, A.M., Forudi, F. and Whitlow, P.L. (1996) Role of intimal hyperplasia and arterial remodeling after balloon angioplasty. An experimental study in the atherosclerotic rabbit model. *Arteriosclerosis, Thrombosis, and Vascular Biology* **16**, 479–487.

Hammond, C.B. (1994) Women's concerns with hormone replacement therapy—compliance issues. *Fertility and Sterility* **62**, 157S–160S.

Hayashi, T., Yamada, K., Esaki, T., Kuzuya, M., Satake, S., Ishikawa, T., Hidaka, H. and Iguchi, A. (1995) Estrogen increases endothelial nitric oxide by a receptor-mediated system. *Biochemical Biophysical Research Communications* **214**, 847–855.

Hishikawa, K., Nakaki, T., Marumo, T., Suzuki, H., Kato, R. and Saruta, T. (1995) Up-regulation of nitric oxide synthesis by estradiol in human aortic endothelial cells. *FEBS Letters* **360**, 291–293.

Karas, R.H., Patterson, B.L. and Mendelsohn, M.E. (1994) Human vascular smooth muscle cells contain functional estrogen receptor. *Circulation* **89**, 1943–1950.

Kauffman, R.F., Bean, J.S. and Bensch, W.R. (1995) Comparative effects of estrogen and raloxifene, a selective estrogen receptor modulator (SERM), upon vascular response to balloon injury. Scientific Conference on Hormonal, Metabolic and Cellular Influences on Cardiovascular Disease in Women San Diego. Abstract.

Kauffman, R.F. and Bryant, H.U. (1995) Selective estrogen receptor modulators. *Drug News & Perspectives* **8**, 531–539.

Kauffman, R.F., Bensch, W.R., Roudebush, R.E., Cole, H.W., Bean, J.S., Phillips, D.L., Monroe, A., Cullinan, G.J., Glasebrook, A.L. and Bryant, H.U. (1997) Hypocholesterolemic activity of raloxifene (LY139481): pharmacological characterization as a selective estrogen receptor modulator. *Journal of Pharmacology and Experimental Therapeutics* **280**, 146–153.

Kim-Schulze, S., McGowan, K.A., Hubchak, S.C., Cid, M.C., Martin, M.B., Kleinman, H.K., Greene, G.L. and Schnaper, H.W. (1996) Expression of an estrogen receptor by human coronary artery and umbilical vein endothelial cells. *Circulation* **94**, 1402–1407.

Knauthe, R., Diel, P., Haegele-Hartung, C., Engelhaupt, A. and Fritzmeier, K.-H. (1996) Sexual dimorphism of steroid hormone receptor messenger ribonucleic acid expression and hormonal regulation in rat vascular tissue. *Endocrinology* **137**, 3220–3227.

Kolodgie, F.D., Jacob, A., Wilson, P.S., Carlson, G.C., Farb, A., Verma, A. and Virmani, R. (1996) Estradiol attenuates directed migration of vascular smooth muscle cells *in vitro*. *American Journal of Pathology* **148**, 969–967.

Kuiper, G.G.J.M., Enmark, E., Pelto-Huikko, M., Nilsson, S. and Gustafsson, J.-A. (1996) Cloning of a novel estrogen receptor expressed in rat prostate and ovary. *Proceedings of the National Academy of Science. U.S.A.* **93**, 5926–5930.

Levine, R.L., Chen, S.-J., Durand, J., Chen, Y.-F. and Oparil, S. (1996) Medroxyprogesterone attenuates estrogen-mediated inhibition of neointima formation after balloon injury of the rat carotid artery. *Circulation* **94**, 2221–2227.

McGill, H.C. and Sheridan, P.J. (1981) Nuclear uptake of sex steroid hormones in the cardiovascular system of the baboon. *Circulation Research* **48**, 238–244.

Morales, D.E., McGowan, K.A., Grant, D.S., Maheshwari, S., Bhartiya, D., Cid, M.C., Kleinman, H.K. and Schnaper, H.W. (1995) Estrogen promotes angiogenic activity in human umbilical vein endothelial cells *in vitro* and in a murine model. *Circulation* **91**, 755–763.

Morey A.K., Pedram, A., Prins, B.A., Hu, R.-M., Biesiada, E. and Levin, E.R. (1996) Estrogen and progesterone inhibit vascular smooth muscle proliferation. 10th International Congress of Endocrinology, San Francisco. Abstract.

Nakai, K., Itoh, C., Hotta, K., Itoh, T., Yoshizumi, M. and Hiramori, K. (1994) Estradiol-17$\beta$ regulates the induction of VCAM-1 mRNA expression by interleukin-1$\beta$ in human umbilical vein endothelial cells. *Life Sciences* **54**, PL221–PL227.

Nishigaki, I., Sasaguri, Y. and Yagi, K. (1995) Anti-proliferative effect of 2-methoxyestradiol on cultured smooth muscle cells from rabbit aorta. *Atherosclerosis* **113**, 167–170.

O'Brien, J.E., Peterson, E.D., Keeler, G.P., Berdan, L.G., Ohman, E.M., Faxon, D.P., Jacobs A.K., Topol, E.J. and Califf, R.M. (1996) Relation between estrogen replacement therapy and restenosis after percutaneous coronary interventions. *Journal of the American College of Cardiology* **28**, 1111–1118.

O'Keefe, J.H., Kim, S.C., Hall, R.R., Cochran, V.C., Lawhorn, S.L. and McCallister, B.D. (1997) Estrogen replacement therapy after coronary angioplasty in women. *Journal of the American College of Cardiology*. **29**, 1–5.

Rasmussen, L.M., Wolf, Y.G. and Ruoslahti, E.K. (1995) Vascular smooth muscle cells from injured rat aortas display elevated matrix production associated with transforming growth factor-$\beta$ activity. *American Journal of Pathology* **147**, 1041–1048.

Rhee, C.Y., Spaet, T.H., Stemerman, M.B., Lajam, F., Shiang, H.H., Caruso, E. and Litwak, R.S. (1977) Estrogen suppression of surgically induced vascular intimal hyperplasia in rabbits. *Journal of Laboratory and Clinical Medicine* **90**, 77–84.

Stampfer, M.J. and Colditz, G.A. (1991) Estrogen replacement therapy and coronary disease: a quantitative assessment of the epidemiological evidence. *Preventive Medicine* **20**, 47–63.

Sullivan, T.R., Karas, R.H., Aronovitz, A., Faller, G.T., Ziar, J.P., Smith, J.J., O'Donnell, T.F., Jr. and Mendelsohn, M.E. (1995) Estrogen inhibits the response-to-injury in a mouse carotid injury model. *Journal of Clinical Investigation* **96**, 2482–2488.

Vargas, R., Wroblewska, B., Rego, A., Hatch, J. and Ramwell, P.W. (1993) Oestradiol inhibits smooth muscle cell proliferation of pig coronary artery. *British Journal of Pharmacology* **109**, 612–617.

Vargas, R., Hewes, B., Rego, A., Farhat, M.Y., Suarez, R. and Ramwell, P.W. (1996) Estradiol effect on proliferation of rat carotid segments: effect of gender and tamoxifen. *Journal of Cardiovascular Pharmacology* **27**, 495–499.

Wade, G.N., Blaustein, J.D., Gray, J.M. and Meredith, J.M. (1993) ICI 182,780: a pure antiestrogen that effects behaviors and energy balance in rats without acting in the brain. *American Journal of Physiology* **265**, R1392–R1398.

Wakeling A.E., Dukes, M. and Bowler, J. (1991) A potent specific pure antiestrogen with clinical potential. *Cancer Research* **51**, 3867–3873.

Yang, N.N., Venugopalan, M., Hardikar, S. and Glasebrook, A. (1996) Identification of an estrogen response element activated by metabolites of 17b-estradiol and raloxifene. *Science* **273**, 1222–1225.

# 15 Estradiol and Accelerated Transplant Arteriosclerosis

**Marie L. Foegh* and Peter W. Ramwell**

*Department of Surgery and Department of Physiology and Biophysics, Georgetown University Medical Center, Washington, DC 20007, USA.*
** Corresponding author, Tel.: 001 202 687 6044, Fax: 001 202 784 3950, 202 687 6573.*

Transplant arteriosclerosis is the cardinal complication limiting long term organ graft survival. The etiology is most likely the alloimmune response that is preceded by the ischemia and reperfusion injury. We find that both physiological and pharmacological doses of estradiol substantially attenuate transplant arteriosclerosis. The mechanisms are associated with supression of upregulation of expression of the major histocompatibility complex (MHC) class II, insulin-like growth factor-I (IGF-I) and interleukin-6 (IL-6). In addition macrophage infiltration of the allograft vascular wall is abrogated in the estrogen treated recipients. Thus a vascular specific estrogen may prevent transplant arteriosclerosis, the major cause of loss of graft function.

## INTRODUCTION

The incidence of cardiovascular disease in young women is low compared to their cohorts (Nabulsi *et al.*, 1993; Manolio *et al.*, 1993; Kafonek, 1994). However, post menopausal women receiving estrogen replacement therapy have a 50% lower incidence of cardiovascular disease when compared to untreated post menopausal women. *In vitro*, estradiol also inhibits proliferation of primary smooth muscle cells and explants of porcine coronary arteries (Vargas *et al.*, 1993). *In vivo* estradiol treatment protects against experimental transplant arteriosclerosis (Foegh *et al.*, 1987; Jacobsson *et al.*, 1992; Cheng *et al.*, 1991; Lou *et al.*, 1996b; Saito *et al.*, in press) and myointimal hyperplasia following balloon injury and inhibits DNA synthesis (Foegh *et al.*, 1993). Female gender of the rat aorta isograft protects against myointimal hyperplasia when implanted in to recipients of either sex (Foegh *et al.*, 1995). Collectively these data suggest that estradiol treatment preserves vascular integrity against immunological injury.

Transplant arteriosclerosis is the cardinal complication limiting long term graft survival (Sarris *et al.*, 1994; Foegh, 1992). Detection is often difficult. Coronary angiography, routinely used for diagnosis of segmental coronary artery atherosclerotic lesions, is not sufficiently sensitive for detecting the diffuse narrowing of the epicardial arteries and intramyocardial branches that is characteristic of transplant arteriosclerosis. Currently, intracoronary ultrasound is a possible alternative but it can only evaluate the larger coronary arteries (Haase *et al.*, 1995; Miller, 1995). Angina is not a symptom of ischemia since the transplant is denervated. Thus the disease often presents as sudden death due to arrhythmia or silent myocardial infarction (Uretsky *et al.*, 1992; Eich *et al.*, 1990). At autopsy, the histology of the graft lesions show that the internal elastic lamina remains largely intact with only a few breaks in its continuity. The intima consists largely of smooth muscle cells with associated monocytes and macrophages; only rarely are lymphocytes seen in the vessel wall (Ip *et al.*, 1990; Kuwahara *et al.*, 1991; Billingham, 1994). Endothelium lines the vessel lumen. Blood lipoproteins, known risk factors for regular atherosclerosis do not relate to the extra ordinary rapid development of transplant arteriosclerosis although hyperlipidemia and other risk factors may interact with the etiology factors for transplant arteriosclerosis to modify the severity of the disease.

Several features of transplant arteriosclerosis suggest that immune activation is central in its development (Russell *et al.*, 1994a,b; Libby *et al.*, 1992; Kuo and Monaco, 1993). Transplant arteriosclerosis is seen in all transplanted organs, independent of age, and can be reproduced by transplanting across immunologically disparate animal strains (Russell *et al.*, 1994a). Further, the tubular nature of transplant arteriosclerosis with smooth muscle cell proliferation that occurs diffusely in the intima throughout the cardiac allograft and limited to the arteries of the graft speaks to a specific immunologic host versus graft reaction (rejection) that occurs in all recipients of organ transplants. No treatment is available.

## THE IMMUNE RESPONSE IN ALLOGRAFT REJECTION

The rejection process occurring in organ transplant recipients is a host versus graft response. It is preceded by a non specific inflammatory response to injury followed by the more specific allo-immune response. The latter depends on clonal expansion of major histocompatibility (MHC) restricted T-lymphocytes in response to foreign antigen (Hosenpud *et al.*, 1992). An important function of the activated T-cells is to elicit a cascade of cytokines. The triggering of $T_{CD4+}$ lymphocytes is a crucial early step. These T-cells are much more efficient than $T_{CD8+}$ lymphocytes in producing important cytokines such as interferon-$\gamma$ (IFN-$\gamma$). The latter is the most effective regulator of both constitutive and inducible MHC class II expression (Pober *et al.*, 1986). The level of expression of MHC class II molecules is a key parameter in T cell activation. Thus MHC class II gene regulation has an essential role in the control of the immune response in allograft rejection.

MHC class II genes are a family of related $\alpha$- and $\beta$-chain molecules (Ie and Ia in mice; HLA-DR, -DQ and -DP in humans). They are expressed in a rigorously controlled fashion. In contrast to MHC class I molecules, the constitutive expression of class II molecules is strictly limited to macrophages, dendritic cells and B-lymphocytes. However, in many MHC class II negative cells and tissues, class II expression is inducible by a variety of stimuli, particularly IFN-$\gamma$ (Amaldi *et al.*, 1989; Steimle *et al.*, 1994). Excessive or ectopic expression of class II molecules is implicated in autoimmune diseases that are characterized by aberrant activation of $T_{CD4+}$ cells.

MHC class II gene expression is controlled primarily at the transcriptional level. The promoter region functions as a single unit in which spacing between the individual cisacting elements is crucial. All the promoters contain conserved cis-acting sequence motifs referred to as the W, X, X2 and Y boxes. Studies of patients with a rare inherited disease (Bare Lymphocyte Syndrome) (Reith *et al*., 1995) have elucidated the molecular defects that lead to lack of class II gene expression. Two crucial regulatory factors were isolated, namely the MHC class II gene specific transcription activator, CIITA and the X-box-binding nuclear protein RFX (Glimcher and Kara, 1992; Emery *et al*., 1996).

CIITA is a 1130 amino acid protein. Its sequence has no homology to other known proteins and contains no obvious structural motifs other than a consensus sequence for an ATP/GTP-binding casette and stretches that are rich in acidic amino acids and in Pro, Ser and Thr. CIITA is differentially displayed like class II. CIITA expression is only detected in MHC class II+ tissues and cells (Glimcher and Kara, 1992). It is present when MHC class II expression is being induced. Thus this pivotal role of the constitutive and inducible CIITA and of the constitutive RFX in expression of class II make them likely molecular targets for the action of estradiol in abrogating MHC class II expression in cardiac allografts.

## ESTROGEN AND THE IMMUNE RESPONSE

Women appear to exhibit enhanced humoral cell mediated immune responses when compared to men. Women have a higher incidence of auto-immune diseases and estradiol is thought to enhance many of these diseases, with the exception of rheumatoid arthritis where estrogen therapy suppresses symptoms (Olsen and Kovacs, 1996; Valentino *et al*., 1993). Aberrant expression of class II molecules is thought to trigger or exacerbate autoimmune disease. This seems to contrast with our finding that estradiol-treatment caused complete repression of class II in the allograft (Lou *et al*., 1996b; Saito *et al*., 1993). It is possible that the aberrant increase in class II expression is due to failure of estradiol to control class II or a change in estradiol metabolism leading to stimulatory metabolites as has been suggested. Premenstrual (lowest estradiol concentration) worsening of asthma is well known. Interestingly, estradiol treatment prevents this. Likewise the estradiol treated recipients in our experimental transplant models showed either the same (Foegh *et al*., 1987; Lou *et al*., 1996b; Saito *et al*., 1997) or a lesser rejection grading score compared to placebo treated, suggesting that the apparent immunostimulatory effects of estrogen are of no consequence for rejection. This may apply to women transplant patients who receive hormone replacement therapy.

## ESTRADIOL AND MHC CLASS II

Recently we discovered that estradiol inhibited vascular expression of MHC class II and this was associated with inhibition of coronary artery transplant arteriosclerosis (Lou *et al*., 1996b). This evidence of reduced graft allogenicity by abolition of class II expression in all the estradiol treated recipients when compared to the placebo treated group suggests it is a potentially important and novel estradiol mechanism. This finding may have implications for autoimmune cell mediated diseases that are know to be improved by estradiol treatment e.g. rheumatoid arthritis. Others, like systemic lupus, a humoral mediated autoimmune disease, respond less clearly to estradiol. Clinically it still remains controversial whether estradiol exacerbates or attenuates systemic lupus.

Cytokines as well as some growth factors also modulate MHC class II expression (Hosenpud *et al*., 1992; Pober *et al*., 1986; Murray *et al*., 1995). For example tumor necrosis factor-α (TNFα) may be an important regulator of the induction of autoimmune heart disease by upregulating MHC class II molecules in the cardiac interstitial cells. Introduction of a TNF receptor null mutation into a mouse strain susceptible to cardiac myosin-induced autoimmune myocarditis failed to express MHC class II in the cardiac interstitial cells and did not show autoimmune myocarditis in spite of production of autoantibody (Bachmaier *et al*., 1997) causing myocarditis. Interestingly, estradiol replacement therapy lowers serum TNFα levels and both unstimulated and stimulated release of TNFα by monocytes (Aune *et al*., 1995). This effect of estradiol was at the level of TNFα mRNA expression (Shanker *et al*., 1994).

## CORTICOSTEROIDS AND MHC CLASS II

It is useful to compare another class of steroids with estradiol. Corticosteroids are the classical treatment of acute rejection in organ transplant recipients. They prevent upregulation of class II expression (Sydner and Unanue, 1982). Corticosteroids bind to the promoter region of the I-κB gene increasing transcription. This results in increased I-κB which prevents translocation of NF-κB from the cytoplasm to the nucleus and prevents upregulation of class II expression. The glucocorticosteroid receptor (GR) is a member of the same superfamily of ligand binding nuclear transcription factors to which the estrogen receptor (ER) belongs. The ER and its chaperone protein (heat shock p90) is located in the nucleus, whereas the GR is located in the cytoplasm and translocated to the nucleus following ligand binding. However, in contrast to corticosteroid, estradiol has little or no effect on I-κB but interferes with the function of the transcription factors NF-κB and C/EBPb (Stein and Yang, 1995). It is possible the liganded ER may reduce the functional competence of CIITA for transactivating MHC class II promoters or interferes with the X-box-binding of RFX.

## ESTRADIOL AND IFN-γ

The release of IFN-γ is a key event in the alloimmune response since treatment with IFN-γ monoclonal antibody prolongs the survival of class II disparate skin grafts and attenuates substantially transplant arteriosclerosis in mouse cardiac allografts (Russell *et al*., 1994c). It is the most widely implicated factor for enhancing both constitutive and inducible expression of MHC class II in all cells including the non-professional antigen presenting cells like smooth muscle cells and endothelial cells. In native vessels the up-regulation of class I and II antigen expression on endothelial cells and smooth muscle cells is linked to a decreased proliferative response of smooth muscle cells to injury (Hansson *et al.*, 1988; Warner *et al.*, 1989). However in the context of transplantation the secretion of IFN-$\gamma$ is counterproductive in that it increases the expression and recognition of donor vascular antigens.

The signal transduction pathway for IFN-$\gamma$ induction and upregulation of MHC class II expression involves activation the JAK-STAT pathway. This pathway is implicated in the signal transduction of a number of other ligands too including IL-2, IL-4 and IL-6 and some ligands using serpentine receptors coupled to heterotrimeric G proteins. However, CIITA like all IFN-$\gamma$ inducible genes cannot be induced in cells deficient in the Janus-family kinase JAK1 or in the signal transducer and activator of transcription 1 (STAT1) (Glimcher and Kara, 1992). Estradiol is believed to regulate IFN-$\gamma$ gene expression through the promoter region of this gene (Fox *et al.*, 1991). Thus estradiol treatment of the cardiac allograft may modify MHC class-II expression by the JAK-STAT signaling pathway.

## OTHER CYTOKINES RELATED TO TRANSPLANT REJECTION

Currently the endothelium is regarded as an important target for estradiol. Endothelial and smooth muscle cells can express MHC class II antigens and express adhesion molecules and may trigger T-cell activation resulting in clonal expansion of cytolytic T-cells. This process results in the release of interleukins from endothelial cells and other antigen presenting cells such as monocytes/macrophages, B lymphocytes and dendritic cells. The activated T-cells release TNF$\alpha$ which in turn augments IL-1 release from the endothelium.

Induction of expression of IL-6 is seen in renal allograft rejection (Vandenbroecke *et al.*, 1991). In experimental cardiac allograft rejection in monkeys both IL-6 and IL-8 mRNA are present prior to rejection. The investigators also found that IL-1, IL-2, TNF$\beta$ and IFN-$\gamma$ were detectable at later times following transplantation (Wu *et al.*, 1992). Interleukin-9, released from T-lymphocytes, is a growth factor for those same cells and possibly other lymphocytes. Interleukin-10 produced by both lymphocytes and macrophages powerfully suppresses many aspects of the cellular immune response by inhibiting monocyte production of cytokines, growth factors and MHC class II antigen (Spits and de Wall Malefyt, 1992). Interleukin-12 is a B-lymphocyte product that induces IFN-$\gamma$ production from T-lymphocytes and acts synergistically with IL-2 to induce lymphokine activated killer cells (LAK cells). It is comitogenic with IL-2 for T-lymphocytes (Scott, 1993).

## ESTRADIOL AND IL-6 MITOGENIC EFFECTS

Some cytokines related to transplant rejection like IL-6 are mitogenic for smooth muscle cells (Porreca *et al.*, 1993) and are associated with induction of expression of one of the early response genes, c-myc (Loppnow and Libby, 1990; Nabata *et al.*,

1990). We find IL-6 is more mitogenic for the allograft vascular explant than for the native vascular explant (Lou *et al.*, 1996a). Estradiol via the ER down-regulates IL-6 transcription. The IL-6 promoter has no estrogen response element (ERE). It seems that the binding sites for NF-$\kappa$B and C/EBPb in the IL-6 promoter mediate repression by estradiol (Stein and Yang, 1995). We find that treatment of allograft recipients with estradiol abolishes the mitogenic effect of IL-6 in the allograft vessel and, in the native vessel estradiol attenuated the IL-6 response. This inhibitory effect is associated with a significant decrease of MHC II antigens and macrophages in coronary arteries of estradiol treated rabbits and a significant decline of myointimal hyperplasia. Thus IL-6 has an important role in myointimal proliferation in transplant arteriosclerosis in that estradiol inhibition of transplant arteriosclerosis may relate to abrogation of the mitogenic effect of IL-6 and other immunoactive cytokines on allograft vessels.

## GROWTH FACTORS—INSULIN-LIKE GROWTH FACTOR-I

The relation of estradiol to growth factors is important since they are expressed by smooth muscle cells during rejection. They include IGF-I and also platelet derived growth factor (PDGF), epidermal growth factor (EGF), basic fibroblast growth factor (bFGF) and transforming growth factor-$\beta$ (TGF-$\beta$) (Banskota *et al.*, 1986; Hongo *et al.*, 1992; Wilson and Lloyd, 1991; Kaufman *et al.*, 1992; Shaddy *et al.*, 1992; Kourembanas and Faller, 1989). The latter suppresses IFN-$\gamma$ upregulation of MHC class II and exerts immunosuppressant activity and is one of the few growth factors that is anti-proliferative.

IGF-I augments the T-cell proliferative response to IL-2. All lymphoid cells and antigen presenting cells, including smooth muscle cells and endothelial cells contain IGF-I receptors that are abundantly expressed in the myointima of the allograft (Lou *et al.*, 1997). IGF-I by itself has a modest mitogenic effect and little expression in vessels under basal conditions. However, extensive IGF-I expression is found in human restenotic coronary atherectomy plaques (Grant *et al.*, 1994). Further, there are substantial mitogenic response to IGF-I in both animal arterial explants (Yamamoto and Yamamoto, 1994) and in human coronary smooth muscle cells (Grant *et al.*, 1994). IGF-I is also mitogenic for arterial explants in a media in which PDGF-AA and PDGF-BB have no mitogenic effect (Schiffers *et al.*, 1994). By itself, PDGF can initiate DNA synthesis but PDGF and IGF-I together stimulated the rate of cell cycle progression in primary culture of rabbit arterial smooth muscle cells (Yamamoto and Yamamoto, 1994). *In vitro*, PDGF promotes IGF-I mRNA and increases IGF-I receptor expression in smooth muscle cells (Delafontaine *et al.*, 1991; Ververis *et al.*, 1993). Delafontaine and collaborators (Delafontaine *et al.*, 1995) found antisense oligonucleotides for IGF-I receptors prevent proliferation induced by both IGF-I and other growth factors like PDGF and FGF. They concluded that IGF-I is an important progression factor for smooth muscle cells.

The activity of IGF-I is complex due to regulation by binding proteins, which not only bind IGF-I but have roles of their own. The IGF-BP's are produced by multiple tissues, including vascular, endothelial and smooth muscle cells and act as local regulators of IGF-I actions. IGF-BP-1 has a binding affinity similar for both IGF-I and IGF-II, while IGF-BP-2 to -6 have higher binding affinities for IGF-II than for IGF-I. IGF-BP-1 and -2 are regulated by insulin, while IGF-BP-3 is regulated by growth hormone and IGF-I. Vascular smooth muscle cells (Kiefer *et al.*, 1993; Ververis *et al.*, 1994) produce only a little IGF-BP-3 (38–42 kDa) and a 31 kDa IGF-BP (possibly IGF-BP-2). These cells synthesize abundant amounts of two glycosylated forms of IGFBP-4 that are regulated by FGF and PDGF (Giannella-Neto *et al.*, 1992; Anwar and Delafontaine, 1994). IGF-BP-3 inhibits growth of fibroblasts with a targeted disruption of the IGF-I receptor gene, so these binding proteins may have a direct effect on genes for both growth factors and their receptors (Valentinis *et al.*, 1995).

Recently we showed IGF-I was substantially increased in the rat aorta allograft at 3 months following transplantation. EGF and PDGF-BB were also increased. In contrast inhibition of myointimal thickening was associated with a decrease in IGF-I as well as the other two growth factors (Hayry *et al.*, 1993). We found increased IGF-I in rabbits in the vessel wall on day 2 and a further increase on day 4 after balloon injury (Howell *et al.*, 1994). A similar finding was observed in the rat aorta following balloon injury by Cercek *et al.* (1990) and by Bornfeldt *et al.* (1992). The latter report also showed that IGF-I infusion stimulated native vascular smooth muscle cell proliferation. As mentioned previously we have confirmed that grafts exposed to IGF-I prior to transplantation exhibit accelerated arterial arteriosclerosis (Motomura, in press).

## ESTRADIOL AND IGF-I

Unfortunately little information is available on the relationship between estradiol and IGF-I in vascular smooth muscle cells. Most studies focus on reproductive organs and other classical estrogen sensitive cells and tissues. In these cells and tissues estradiol is often related to an increase in the effect of IGF-I, as shown in breast cancer cells (Ruan *et al.*, 1995) and *in vivo* in the uterus (Sahlin *et al.*, 1994). However inhibitory effects of estradiol on serum IGF-I is seen following oral estrogen treatment in post-menopausal women (Weissberger *et al.*, 1991). Recently we showed allograft vessels exhibit increased ER expression and that inhibition of transplant arterioscleroses induced by chronic estradiol-treatment is associated with inhibition of IGF-I and IGF-I receptor expression in the allograft vessels. Recently we also documented the inhibitory effect of the chronic E2-treatment on the proliferative response of the allograft vessel to IGF-I (Lou *et al.*, 1997). This anti-proliferative effect contrasts with the stimulatory effects of estradiol on cell proliferation in the female reproductive organs as well as in endometrial and breast cancer. The IGF-I promoter is a target of estradiol regulation. The investigator did not identify any conventional estrogen-responsive element within the promoter region but the AP-1 motif located therein was shown to be essential. The estradiol-responsive enhancement of the Fos-Jun binding to the AP-1 motif mediated the estrogen response. Estradiol receptor binding to the target DNA was not involved, but a unique function of the DNA binding domain of the estrogen receptor is to facilitate protein–protein interaction (Umayahara *et al.*,

1994). Thus estradiol can exert both positive and negative regulatory roles. The mechanisms of estradiol inhibition of myointimal hyperplasia in transplanted organs may involve many genes with estrogen responsive elements that either may activate or silence a specific gene. In addition ER activation can affect genes that do not have ER response elements by physical interaction with binding sites for transcription factors as shown for the IL-6 gene (Stein and Yang, 1995).

## ESTRADIOL AND MACROPHAGES

Macrophages are present in vascular grafts exhibiting transplant arteriosclerosis. And as early as 3 days post-transplantation in a non-immunosuppressed transplant model. As stated previously these cells are antigen presenting cells and as such upregulate MHC class II antigen expression in response to IFN-$\gamma$ released from activated T-cells. Estradiol activates macrophages, thereby promoting the production of IL-1; estradiol also increases complement activation and macrophage phagocytic activity. During pregnancy the number of macrophages and their phagocytic capability increases with rising concentrations of estrogen (Vernon-Roberts, 1969; Barano, 1992). In the cardiac and aorta transplant models we consistently find estradiol treatment to prevent macrophage infiltration of the intimal and medial layers of the allograft. Four different doses of estradiol preserved the normal ultrastructure of the endothelium, up to the time of sacrifice (Cheng *et al.*, 1991). In contrast, the endothelium of the control animals underwent severe degenerative changes. Whether the preservation of the normal ultrastructure of the endothelium may be synonymous with lack of activation and lack of release of cytokines and growth factors remains to be shown.

## CONCLUSION

We consistently find chronic estradiol treatment of transplant recipients to inhibits arteriosclerosis in both cardiac and aorta transplant models with and without concomitant administration of the immunosuppressant, cyclosporin A. This inhibition occurs with physiological as well as pharmacological doses of estradiol. Treatment with estradiol does not accentuate acute rejection in spite of its well known general stimulation of the immune system. The inhibition of transplant arteriosclerosis is most likely due to estradiol negatively regulating at least three genes, namely MHC class II, IL-6 and IGF-I. This leads to inhibition of chronic rejection and smooth muscle cell proliferation in response to the mitogens IL-6 and IGF-I. In addition the prevention of macrophage infiltration of the subintima and media of the vascular wall reduces the exposure of the vascular smooth muscle cell to chemoattractants and mitogens. Thus a non-feminizing estrogen seems an ideal drug for prevention of one of the major causes of loss of transplant function.

## REFERENCES

Amaldi I, Reith W, Berte C and Mach B. (1989). Induction of HLA genes by IFN-g is transcriptional and requires a trans-acting protein. *J. Immunol.* **142**: 999–1004.

Anwar A and Delafontaine P. (1994). Hypertension increases insulin-like growth factor binding protein-4 mRNA levels in rat aorta. *Hypertension* **24**: 679–685.

Aune B, Oian P, Omsjo I and Osterud B. (1995). Hormone replacement therapy reduces the reactivity of monocytes and platelets in whole blood—a beneficial effect on atherogenesis and thrombus formation? *Am. J. Obstet. Gynecol.* **173**: 1816–1820.

Bachmaier K, Pummerer C, Kozieradzki I, Pfeffer K, Mak TW, Neu N and Penninger JM. (1997). Low-molecular-weight tumor necrosis factor receptor p55 controls induction of autoimmune heart disease. *Circulation* **95**: 655–661.

Banskota NK, Carpentier JL and King GL. (1986). Processing and release of insulin and insulin-like growth factor I by macro- and microvascular EC. *Endocrine* **119**: 1904–1913.

Barano RI, Tenenbaum A, Sales ME and Rumi LS. (1992). Functional alterations of murine peritoneal macrophages during pregnancy. *Am. J. Reprod. Immunol.* **27**: 82–86.

Billingham ME. (1994). Pathology and etiology of chronic rejection of the heart. *Clin. Transplantation* **8**: 289–292.

Bornfeldt KE and Arnquist HJ. (1992). Proliferation of vascular smooth muscle cells and regulation by insulin like growth factor I and insulin. *Diabetogia* **35**: 104–108.

Cercek B, Fishbein MC, Forrester JS, Helfant RH and Fagin JA. (1990). Induction of insulin like growth factor I messenger RNA in rat aorta after balloon denudation. *Circ. Res.* **66**: 1755.

Cheng LP, Kuwahara M, Jacobsson J and Foegh ML. (1991). Inhibition of myointimal hyperplasia and macrophage infiltration by estradiol in aorta allografts. *Transplantation* **52**: 967–972.

Delafontaine P, Lou H and Alexander RW. (1991). Regulation of insulin-like growth factor I messenger RNA levels in vascular smooth muscle cells. *Hypertension* **18**: 742–747.

Delafontaine P, Meng XP and Du J. (1995). Regulation of vascular smooth muscle cell insulin like growth factor-I receptors by phosphorthioate oligonucleotides: effects on cell growth and evidence that sense targeting at the ATG site increases receptor expression. *J. Biol. Chem.* **270**: 14383–14388.

Eich DM, Johnson DE, Hastillo A, Thompson JA, Barnhart GR, Ko DF, Lower RR and Hess ML. (1990). Accelerated coronary atherosclerosis in cardiac transplantation. *Cardiovasc. Clin.* **20**: 199–211.

Emery P, Durand B, Mach B and Reith W. (1996). RFX proteins, a novel family of DNA binding proteins conserved in the eukaryotic kingdom. *Nucleic Acids Res.* **24**: 803–707.

Foegh M, Rego A, Lou H, Katz N and Ramwell P. (1995). Gender effects on Myointimal Hyperplasia. *Transpl. Proc.* **27**: 2070–2072.

Foegh ML. (1992). Accelerated atherosclerosis following transplantation angioplasty and bypass surgery and inhibition by Angiopeptin. In: Strategies in primary and secondary prevention of coronary artery disease. Eds. Ischinger TH and Gohlke H. W. Zuckschwerdt Verlag, pp. 175–180.

Foegh ML, Asotra S, Howell MH and Ramwell PW. (1993). Estradiol inhibition of arterial neointimal hyperplasia after balloon injury. *J. Vasc. Surg.* **19**: 722–726.

Foegh ML, Khirabadi BS, Nakanishi T, Vargas R and Ramwell PW. (1987). Estradiol protects against experimental cardiac transplant atherosclerosis. *Transplant. Proc.* **19**(suppl. 5): 90.

Fox HS, Bond BL and Parslow TG. (1991). Estrogen regulates the IFN-gamma promoter. *J. Immunology* **146**: 4362–4367.

Giannella-Neto D, Kamyar A, Sharifi B, Pirola CJ, Kupfer J, Rosenfeld RG, Forrester JS and Fagin JA. (1992). Platelet-derived growth factor isoforms decrease insulin-like growth factor I gene expression in rat vascular smooth muscle cells and selectively stimulate the biosynthesis of insulin-like growth factor binding protein 4. *Circ. Res.* **71**: 646–656.

Glimcher LH and Kara CJ. (1992). Sequences and Factors: A guide to MHC class-II transcription. *Annu. Rev. Immunol.* **10**: 13–49.

Grant MB, Wargovich TJ, Ellis EA, Caballero S, Mansour M and Pepine CJ. (1994). Localization of insulin like growth factor I and inhibition of coronary smooth muscle cell growth by somatostatin analogues in human coronary smooth muscle cells. *Circ.* **89**: 1511–1517.

Haase J, Ozaki Y, DiMario C, Escaned J, De Feyter PJ, Roelandt JRTC and Serruys PW. (1995). Can intracoronary ultrasound correctly assess the luminal dimensions of coronary artery lesions? A comparison with quantitative angiography. *Europ. Heart J.* **16**: 112–119.

Hansson GK, Jonasson L, Holm J, Clowes MM and Clowes AW. (1988). Gamma-interferon regulates vascular smooth muscle proliferation and Ia antigen expression *in vivo* and *in vitro*. *Circulation Res.* **63**(4): 712–719.

Hayry P, Raisanen A, Ustinov J, Mannander A and Paavonen T. (1993). Somatostatin analog, lanreotide, inhibits myocyte replication and several growth factors in allograft arteriosclerosis. *FASEB J.* **7**: 1055–1060

Hongo M, Itoi M, Yamaguchi N and Imanishi J. (1992). Distribution of epidermal growth factors (EGF) receptors in rabbit corneal epithelial cells, keratocytes and EC and the changes induced by transforming growth factor-beta 1. *Exp. Eye Res.* **54**: 9–16.

Hosenpud JD, Shipley DG and Wagner CR. (1992). Cardiac allograft vasculopathy: Current concepts, recent developments, and future directions. *J. Heart Lung Transplant.* **11**: 9–23.

Howell M, Orskov H, Frystyk J, Flyvbjerg A, Gronbaek H and Foegh M. (1994). Lanreotide, a somatostatin analogue, reduces insulin-like growth factor I accumulation in proliferating aortic tissue in rabbits *in vivo.* A preliminary study. *Eur. J. Endocrinol.* **130**: 422–425.

Ip JH, Fuster V, Badimon J, Taubman MB and Chesebro JH. (1990). Syndromes of Accelerated Atherosclerosis: Role of Vascular Injury and Smooth Muscle Cell Proliferation. *J. Am. Col. Cardiol.* **15**: 1667.

Jacobsson J, Cheng L, Lyke K, Kuwahara M, Kagan E, Ramwell PW and Foegh ML. (1992). Effect of estradiol on accelerated atherosclerosis in rabbit heterotopic aortic allografts. *J. Heart Lung Transplant.* **11**: 1188–1193.

Kafonek SD. (1994). Postmenopausal hormone replacement therapy and cardiovascular risk reduction. *Drugs* **47** (suppl. 2): 16–24.

Kaufman BR, Fox PL and Graham LM. (1992). Platelet-derived growth factor production by canine aortic grafts seeded with EC. *J. Vascular Surgery* **15**: 699–706.

Kiefer MC, Schmid C, Waldvogel M, Schlapfer I, Futo E, Masiarz FR, Green K, Barr PJ and Zapf J. (1993). Recombinant human insulin-like growth factor binding proteins 4, 5 and 6: biological and physiochemical Characterization. *Growth Regulation* **3**: 56–59.

Kourembanas S and Faller DV. (1989). Platelet-derived growth factor production by human umbilical vein EC is regulated by basic fibroblast growth factor. *J. Biol. Chem.* **264**: 4456–4459.

Kuo P and Monaco AP. (1993). Chronic rejection and suboptimal immunosuppression. *Transpl. Proc.* **25**: 2082–2084.

Kuwahara M, Jacobsson J, Kuwahara M, Kagan E, Ramwell PW and Foegh M. (1991). Coronary artery ultrastructural changes in cardiac transplant atherosclerosis in the rabbit. *Transplantation* **52**: 759–765.

Libby P, Swanson SJ, Hiroyuki T, Murray A, Schoen FJ and Pober J. (1992). Immunopathology of coronary arteriosclerosis in transplanted hearts. *J. Heart Lung Transplant.* **11**(3, part 2): S5–6.

Loppnow H and Libby P. (1990). Proliferating or interleukin 1-activated human vascular smooth muscle cells secrete copious interleukin 6. *J. Clin. Invest.* **85**: 731–738.

Lou H, Kodama T, Wang YN, Katz NM, Ramwell PW and Foegh ML. (1996a). L-Arginine prevents cardiac transplant arteriosclerosis by modulating the vascular cell proliferative response to insulin-like growth factor-I and interleukin 6. *J. Heart Lung Transpl.* **15**: 1248–1257.

Lou H, Kodama T, Zhao YJ, Maurice P, Wang YN, Katz N and Foegh ML. (1996b). Inhibition of transplant arteriosclerosis in rabbits by chronic estradiol treatment is associated with abolition of MHC class II antigen expression. *Circulation* **94**: 3355–3361.

Lou H, Zhao Y, Delafontaine P, Kodama T, Katz NM, Ramwell PW and Foegh ML. (1997). Estrogen effects on insulin-like growth factor-I (IGF-I) induced cell proliferation and IGF-I expression in native and allograft vessels. *Circulation* **96**: 927–933.

Manolio TA, Furberg CD, Shemanski L, Psaty BM, OLeary DH, Tracy RP and Bush TL. (1993). Associations of postmenopausal estrogen use with cardiovascular disease and its risk factors in older women. *Circ.* **88**: 2163–2171.

Miller L. (1995). The role of intracoronary ultrasound for diagnosis of graft vessel disease. *Transpl. Proc.* **17**: 1989–1992.

Motomura N, Lou H, Maurice P and Foegh ML. (1997). Transplant arteriosclerosis of rat aorta is accelerated by insulin-like growth factor-I. *Transplantation* **63**: 932–936.

Murray AG, Libby P and Pober JS. (1995). Human vascular smooth muscle cells poorly costimulate and actively inhibit allogeneic CD4+ T cell proliferation *in vitro. J. Immunol.* **154**: 151–161.

Nabata T, Morimoto S, Koh E, Shiraishi T and Ogihara T. (1990). Interleukin 6 stimulates c-myc expression and proliferation of cultured vascular smooth muscle cell. *Biochem. Int.* **20**: 445–453.

Nabulsi AA, Folsom AR, White A, Patsch W, Heiss G, Wu KK and Szklo M. (1993). Association of hormone-replacement therapy with various cardiovascular risk factors in post menopausal women. *N. Engl. J. Med.* **328**: 1069–1075.

Olsen NJ and Kovacs WJ. (1996). Gonadal steroids and immunity. *Endocr. Rev.* **17**: 369–384.

Pober JS, Collins T, Gimbrone MA. Jr, Libby P and Reiss CS. (1986). Inducible expression of class II major histocompatibility complex antigens and the immunogenicity of vascular endothelium. *Transplantation* **41**: 141–146.

Porreca E, DiFebbo C, Barbacane RC, Panara MR, Cuccurullo F and Conti P. (1993). Effect of interleukin-1 Receptor antagonist on vascular smooth muscle cell proliferation. *Atherosclerosis* **99**: 71–78.

Reith W, Steimle V and Mach B. (1995). Molecular defects in the bare lymphocyte syndrome and regulation of MHC class II genes. *Immunol. Today* **16**: 539–546.

Ruan W, Catanese V, Wieczorek R, Feldman M and Kleinberg DL. (1995). Estradiol enhances the stimulatory effect of insulin-like growth factor-I (IGF-I) on mammary development an growth hormone induced IGF-I messenger ribonucleic acid. *Endocrinology* **136**: 1296–1302.

Russell PS, Chase CM, Winn HJ and Colvin RB. (1994a). Coronary atherosclerosis in transplanted mouse hearts: Time course and immunogenetic and immunopathological considerations. *Am. J. Pathol.* **144**: 260–274.

Russell PS, Chase CM, Winn HJ and Colvin RB. (1994b). Coronary atherosclerosis in transplanted mouse hearts: Importance of humoral immunity. *J. Immunol.* **152**: 5135–5141.

Russell PS, Chase CM, Winn HJ and Colvin RB. (1994c). Coronary atherosclerosis in transplanted mouse hearts: Effects of recipient treatment with a monoclonal antibody to interferon-g. *Transplantation* **57**: 1367–1371.

Sahlin L, Norstedt G and Erikson H. (1994). Estrogen regulation of the estrogen receptor and insulin-like growth factor-I in the rat uterus: a potential coupling between effects of estrogen and IGF-I. *Steroids* **59**: 421–430.

Saito S, Motomura N, Lou H, Ramwell PW and Foegh ML. Specific effects of estrogen on growth factor and major histocompatibility (MHC) class II expression in rat aorta allograft. *J. Thorac. Cardiovasc. Surg.* in press.

Sarris GE, Moore KA, Schroeder JS, Hunt SA, Fowler MB, Valantine HB, Vagelos RH, Billingham ME *et al.* (1994). Cardiac transplantation: The Stanford experience in the cyclosporine era. *J. Thorac. Cardiovasc. Surg.* **108**: 240–252.

Schiffers PMH, Fazzi GE, van Ingen Schenau D and De Mey JGR. (1994). Effects of candidate autocrine and paracrine mediators on growth responses in isolated rat arteries. *Arterioscler. Thromb.* **14**: 420–426.

Scott P. (1993). IL-12: Initiation cytokine for cell-mediated immunity. *Science* **260**: 496–497.

Shaddy RE, Hansen CJ and Cowley CG. (1992). Effects of T cell on platelet-derived growth factor-like protein secretion from EC. *J. Heart Lung Transplant.* **11**: 48–57.

Shanker G, Sorci-Thomas M and Adams MR. (1994). Estrogen modulates the expression of tumor necrosis factor alpha mRNA in phorbol ester-stimulated human monocytic THP-1 cells. *Lymphokine Cytokine Res.* **13**: 377–382.

Spits H and de Wall Malefyt R. (1992). Functional characterization of human IL-10. *Int. Arch. allergy. Immunol.* **99**: 8–15.

Steimle V, Siegrist C-A, Mottet A, Lisowska-Grospierre B and Mach B. (1994). Regulationof MHC class II expression by interferon-g mediated by the transactivator gene CIITA. *Science* **264**: 1920–1921.

Stein B and Yang MX. (1995). Repression of the interleukin-6 promoter by estrogen receptor is mediated by NF-$\kappa$B and C/EBP$\beta$. *Mol. Cell. Biol.* **15**: 4971–4979.

Sydner DS and Unanue ER. (1982). Corticosteroids inhibits murine macrophage Ia expression and interleukin I production. *J. Immunol.* **129**: 1803–1808.

Umayahara Y, Kawamori R, Watada H, Imano E, Iwama N, Morishima T, Yamasaki Y, Kasjimoto Y and Kamada T. (1994). Estrogen regulation of the insulin-like growth factor-I gene transcription involves an AP-l. *J. Biol. Chem.* **269**: 16433–16442

Uretsky BF, Kormos RL, Zerbe TR, Lee A, Tokarczyk TR *et al.* (1992). Cardiac events after heart transplantation: incidence and predictive value of coronary arteriography. *J. Heart Lung Transplant.* **11**: S45–51.

Valentinis B, Bhala A, DeAngelis T, Baserga R and Cohen P. (1995). The human insulin-like growth factor (IGF) binding protein-3 inhibits the growth of fibroblasts with a targeted disruption of the IGF-I receptor gene. *Molecular Endocrinol.* **9**: 361–367.

Valentino R, Savastano S, Tommaselli AP, Riccio A, Mariniello P, Pronesti G, De Divitiis PM and Lombardi G. (1993). Hormonal pattern in women affected by rheumatoid arthritis. *J. Endocrinol. Invest.* **16**: 619–624.

Vandenbroecke C, Caillat-Zucman S, Legendre C, Noel LH, Kreis H, Woodrow D, Bach JF and Tovey MG. (1991). Differential *in situ* expression of cytokines in renal allograft rejection. *Transplantation* **51**: 602–609.

Vargas R, Wroblewska B, Rego A, Hatch J and Ramwell PW. (1993). Oestradiol inhibits smooth muscle cells proliferation of pig coronary artery. *Br. J. Pharmacol.* **109**: 612–617.

Vernon-Roberts B. (1996). The effect of steroid hormones on macrophage activity. *Int. Rev. Cytol.* **25**: 131–159.

Ververis JJ, Ku L and Delafontaine P. (1994). Fibroblast growth factor regulates insulin-like growth factor binding protein production by vascular smooth muscle cells. *Am. J. Med. Sci.* **307**: 222–227.

Ververis JJ, Ku L and Delafontaine P. (1993). Regulation of insulin-like growth factor I receptors on vascular smooth muscle cells by growth factors and phorbol esters. *Circ. Res.* **72**: 1285–1292.

Warner SJ, Friedman GB and Libby P. (1989). Immune interferon inhibits proliferation and induces 2'-5'-oligoadenylate synthetase gene expression in human vascular smooth muscle cells. *J. Clin. Invest.* **83**: 1174–1182.

Weissberger AJ, Ho KK and Lazarus L. (1991). Contrasting effects of oral and transdermal routes of estrogen replacement therapy on 24-hour growth hormone (GH) secretion, insulin like growth factor I, and GH-binding protein in postmenopausal women. *J. Clin. Endocrinol. Metab.* **72**: 374–381.

Wilson SE and Lloyd SA. (1991). Epidermal growth factor and its receptor, basic fibroblast growth factors, transforming growth factor beta-1, and interleukin-1 alpha mRNA production in human corneal EC. *Invest. Ophthalmo. Vi. Sci.* **32**: 2747–2756.

Wu CJ, Lovett M, Wong-Lee J, Moeller F, Kitamura M, Goralski TJ, Billingham ME, Starnes VA and Clayberger C. (1992). Cytokine gene expression in rejecting cardiac allografts. *Transplantation* **54**: 326–332.

Yamamoto M and Yamamoto K. (1994). Growth regulation in primary culture of rabbit arterial smooth muscle cells by platelet derived growth factor, insulin-like growth factor-I and epidermal growth factor. *Exp. Cell Research* **212**: 62–68.

# 16 Estrogens and Coronary Artery Reactivity in Atherosclerotic Monkeys

**J. Koudy Williams *, Michael R. Adams, Erika K. Honoré and Thomas B. Clarkson**

*Comparative Medicine Clinical Research Center, Department of Comparative Medicine, Bowman Gray School of Medicine of Wake Forest University, Winston-Salem, NC 27157-1040, USA.*
** Corresponding author, Tel.: 001 910 716 1631, Fax: 001 910 716 1515, Email: kwilliams@cpm.bgsm.edu.*

It has been known for many years that sex hormones modulate vasodilator responses of arteries supplying the uterus with blood. Recently, it has been shown that sex hormones such as estrogen modulate vasomotor responses of other arteries, including coronary arteries. It is thought that modulation of vasodilator and constrictor responses of coronary arteries may be one mechanism by which estrogen affects the risk of coronary heart disease. Although several studies have examined the effects (and potential mechanisms) of estrogen on vasodilator responses of nonatherosclerotic arteries, few have focused on estrogen's effects on atherosclerotic coronary arteries. In studies of ovariectomized atherosclerotic female cynomolgus monkeys, both long-term (2 years) and short-term (20 minutes) estradiol treatment augments dilator responses to acetylcholine, but not nitroglycerin. Presumably, this indicates an effect of estradiol on endothelium-mediated dilator responses of coronary arteries. Addition of the progestin medroxyprogesterone acetate diminishes the beneficial effect of conjugated equine estrogens on these dilator responses. This is significant because a progestin is usually added to estrogen replacement to reduce the risk of endometrial and breast cancer associated with unopposed estrogen therapy. Studies in premenopausal female monkeys indicate that the effective plasma dose of estradiol is approximately 60–70 pg/ml; below that, estradiol does not affect vasodilation. When atherosclerotic, ovariectomized monkeys receive dietary lowering of cholesterol with and without hormone replacement therapy, vasodilator response improves little beyond that seen with lipid lowering alone. Furthermore, hormone status (being female and being premenopausal) has an effect on dilator responses of coronary arteries only up to a certain complexity of atherosclerotic lesion. The greatest effect of hormones are on moderately sized lesions with little calcification and necrosis. These last two studies may indicate some limitations of hormone therapy in affecting coronary vasomotor responses. However, estrogens remain an important modulator of arterial vasomotor responses and may play an important role in estrogen's effects on coronary heart disease risk.

KEYWORDS: Estrogen, progestins, atherosclerosis, coronary arteries, monkeys.

## INTRODUCTION

It has been known since the 1800s that estrogens modulate vascular tone and vascular reactivity. Over a century ago, MacKenzie (1884) reported that menstrual cycle and pregnancy could cause changes in the degree of hyperemia and vascularity of mucous membranes in women. Sixty years later, Reynolds and Foster (1940) reported that estrogen induced dilation of small vessels in the ear of ovariectomized rabbits. In the 1960s, Ueland and Parer (1966) reported a decrease in systemic vascular resistance during pregnancy. These early studies laid the ground work for a steadily increasing amount of work being done to determine estrogen's direct effects on vasomotor tone.

## VASCULAR CELLS AFFECTED BY ESTROGEN

Silva de Sa and Meirelles (1977) reported that estradiol cause relaxation of isolated, perfused human umbilical arteries suggesting a direct effect of estrogens on arteries. Although intriguing, many of these early studies were done using estrogen concentrations several fold higher than those produced endogenously or given as hormone replacement therapy exogenously. Additionally, these early studies did not identify which vascular cells were being affected. Harder and Coulson (1979) first showed that vascular smooth muscle cells can respond to estrogen. In their studies, the estrogen diethylstilbestrol (DES) caused a dose-dependent hyperpolarization of isolated smooth muscle cells. In 1980, Chang *et al.* reported that estrogen might enhance basal levels of prostacyclin secretion thereby affecting vascular dilator responses. Later, Jiang *et al.* (1991) reported that estrogen promoted endothelium-independent, smooth muscle cell-mediated dilation of arteries, possible through regulation of ion channels. Additionally, Jiang *et al.* (1992) reported that estradiol may also modulate vasoconstrictor responses to endothelin-1.

There is also evidence that estrogens may affect vascular reactivity through actions on endothelial cells. Gisclard *et al.* (1987, 1988) and Miller *et al.* (1988) studied the effects of estradiol on both endothelium-mediated and smooth muscle-mediated vascular reactivity in vascular rings from ovariectomized rabbits treated or not treated with intramuscular estradiol. The results of their studies indicate direct effects of estrogen on selected endothelium-mediated dilator responses. Miller *et al.* (1988) hypothesized that estrogen may modulate endothelium function through its effects on endothelium-produced nitric oxide. Potential mechanisms suggested are estrogen's effects on nitric oxide (Hayashi *et al.*, 1992), endothelin (Miller *et al.*, 1988; Jiang *et al.*, 1992), and ion channels (Jiang *et al.*, 1991).

## LONG-TERM EFFECTS OF ESTROGEN

We have focused on the effects of estrogens and progestins on vascular reactivity in atherosclerotic coronary arteries since that is the most clinically relevant model. In the first study, cynomolgus monkeys were ovariectomized, fed an atherogenic diet, and given either no hormone replacement or subcutaneous estradiol (to produce plasma estradiol concentrations of 350 pg/ml) for 3 years (Williams *et al.*, 1990) (Figure 16.1). The coronary arteries of untreated monkeys constricted in response to intracoronary infusion of acetylcholine. However, the coronary arteries of the estrogen-treated monkeys did not constrict in response to acetylcholine. Coronary

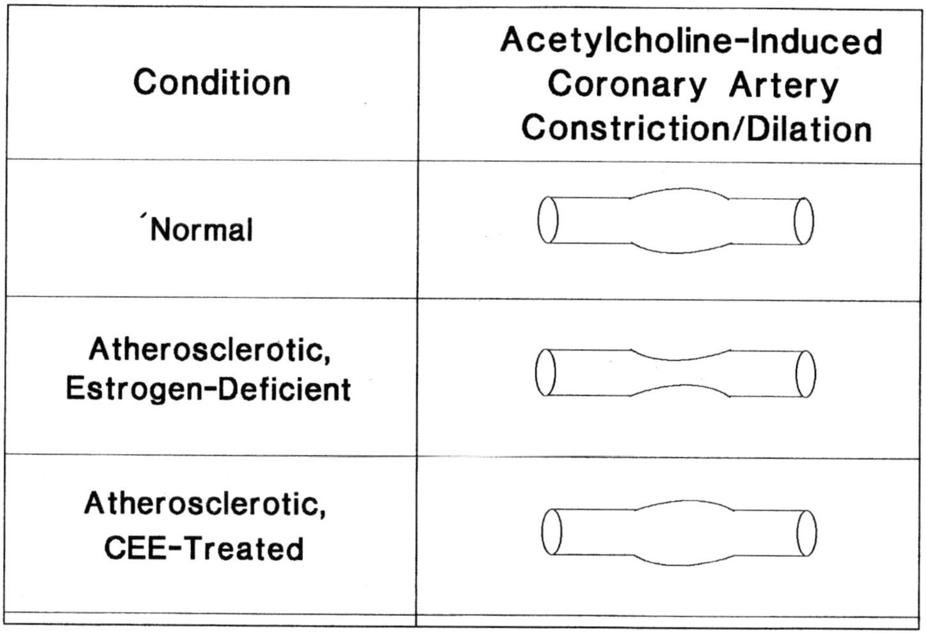


**Figure 16.1.** Normal coronary arteries dilate in response to intracoronary infusion of acetylcholine (top). When coronary arteries develop atherosclerosis, dilation changes to constriction (middle). When atherosclerotic monkeys are treated with estrogens, the constrictor response reverts to a more normal dilator response (bottom).

arteries from both groups of monkeys dilated in response to nitroglycerin. Although the estrogen-treated monkeys had less coronary artery atherosclerosis than their untreated counterparts, the correlation between plaque size and vascular response was low. We concluded that estrogen treatment protects atherosclerotic coronary arteries from abnormal constrictor responses to acetylcholine. We speculated that estrogen preserved normal endothelium-mediated dilation of coronary arteries, and may be one mechanism by which estrogen reduces coronary heart disease risk in postmenopausal women. Similar studies have been done in women (Herrington *et al.*, 1994) replicating this phenomenon.

## SHORT-TERM EFFECTS OF ESTROGEN

In the next study, coronary artery dilator responses of atherosclerotic ovariectomized cynomolgus monkeys were measured before and 20 minutes after intravenous infusion of ethinyl estradiol (to produce plasma estradiol concentrations of 350 pg/ml) (Williams *et al.*, 1992). As observed earlier, the coronary arteries of ovariectomized monkeys constricted in response to acetylcholine. However, the constrictor response to acetylcholine was changed to a dilator response within 20 minutes of estradiol infusion (Figure 16.2). Arteries dilated in response to nitroglycerin both before and after

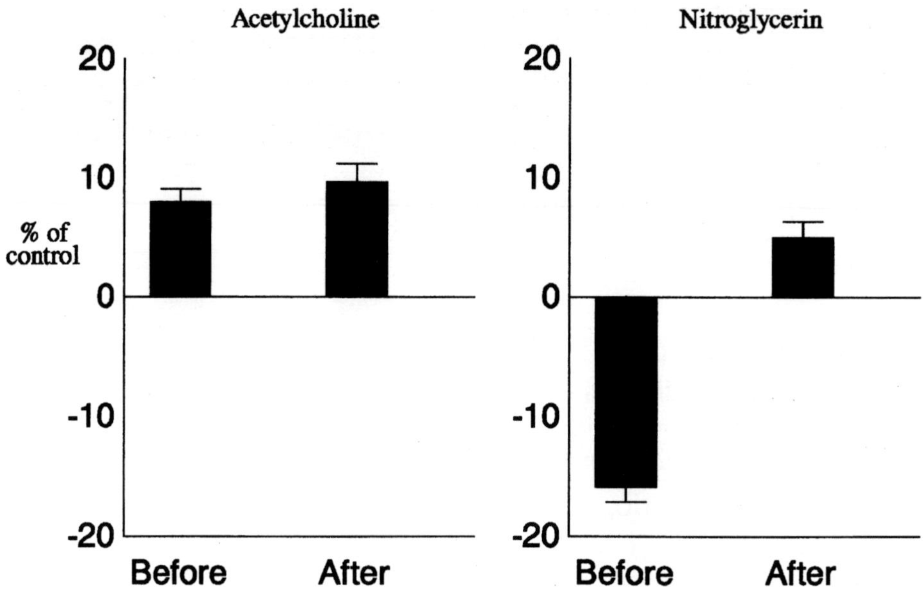


**Figure 16.2.** Percent change in coronary artery diameter in response to nitroglycerin (right panel) and acetylcholine (left panel). Responses were done before short-term infusion of ethinyl estradiol (Before) and 20 minutes after the infusion (After). Constrictor response to acetylcholine diminished within 20 minutes of estradiol infusion.

estradiol infusion. These results confirm that the estrogen response is independent of plaque extent and plasma lipid concentrations. An acute effect of estrogen on the dilator responses of atherosclerotic coronary arteries to chronic and acute estradiol administration also has been shown in women (Reis *et al.*, 1994; Herrington *et al.*, 1994).

## POTENTIAL MECHANISMS

Although the results of these studies establish a clear effect of estrogen on vascular tone and (because of its rapidity) indicate that the response is nongenomic, there remains a controversy about whether estrogen modulates vascular reactivity through actions on endothelial cells, smooth muscle cells, or both. The answer appears to vary depending on the dose and the disease state of the artery. As noted above, there is good evidence that estrogen affects reactivity through smooth muscle-mediated mechanisms. This is probably true when estrogen is given in large (pharmacologic doses) to normal or non-diseased arteries. However, when estrogen is given at physiologic doses to atherosclerotic arteries, the endothelium-mediated responses seem to predominate. This seems logical, since endothelial function is significantly impaired in atherosclerotic arteries (Ludmer *et al.*, 1987), but does not rule out the possibility that smooth muscle function also is improved by estrogen in atherosclerotic arteries (perhaps via a more subtle mechanism).

The exact mechanism(s) of action is currently not known, but again appears to differ with dose and disease state of the artery. As mentioned previously, large doses of estrogen may affect smooth muscle function through calcium channels (Zhang *et al.*, 1994), responses to endothelin (Jiang *et al.*, 1992), or direct hyperpolarization (Jiang *et al.*, 1991). A common theme throughout most of the literature is estrogen's effects on nitric oxide. This could merely be because nitric oxide is a widely studied molecule. Nonetheless, results reported by Williams *et al.* (1990, 1992) and Reis *et al.* (1994) suggest a beneficial effect of estrogen on either the enhanced production or reduced breakdown of nitric oxide. This is supported by the studies of Hayashi *et al.* (1992), which show that basal release of nitric oxide from endothelium-intact rabbit aortic rings is greater in intact females than ovariectomized rabbits or males. Interestingly, no differences were detected in the stimulated release of nitric oxide among groups of rabbits. Nitric oxide release is increased in pregnant rats (Conrad *et al.*, 1993) and calcium-dependent nitric oxide synthase (eNOS) activity, as well as messenger RNA levels, are increased in pregnant and estrogen-treated guinea pigs (Weiner *et al.*, 1994). Thus, estrogen's effects on nitric oxide may be one important mechanism by which this hormone modulates vasomotor tone. Estrogen's effects on nitric oxide may be particularly important in atherosclerotic arteries.

## VASCULAR REACTIVITY AND CORONARY HEART DISEASE

It is well established that hormone replacement therapy reduces the risk of coronary heart disease (CHD) in postmenopausal women. What is not known is the mechanism by which these hormones act on the artery to reduce CHD risk. Studies by Adams *et al.* (1990) indicate that estrogen therapy may, in part, reduce the CHD risk by inhibiting the progression of coronary artery atherosclerosis. Another beneficial effect of hormone replacement therapy may be an overall improvement in coronary vasomotor tone, thereby reducing the risk of vasospasm and angina pectoris. It also is possible that impaired dilator responses and augmented constrictor responses of coronary arteries promote atherogenesis by altering shear stresses and wall tension.

As mentioned previously, it has been shown that both chronic and acute administration of estrogen improve the impaired dilator responses of coronary arteries to acetylcholine in ovariectomized monkeys (Williams *et al.*, 1990; 1992) (Figures 16.1 and 16.2) and postmenopausal women (Reis *et al.*, 1994; Herrington *et al.*, 1994). Rosano *et al.* (1993) have shown that treadmill tolerance is improved in women within 10 minutes of sublingual administration of estradiol. These results indicate that estrogen may promote dilation (or reduce peripheral resistance), thereby reducing the incidence of angina pectoris in women. They also lead to interesting speculation about effects of estrogen on the coronary microvasculature (which regulates coronary resistancc) and microvascular angina in women.

## EFFECTIVE DOSE OF ESTROGENS

For the most part, studies exploring the effects of estrogen on vascular reactivity have used fairly high doses, ranging from 300 pg/ml plasma (the concentration achieved during at peak estrus) to several fold higher than that (pharmacologic

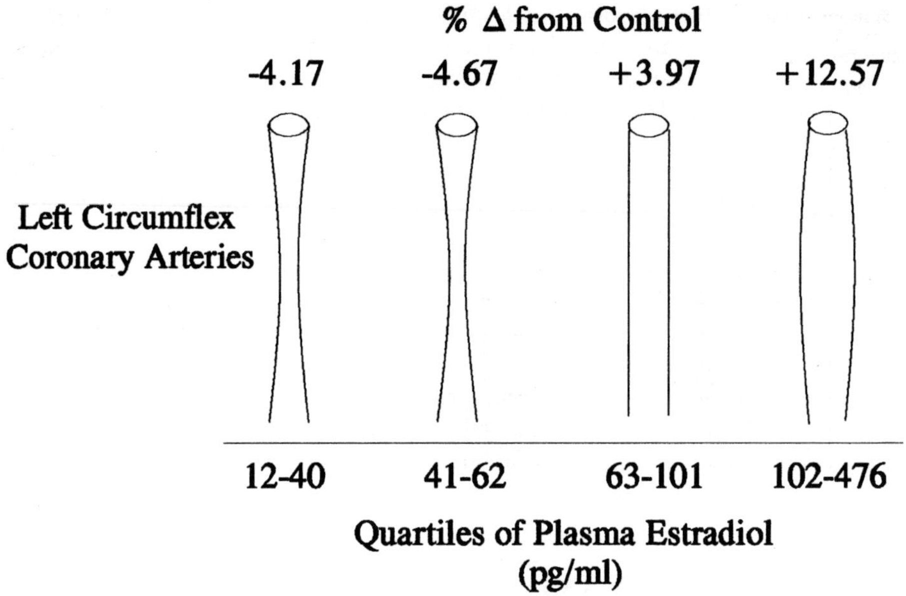


**Figure 16.3.** Percent change in coronary artery diameter in response to acetylcholine in premenopausal atherosclerotic female monkeys with plasma estradiol concentrations of 12–476 pg/ml on the day of angiography. Responses divided into quartiles based on plasma estradiol concentrations. Constrictor response to acetylcholine changes to dilation at plasma estradiol concentrations of about 60–70 pg/ml. Adapted from Williams *et al.*, 1994a.

doses). Although valuable in establishing a potential role of estrogens in the control of vasomotion, these studies do not address whether estrogen affects vasomotion *in vivo*, at doses commonly achieved by endogenous production or exogenous administration of estrogen. Indeed, few studies have concentrated on the lower doses of estrogen (20–100 pg/ml) commonly found in normally cycling women or during hormone replacement. In one study (Williams *et al.*, 1994a), premenopausal cynomolgus monkeys with diet-induced atherosclerosis underwent coronary angiography at random times during their menstrual cycle. Results indicate that plasma estradiol concentrations above 70 pg/ml were "protective" against the paradoxical constrictor responses of atherosclerotic coronary arteries to acetylcholine, whereas concentrations below this value were not (Figure 16.3). It is unclear if this result indicates a critical concentration of estradiol needed to affect reactivity, or a linear dose-response curve. Further studies on effective estrogen doses are greatly needed.

## MODULATION OF VASCULAR REACTIVITY BY PROGESTINS

A progestin is usually given in conjunction with estrogen replacement therapy to reduce the risk of breast and uterine cancer. However, added progestins may have a harmful effect on plasma lipid concentrations and have been speculated to diminish the beneficial effects of estrogens on coronary heart disease risk. Relevant to this

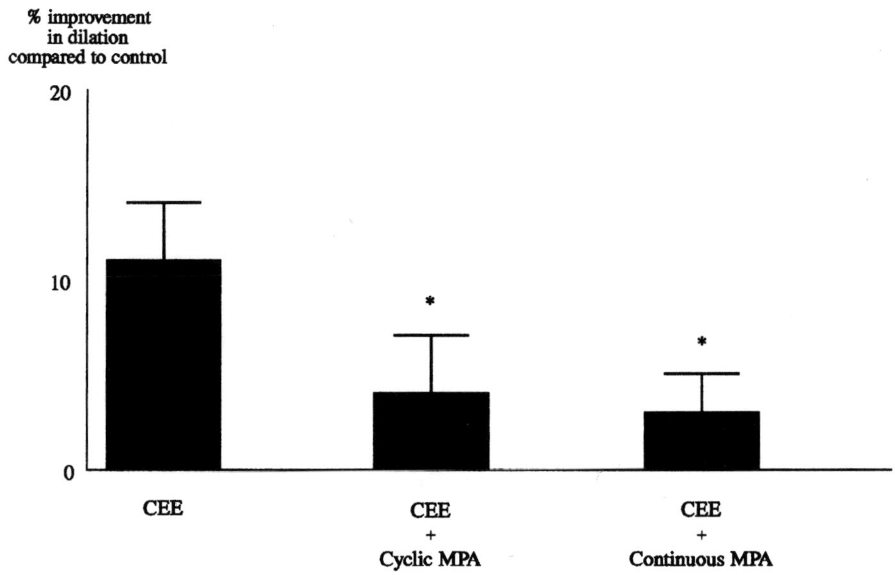


**Figure 16.4.** Effects of added cyclic high-dose or continuous low-dose medroxyprogesterone acetate (MPA) and conjugated equine estrogens (CEE) on coronary artery reactivity to acetylcholine. Results are expressed as % change in diameter from control. * $= p < 0.05$ vs CEE. Reprinted from Williams *et al.*, 1994b.

discussion is whether added progestin therapy affects the beneficial effects of estrogen on coronary vascular reactivity. Early studies by Jiang *et al.* (1992) indicated that progesterone may promote dilation of arteries. These studies were done *in vitro* using large doses of progesterone. The progestin medroxyprogesterone acetate (MPA) appears to do the opposite when given in doses similar to those given to postmenopausal women (Williams *et al.*, 1994b). In this study, MPA was given to ovariectomized cynomolgus monkeys with diet-induced atherosclerosis either in a continuous, low-dose regimen (equivalent to 2 mg/day MPA) or a cyclic, high-dose regimen (10 mg/day). In either case, MPA diminished the beneficial effects of conjugated equine estrogens on dilator responses to acetylcholine (Figure 16.4). It should be noted that progesterone and MPA are different compounds, and these studies were done with widely differing concentrations of hormones. Therefore, different progestins may have different effects on the coronary vasculature.

## EFFECTS OF LIPID LOWERING AND HORMONE REPLACEMENT ON VASCULAR REACTIVITY

The National Cholesterol Education Program guidelines (1994) emphasize dietary and pharmacological interventions for lowering LDL cholesterol concentrations in patients with established CHD. The guidelines also endorse the use of estrogen replacement therapy as an alternative or supplement to conventional lipid-lowering

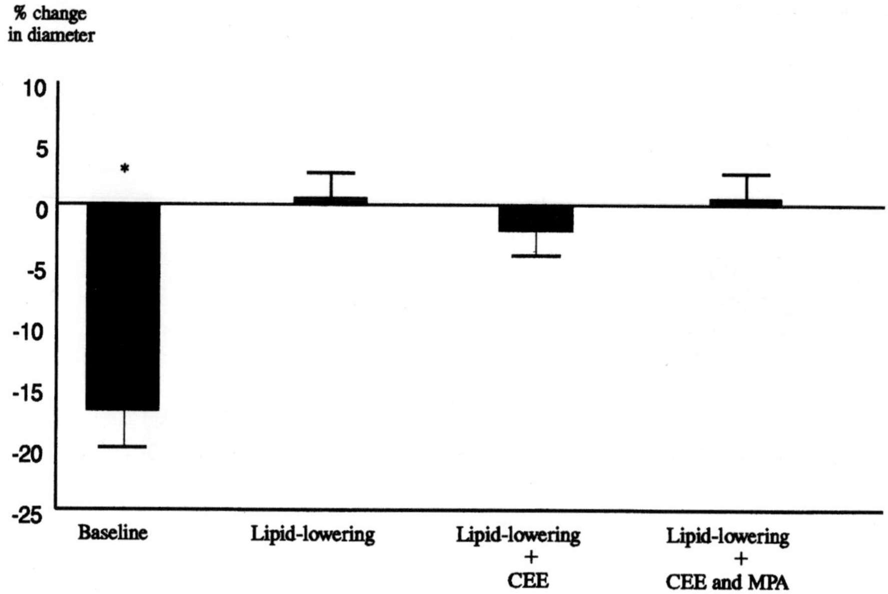


**Figure 16.5.** Percent change in coronary artery diameter from control in response to acetylcholine. Responses measured in atherosclerotic monkeys (Baseline), in monkeys after lipid-lowering diet only (Lipid-lowering), in monkeys receiving lipid-lowering diet +CEE, and in monkeys receiving a lipid-lowering diet +CEE +MPA. * $= p < 0.05$ vs treatment groups. Reprinted from Williams *et al.*, 1995.

interventions to reduce the risk of CHD in postmenopausal women. One of the primary goals of these therapies is to produce regression of atherosclerotic lesions in coronary arteries. Traditionally, regression was used to describe reduction in lesion size after lipid-lowering therapy. More recently, regression has come to mean the structural, chemical, and functional improvements that occur after treatment (Brown *et al.*, 1993). It has been shown that both lipid lowering (Williams *et al.*, 1991) and estrogen treatment independently improve dilator responses to acetylcholine in atherosclerotic coronary arteries. However, the question remains whether estrogen therapy and lipid lowering interact to improve coronary artery reactivity. In a recent study by Williams *et al.* (1995), female cynomolgus monkeys undergoing nutritionally induced lipid lowering were either given, or not given, estrogen replacement therapy. The results indicate that estrogen replacement therapy plus lipid lowering did not further improve dilator responses to acetylcholine in large epicardial coronary arteries, compared to those of monkeys undergoing lipid lowering alone (Figure 16.5). However, coronary flow reserve (a measure of vascular dilator capacity of smaller, resistance-size coronary arteries) was enhanced by estrogen replacement therapy. These results indicate that, except for a modest improvement in small artery reactivity, estrogen and dietary lipid lowering do not have interactive effects on coronary artery reactivity.

## ALTERNATIVE "ESTROGENS"

As mentioned earlier, current hormone replacement regimens have a poor compliance rate due to fear of cancer and side effects such as bleeding and malaise. Therefore, other "estrogens" have been considered as potential new therapies for postmenopausal hormone replacement. A series of estrogen agonist/antagonists are being considered as candidates. Probably the best known of these agents is tamoxifen, which is commonly used to treat estrogen receptor-positive breast cancer. In a recent study (Honoré *et al.*, 1997), we compared the effects of conjugated equine estrogens and tamoxifen on coronary artery reactivity among ovariectomized cynomolgus monkeys with diet-induced atherosclerosis. Similar to what was seen before, coronary arteries of ovariectomized monkeys constricted, whereas those from estrogen-treated animals dilated in response to acetylcholine. The coronary arteries of tamoxifen-treated monkeys constricted in response to acetylcholine, similar to untreated controls, suggesting that tamoxifen does not improve coronary artery dilator responses. These results indicate that tamoxifen (and possibly other tamoxifen-like compounds) may have anti-estrogenic effects on coronary artery reactivity, and indirectly implicate the estrogen receptor as an important modulator of this response.

Estrogen replacement regimens use estradiol, conjugated equine estrogens, synthetic estrogens, and other mammalian-based estrogens. Plants make estrogenic-like compounds (isoflavonoids) with structural similarities to estradiol that bind weakly to estrogen receptors. Thus, they have been called plant estrogens, although the term "isoflavones" is more accurate. Many plants make isoflavones, but soy protein is especially rich in them (particularly genistein and daidzein). Genistein is well known for its tyrosine kinase-inhibitory properties, which has increased interest in genistein as an antiproliferative agent. Since atherosclerosis is a proliferative disease in a sense, there is interest in the antiatherogenic effects of genistein.

Genistein and other isoflavones may have "estrogenic" effects on coronary artery reactivity. Therefore, we examined the long-term and short-term effects of isoflavone treatment on coronary artery reactivity of female cynomolgus monkeys with diet-induced atherosclerosis. The coronary arteries of monkeys fed a diet low in soy-based isoflavones constricted in response to acetylcholine. In monkeys were fed a high-isoflavone diet, coronary arteries dilated in response to acetylcholine (Honoré *et al.*, 1997), similar to responses in monkeys treated with mammalian estrogens (Williams *et al.*, 1992; 1994b). Similarly, short-term infusion of genistein changed the constrictor response to acetylcholine to dilation within 20 minutes of administration. Therefore, isoflavones appear to have estrogenic effects on coronary artery reactivity.

## PLAQUE SEVERITY AND ESTROGEN'S EFFECTS ON CORONARY ARTERY REACTIVITY

It is fairly certain that estrogens modulate coronary artery reactivity in normal and atherosclerotic coronary arteries. However, studies have been done in women and monkeys with only mild to moderate amounts of atherosclerosis. It is uncertain whether estrogen's effect on vascular reactivity persists in arteries with advanced lesions. We examined this question in female and male rhesus monkeys with varying amounts and severity of atherosclerotic lesions. The three groups of monkeys were premenopausal female monkeys, ovariectomized monkeys, and male monkeys. They had been fed atherogenic diets for 2–14 years and had coronary lesions that ranged from fatty streaks to advanced, calcified atherosclerosis. These advanced lesions had large areas of necrosis, calcification, neovascularization, and fibrous caps. The arteries were evaluated on a basis of a score assigned to the amount of lesion (intimal area), the amount of necrosis (area of intima taken up by necrosis), and the amount of calcification. As expected, coronary arteries from male and ovariectomized females constricted in response to acetylcholine, whereas the arteries from premenopausal monkeys (with normal concentrations of circulating estrogen) dilated (Honoré *et al.*, 1997). The most interesting part of the experiment, however, was that most of the variability in response (dilation or constriction) occurred in arteries with minimal lesions. The arteries with advanced atherosclerosis neither dilated nor constricted to any great degree. These results suggest that estrogen's effects on vascular reactivity may have a finite window of opportunity in which to work. If lesions become too advanced, estrogen may not be an effective modulator of vascular reactivity.

## CONCLUSIONS

It is now clear that sex hormones (specifically estrogens and progestins) modulate vascular tone and reactivity. The implications of this depend on the arterial site of question. For instance, estrogen's effects on uterine blood flow are important in normal uterine function during the menstrual cycle and during pregnancy. Recently, attention has turned to estrogen's effects on coronary arteries. These effects may help explain, in part, estrogen's protective effects on risk of coronary heart disease, and may point the way for identifying mechanisms by which sex hormones affect the vasculature. Conversely, other sex hormones (progestins) and estrogen agonist/antagonists (tamoxifen) may have different effects on vascular reactivity. Studies are needed to help identify the mechanism(s) by which sex hormones modulate vascular reactivity. Identification of these mechanisms may help increase understanding of how hormones affect the pathogenesis of coronary heart disease.

## REFERENCES

Adams, M.R., Kaplan, J.R., Manuck, S.B., Koritnik, D.R., Parks, J.S., Wolfe, M.S. *et al.* (1990) Inhibition of coronary artery atherosclerosis by 17 beta-estradiol in ovariectomized monkeys: Lack of an effect of added progesterone. *Arteriosclerosis*, **10**, 1051–1057.

Brown, B.G., Zhao, X.-Q., Sacco, D.E., and Albers, J.J. (1993) Lipid lowering and plaque regression: new insights into prevention of plaque disruption and clinical events in coronary disease. *Circulation*, **87**, 1781–1791.

Chang, W.C., Nakao, J., Orimo, H., and Murota, S.L. (1980) Stimulation of prostaglandin cyclooxygenase and prostacyclin synthetase activities by estradiol in rat aortic smooth muscle cells. *Biochimica Biophysica Acta*, **620**, 472–482.

Conrad, K.P., Joffe, G.M., Kruszyna, H., Kruszyna, R., Rochelle, L.G., Smith, R.P. *et al.* (1993) Identification of increased nitric oxide biosynthesis during pregnancy in rats. *Federation of American Societies for Experimental Biology Journal*, **7**, 566–571.

Gisclard, V., Flavahan, N.A., and Vanhoutte, P.M. (1987) Alpha adrenergic responses of blood vessels of rabbits after ovariectomy and administration of 17 beta-estradiol. *Journal of Pharmacology and Experimental Therapeutics*, **240**, 466–470.

Gisclard, V., Miller, V.M., and Vanhoutte, P.M. (1988) Effect of 17 beta-estradiol on endothelium-dependent responses in rabbits. *Journal of Pharmacology and Experimental Therapeutics*, **244**, 19–22.

Harder, D.R. and Coulson, P.B. (1979) Estrogen receptors and effects of estrogen on membrane electrical properties of coronary vascular smooth muscle. *Journal of Cell Physiology*, **100**, 375–382.

Hayashi, T., Fukuto, J.M., Ignarro, L.J., and Chaudhuri, G. (1992) Basal release of nitric oxide from aortic rings is greater in female rabbits than male rabbits: Implications for atherosclerosis. *Proceedings of the National Academy of Sciences of the United States of America*, **89**, 11259–11263.

Herrington, D.M., Braden, G.A., Williams, J.K., and Morgan T.M. (1994) Estrogen modulates coronary vasomotor responses in postmenopausal women with early atherosclerosis. *American Journal of Cardiology*, **73**, 951–952.

Honoré, E.K., Williams, J.K., Anthony, M.S., and Clarkson T.B. (1997) Soy isoflavones enhance coronary vascular reactivity in atherosclerotic female macaques. *Fertility and Sterility*, **67**, 148–154.

Jiang, C., Sarrel, P.M., Lindsay, D.C., Poole-Wilson, P.A., and Collins, P. (1991) Endothelium-independent relaxation of rabbit coronary artery by 17-beta estradiol. *British Journal of Pharmacology*, **104**, 1033–1037.

Jiang, C., Sarrel, P.M., Lindsay, D.C., Poole-Wilson, P.A., and Collins, P. (1992) Progesterone induces enothelium-independent relaxation of rabbit coronary artery *in vitro*. *European Journal of Pharmacology*, **211**, 163–167.

Jiang, C., Sarrel, P.M., Poole-Wilson, P.A., and Collins P. (1992) Acute effect of 17 beta-estradiol on rabbit coronary artery contractile responses to endothelin-1. *American Journal of Physiology*, **263**, H271–H275.

Ludmer, P.L., Selwyn, A.P., Shook, T.L., Wayne, R.R., Mudge, G.H., Alexander, R.W. *et al.* (1986) Paradoxical vasoconstriction induced by acetylcholine in atherosclerotic coronary arteries. *New England Journal of Medicine*, **315**, 1046–1051.

MacKenzie, J.N. (1884) Irritation of the sexual apparatus. *American Journal of Medical Science*, **87**, 360–365.

Miller, V.M., Gisclard, V., and Vanhoutte, P.M. (1988) Modulation of endothelium-dependent and vascular smooth muscle responses by oestrogens. *Phlebology*, **3**(Suppl 1), 63–69.

National Cholesterol Education Program Expert Panel. (1994) Second Report of the National Cholesterol Education Program Expert Panel of Detection, Evaluation, and Treatment of High Blood Cholesterol in Adults (Adult Treatment Panel II). *Circulation*, **89**, 1329–1345.

Reis, S.E., Gloth, S.T., Blumenthal, R.S., Zacur, H.A., Gerstenblith G. *et al.* (1994) Ethinyl estradiol acutely attenuates abnormal coronary vasomotor responses to acetylcholine in postmenopausal women. *Circulation*, **89**, 52–60.

Reynolds, S.R.M. and Foster, F.I. (1940) Peripheral vascular action of estrogen observed in the ear of the rabbit. *Journal of Pharmacology and Experimental Therapeutics*, **68**, 173–177.

Rosano, G.M., Sarrel, P.M., Poole-Wilson, P.A., and Collins, P. (1993) Beneficial effect of oestrogen on exercise-induced myocardial ischemia in women with coronary artery disease. *Lancet*, **342**, 133–136.

Silva de Sa, M.F. and Meirelles, R.S. (1977) Vasodilation effect of estrogen on the human umbilical artery. *Gynecologic Investigation*, **8**, 307–313.

Ueland, K. and Parer, J.T. (1966) Effects of estrogens on the cardiovascular system of the ewe. *American Journal of Obstetrics and Gynecology*, **96**, 400–406.

Weiner, C.P., Lizasoian, L., Baylis, S.A., Knowles, R.G., Charles, I.G., and Moncada, S. (1994) Induction of calcium-dependent nitric oxide synthesis by sex hormones. *Proceedings of the National Academy of Sciences of the United States of America*, **91**, 5212–5216.

Williams, J.K., Adams, M.R., and Klopfenstein, H.S. (1990) Estrogen modulates responses of atherosclerotic coronary arteries. *Circulation*, **81**, 1680–1687.

Williams, J.K., Adams, M.R., Herrington, D.M., and Clarkson, T.B. (1992) Short-term administration of estrogen and vascular responses of atherosclerotic coronary arteries. *Journal of the American College of Cardiology*, **20**, 452–457.

Williams, J.K., Anthony, M.S., Honoré, E.K., Herrington, D.M., Morgan, T.M., Register, T.C. *et al.* (1995) Regression of atherosclerosis in female monkeys. *Arteriosclerosis, Thrombosis and Vascular Biology*, **15**, 827–834.

Williams, J.K., Honoré, E.K., Washburn, S.A., and Clarkson, T.B. (1994b) Effects of hormone replacement therapy on reactivity of atherosclerotic coronary arteries in cynomolgus monkeys. *Journal of the American College of Cardiology*, **24**, 1757–1761.

Williams, J.K., Shively, C.A., and Clarkson, T.B. (1994a) Determinants of coronary artery reactivity in premenopausal female monkeys with diet-induced atherosclerosis. *Circulation*, **90**, 983–987.

Williams, J.K., Vita, J.A., Manuck, S.B., Selwyn, A.P., and Kaplan, J.R. (1991) Psychosocial factors impair vascular responses of coronary arteries. *Circulation*, **84**, 2146–2153.

Zhang, R., Ram, J.L., Standley, P.R., and Sowers, J.R. (1994) 17 beta-estradiol attenuates voltage-dependent Ca currents in A7r5 vascular smooth muscle line. *American Journal of Physiology*, **266**, C975–C980.

# 17 Estrogen and Progesterone Protective Actions on Coronary Arteries in Rhesus Monkeys

**Kent Hermsmeyer** *, **Richard D. Minshall, Koichi Miyagawa and Jaclyn Vidgoff**

*Oregon Regional Primate Research Center, The Charles Dotter Institute for Interventional Radiology and The Departments of Medicine and Cell and Developmental Biology, Oregon Health Sciences University, Portland, OR 97201, USA.*

* *Corresponding author: Kent Hermsmeyer, Oregon Regional Primate Research Center, 505 NW 185th Avenue Beaverton, OR 97006, USA, Tel.: 001 503 690 5580, Fax: 001 503 690 5563.*

Recent studies on coronary reactivity have shown the importance of modulation by ovarian steroid hormones, as summarized in this review. Our studies were designed to determine how coronary vasospasms, produced by a hyperreactive segment of a coronary artery in adult ovariectomized rhesus monkeys, are affected by the absence and replacement of estrogen. We examined the influence of *in vivo* treatment with estrogen in combination with natural progesterone, on the susceptibility to drug-induced coronary vasospasms. In addition, the reactivity of isolated single coronary artery muscle cells was assessed from the same animals after the vasospasm challenge by measuring agonist stimulated changes in intracellular $Ca^{2+}$ and protein kinase C (PKC). The experiments were designed to (1) determine conditions for provoking drug-induced coronary vasospasms in the absence of pathology, (2) determine whether exposure of ovariectomized rhesus monkeys to physiological concentrations of estrogen and progesterone would normalize coronary artery reactivity, and (3) investigate the coronary artery muscle cell basis for increased reactivity.

KEYWORDS: Calcium, $Ca^{2+}$, protein kinase C, postmenopausal, vascular muscle cell, hyperreactivity.

## INTRODUCTION

Cardiovascular disease is the leading cause of death in postmenopausal women; slightly more than 1 of every 2 women in this age group will die from heart disease (Gura, 1995). Sudden cardiac death is now recognized to be more prevalent among postmenopausal women than any other segment of the population, a statistic that has been linked to the hypoestrogenic state. The risk of death due to cardiovascular illnesses in women in the over 50 age group may be up to four times the combined death rates for breast and endometrial cancer, and replacement of ovarian steroid hormones significantly reduces this risk (Wenger *et al.*, 1993). The reluctance for increased use of estrogen replacement therapy is based on the increased risk unopposed estrogen places on developing endometrial cancer (Gorodeski and Utian, 1994). However, addition of progesterone to the regimen provides the balance and eliminates the harmful effects of unopposed estrogen (Gorodeski and Utian, 1994; Miller, for The Writing Group for the PEPI Trial, 1995). Therefore, the observations made thus far suggest that estrogen replacement therapy for postmenopausal women will decrease the incidence of coronary atherosclerosis, angina, myocardial infarction and stroke, with a total risk reduction of developing coronary heart disease hypothesized to be up to 50% (Belchetz, 1994; Miller, 1995; Healy, 1995), prompting further investigations into their cardiovascular effects.

Our hypothesis is that estrogen acts via $Ca^{2+}$ and other intracellular signals to protect coronary arteries against the irreversible contraction known as vasospasm, a common pathogenic element in acute myocardial infarction and sudden cardiac death. There is increasing evidence that the sex steroid hormones (SSH) act directly on the cardiovascular system. Estrogen and progesterone regulate gene expression, gaining access to the nucleus by a receptor-mediated transport process (Sutherland *et al.*, 1988). In the cardiovascular system, progesterone directly lowers forearm blood pressure in humans (Rylance *et al.*, 1985). Progesterone receptors were localized in human thoracic aorta, internal carotid and coronary arteries, and left atria using a specific monoclonal antibody to the primate progesterone receptor (Ingegno *et al.*, 1988; Lin *et al.*, 1986), although this point is controversial (Ravalli *et al.*, 1995). In the carotid and coronary arteries, receptor staining could be localized to the endothelial nuclei of intima and in vascular channels within atherosclerotic plaques and uninvolved vascular muscle cells of the media and intima. In the myocardium, staining was localized to the nuclei of myocardial fibers (Ingegno *et al.*, 1988). Vascular effects of estrogen were recently described in women, supporting the idea that estrogen therapy reduces cardiovascular events in postmenopausal women. While no direct effects of estrogen on blood pressure were detected, estrogen potentiated forearm vasodilation in response to acetylcholine and sodium nitroprusside (Gilligan *et al.*, 1994) and restored normal dilator responses to acetylcholine in ovariectomized atherosclerotic monkeys (Williams *et al.*, 1994). In the baboon, nuclear uptake of estrogen was detected mainly in interstitial and adventitial cells and to a much lesser extent by atrial and ventricular myocardial fibers, arterial endothelial cells and vascular muscle cells of the arterial media (McGill and Sheridan, 1981).

While structural occlusion provides an explanation for a major fraction of fatal ischemic events, coronary vasospasm (focal constrictions and downstream dilations in an hour-glass shaped pattern) (Hermsmeyer *et al.*, 1997) is characteristically localized and episodic, and not always correlated with atherosclerotic plaques (Bertrand *et al.*, 1982; Maseri *et al.*, 1990; Bogaty *et al.*, 1994). Adverse coronary vascular muscle reactivity, or spasm, may not be a peculiar occurrence limited to rare forms of angina, but a common pathogenic element in acute myocardial infarction and in sudden cardiac death (Maseri *et al.*, 1990; Kalsner, 1995). Locally released vasoconstrictors are obligatory to an explanation of the segmental occlusion pattern and probably initiate vasospasm, particularly in cases where plaques are stable or absent (Hermsmeyer *et al.*, 1997). Variant angina in the absence of plaques may be due to nonspecific supersensitivity of coronary artery segments to multiple stimuli rather than a specific abnormal agonist-receptor interaction (Kaski *et al.*, 1986). Several vasoactive agents, including adrenergic and cholinergic neurotransmitters, arachidonic acid metabolites such as thromboxane $A_2$, prostaglandins and leukotrienes, as well as serotonin, histamine, endothelin and angiotensin II have been hypothesized to produce vasospasms (Kaski *et al.*, 1986; Maseri *et al.*, 1990; Bogaty *et al.*, 1994; Willerson, 1995; Kalsner, 1995). We have recently developed an ovarian steroid sensitive model of coronary vasospasm in rhesus monkeys (Hermsmeyer *et al.*, 1997) in which spasms can be reliably induced with a combination of serotonin and thromboxane $A_2$. In addition, we found that the coronary artery muscle cells isolated from ovariectomized monkeys retained more reactive responses to a given stimulus of thromboxane and serotonin (Miyagawa *et al.*, 1997b). This manuscript summarizes our findings regarding the effects of estrogen ($E_2$) and progesterone (P) treatment of ovariectomized rhesus monkeys on agonist-stimulated coronary artery reactivity and underlying single vascular muscle cell $Ca^{2+}$ and PKC signals from the same coronary arteries.

## EXPERIMENTAL DESIGN AND METHODS

The intention of these experiments was to understand the vascular muscle cell actions of physiological concentrations of estradiol $17\beta$ ($E_2$) and natural progesterone (P) on coronary artery reactivity and the calcium $Ca^{2+}$ and PKC signals on which the strong contractions and vasospasm depend. Effects of 4 week $E_2$ and P replacement therapy were examined by comparing coronary artery reactivity of ovariectomized rhesus macaques in the catheterization laboratory. The coronary artery vasospasm protocol used the combination of serotonin and U46619 (the stable thromboxane $A_2$ mimetic), which are a putative pathophysiological combination, to determine coronary reactivity (Hermsmeyer *et al.*, 1997; Miyagawa *et al.*, 1997a). The effect of the thromboxane $A_2$ and serotonin combination were further studied at the coronary vascular muscle cell level measured by $Ca^{2+}$ mobilization and PKC translocation (Miyagawa *et al.*, 1997b). Freshly dispersed or primary cultured vascular muscle cells were derived from the left anterior descending, circumflex, and right coronary arteries of the same monkeys studied in the vasospasm challenge.

### Coronary Vasospasm Induced by Platelet Release Substances

Female rhesus monkeys (Macaca mulatta), free of metabolic or cardiovascular disease, were ovariectomized (ovx) at least 3 months prior to entering the study. All experimental procedures were carried out according to the *Guide for the Care and Use of Laboratory Animals* (NIH Publication No. 86-23, 1985), and approved by the ORPRC institutional animal care and use committee. As a pre-treatment, subdermal silastic tubing containing 200 mg $E_2$ was implanted. After two weeks of $E_2$ alone, tubing with 400 mg of P was additionally implanted ($E_2$+P), or the $E_2$ implant was withdrawn (W), and the animals were maintained for another 4 weeks (treatment period). The implants established physiological plasma steroid levels, measured as $104 \pm 10$ pg/ml ($n=6$) for $E_2$ and $6670 \pm 676$ pg/ml ($n=6$) for P in the $E_2$+P group (Miyagawa *et al.*, 1997a).

Each monkey was studied in the coronary catheterization laboratory after the treatment period. The vasospasm model is based on the catheterization procedures

**TABLE 17.1** Drug-induced protocol for coronary vasospasm

1. IC injection of contrast media to establish artery diameters before drugs (control)
2. IC injection of 100 nM (0.018 μg) acetylcholine to show dilator capacity
3. IC injection of 1 μM acetylcholine (0.18 mg) to show dilator capacity
4. IC injection of 100 μM (17.6 μg) serotonin
5. IC injection of 1 mM (176 μg) serotonin
6. IC injection of 100 nM (0.035 μg) U46619 (stable thromboxane $A_2$ mimetic)
7. IC injection of 1 μM (0.35 μg) U46619
8. Challenge 1 with combined 100 μM (17.6 μg) serotonin and 1 μM (0.35 μg) U46619
9. Challenge 2 with combined serotonin-U46619 stimulation
10. Challenge 3 with combined serotonin-U46619 stimulation
11. Triple combination with S + U + 1 nM endothelin 1, or + 1 μM angiotensin II
12. Dilation 15 min later with 1 or 10 μM nitrendipine or mibefradil

All concentrations are expressed as the undiluted solution in the injection syringe, but would have been diluted 15 times by coronary blood flow as the 1 ml was injected over 30 sec; coronary blood flow under these conditions was estimated at 30 ml/min (allowing for 50% flow obstruction by the catheter), as explained in the Hermsmeyer *et al.* (1997). This protocol was carried on 7 ovx rhesus monkeys. (Reproduced with permission of the *Journal of the American College of Cardiology*, 1997.)

and stimuli similar to those used for angiographic diagnosis of human coronary artery vasospastic disease (Maseri *et al.*, 1990). Intracoronary injections of 0.1 mM serotonin together with 1 μM U46619 were used to induce coronary vasospasms, visualized by intracoronary injection of contrast media and simultaneous angiographic recordings as previously described (Hermsmeyer *et al.*, 1997). Vasospasms were defined as contractions of epicardial coronary arteries that include focal or diffuse areas of constriction to less than 25% of control diameter, followed by adjacent downstream dilation, which persist for at least 5 min. After angiography, monkeys were taken to necropsy for dissection of coronary arteries and subsequent isolation of vascular muscle cells (VMC) for single cell experiments. The injection sequence that produced vasospasm is given in Table 17.1, and the tests of candidate agonists in Table 17.2. More details can be found in Hermsmeyer *et al.* (1997).

## VMC Preparation

Single VMCs from the left anterior descending, circumflex, and right coronary arteries of adult rhesus monkeys were isolated and studied both as freshly dispersed and as primary cultures as described previously (Miyagawa *et al.*, 1997b). The short-term primary cultured cells (never passaged) maintain the characteristics of the source tissue for 2–3 weeks, including contraction, relaxation, receptors, and membrane electrical properties. VMCs were dissociated with collagenase and protease enzymes in a potassium glutamate solution (KG) that prevents loading with $Na^+$, $Ca^{2+}$, or $Cl^-$ and results in a high proportion of viable, contracting cells (Self *et al.*, 1994; Rusch *et al.*, 1995). The cells prepared for culture were seeded at low density in cardiovascular culture solution for mammals, fifth generation (CV5M) on glass coverslips to facilitate selection of individual cells. VMC were used for experiments 7–14 days after attaching to coverslips. Further details can be found in Miyagawa *et al.* (1997b).

**Table 17.2** Sensitivity to vasospasm stimuli

| | $ED_{50}$ $(-\log M)$ | $EC_{50}$ $(-\log M)$ | *Max dose* $(-\log M)$ | *Spasms/N* | *Protocol* |
|---|---|---|---|---|---|
| *Single stimuli* | | | | | |
| Serotonin | 2.5 | 3.8 | 2 | 0/8 | 1* |
| U46619 | 5.5 | 6.8 | 5 | 0/8 | 1* |
| Endothelin 1 | 8.5 | 9.8 | 8 | 0/3 | 2 |
| Angiotensin II | 5.5 | 6.8 | 5 | 0/3 | 2 |
| *Combined stimuli* | | | | | |
| Endothelin 1+S | 9 | 10.3 | 8 | 0/3 | 2 |
| Angiotensin II+S | 6 | 7.3 | 5 | 0/3 | 2 |
| Mechanical injury+S | na | na | na | 3/3 | 2 |
| U46619+S | 6 | 7.3 | 5 | 5/7 | 1 |

$ED_{50}$ = the dose in μM (as the concentration in the syringe, without allowing for any dilution due to coronary blood flow) of the named vasoconstrictor that caused a 50% reduction of diameter from control; S in the combined stimuli means 100 μM serotonin, except in the case of mechanical injury vasospasm which required 1 mM serotonin; na = not applicable.

$EC_{50}$ = the estimated actual coronary artery blood concentration in − log molar based on a 15X dilution (Hermsmeyer *et al.*, 1997).

Max dose = the maximum dose tried (syringe concentration) in − log molar for each vasoconstrictor; Spasms = the number of animals with vasospasm; $N$ = the number of animals in a protocol.

* One of the Protocol 2 animals also was tested with serotonin alone and U46619 alone, after the angiotensin II step, accounting for the total of 8 in the first 2 rows.

Vasospasms all consisted of reduction to ⩽25% of control diameter with downstream dilation (hour-glass shape) that persisted for ⩾15 min. (Reproduced with permission of the *Journal of the American College of Cardiology*, 1997.)

## Subcellular $Ca^{2+}$ and PKC Localization

Freshly dispersed or primary cultured VMCs on glass coverslips were placed in a chamber of laminar flow design and observed with a Zeiss Plan Neofluar 25X/0.80 water immersion objective (Miyagawa, *et al.*, 1997b). Ionic solution for mammals version 2 (ISM2) was continuously pumped through the chamber (at 1 ml/min) to provide continuous equilibration and washout of drugs. After a 15 min equilibration period, VMC were loaded for 15 min at room temperature with 3 μM fluo 3 (Molecular Probes, Inc) for sensing $Ca^{2+}$ and for 10 min with 30–100 nM hypericin (LC Laboratories) for reporting PKC. Individual VMC were stimulated by adding 20 μl of 10 μM serotonin plus 100 nM U46619 over the individual cell. After 15 sec under no-flow conditions, continuous flow of ISM2 was reinstated and a chamber volume of approximately 300 μl was maintained. Fluorescent images were taken at 1, 2, 5, 10, 15, 20 and 30 min after stimulation for $Ca^{2+}$ and at 3, 4, 9, 16, 21 and 31 min for PKC to evaluate subcellular distribution of these signals, and to determine VMC contraction. Other details can be found in Miyagawa *et al.* (1997b).

## Drugs and Solutions

ISM2 contains (in mM): 100 NaCl, 4 $NaHCO_3$, 0.5 $NaH_2PO_4$, 4.7 KCl, 1.8 $CaCl_2$, 0.41 $MgCl_2$, 0.41 $MgSO_4$, 50 HEPES (pH 7.37), and 5.5 dextrose. KG solution consists of (in mM): 140 K-glutamate, 16 $NaHCO_3$, 0.5 $NaH_2PO_4$, 30 HEPES, and 16.5 dextrose at pH 7.3. CV5M consists of: 4.0 mM L-glutamine, 10 µg/ml ciprofloxacin (Miles Pentex/Bayer Corporation), 50 mM HEPES (pH 7.3), and 16 mM $NaHCO_3$ dissolved in 85% MEM-Earle's salts and 15% horse serum. Phenol red, known to have estrogenic activity, was omitted from all solutions.

## RESULTS

### Coronary Catheterizaton

In the absence of injury or plaques, intracoronary stimulation with serotonin and U46619 evoked coronary vasospasm in 5 of 7 ovariectomized rhesus monkeys of the form shown in Figures 17.1–17.3 and summarized in Table 17.2 (Hermsmeyer, 1997). The radiological appearance and pharmacological profile closely mimic human coronary artery vasospasm induced by intracoronary injections of ergonovine, histamine, or high dose acetylcholine (Maseri, 1979; van den Berg *et al.*, 1989; Golino *et al.*, 1991; Egashira *et al.*, 1992; Bogarty *et al.*, 1994). Based on initiating stimulus, characteristic hour-glass pattern of focal constriction followed by downstream dilation (Figure 17.1B and D, Figure 17.2C, and Figure 17.3C), extended time course of intense vasoconstriction, and pharmacology of vasospasm reversal, this rhesus monkey coronary vasospasm phenomenon closely approximates human coronary vasospasm. Vasospasm induced by mechanical injury in Figure 17.1 resulted in a focal constriction and downstream dilation pattern, to which Figures 17.2 and 17.3 can be compared. The multiple examples of hourglass patterns show that without injury or pathology, and only dependent on ovariectomy, hyperreactivity to the synergistic combination of serotonin and U46619 (the thromboxane $A_2$ mimetic) is sufficient to account for the vasospasms in Figures 17.2 and 17.3 (Hermsmeyer *et al.*, 1997). This combination of platelet release products, suggested as a pathophysiological mechanism and supported by effective block with a monoclonal antibody (7e3) directed against the glycoprotein IIb/IIIa receptor (Willerson, 1995), appears sufficient to explain vasospasm in primate coronary arteries. In this respect, the lack of epicardial vasospasm in response to the individual agonists is remarkable, and probably accounts for the difficulty in provoking vasospasms in non-ovx monkeys. These results support the hyperreactivity hypothesis for human coronary vasospasm (Kaski *et al.*, 1986; 1989).

This vasospasm model was developed to provide an animal analogy which can be useful for study of increasing cardiovascular disease in postmenopausal women. In the ovx monkey, $E_2+P$ replacement for 4 weeks protected against the drug-induced vasospasm in 5 of 7 monkeys (Miyagawa *et al.*, 1997b). This modulation of reactivity by ovarian steroid hormones emphasizes that coronary arteries represent an important target organ that has been only partially accepted, as evidenced by the controversy surrounding hormone replacement therapy in post-menopausal women. In rhesus, as in women, loss of ovarian function is accompanied by a dramatically increased risk of coronary vasospasm (Gorodeski and Utian, 1994; Belchetz, 1994). The data summarized here suggest that the explanation may be modulation of coronary reactivity by SSH.

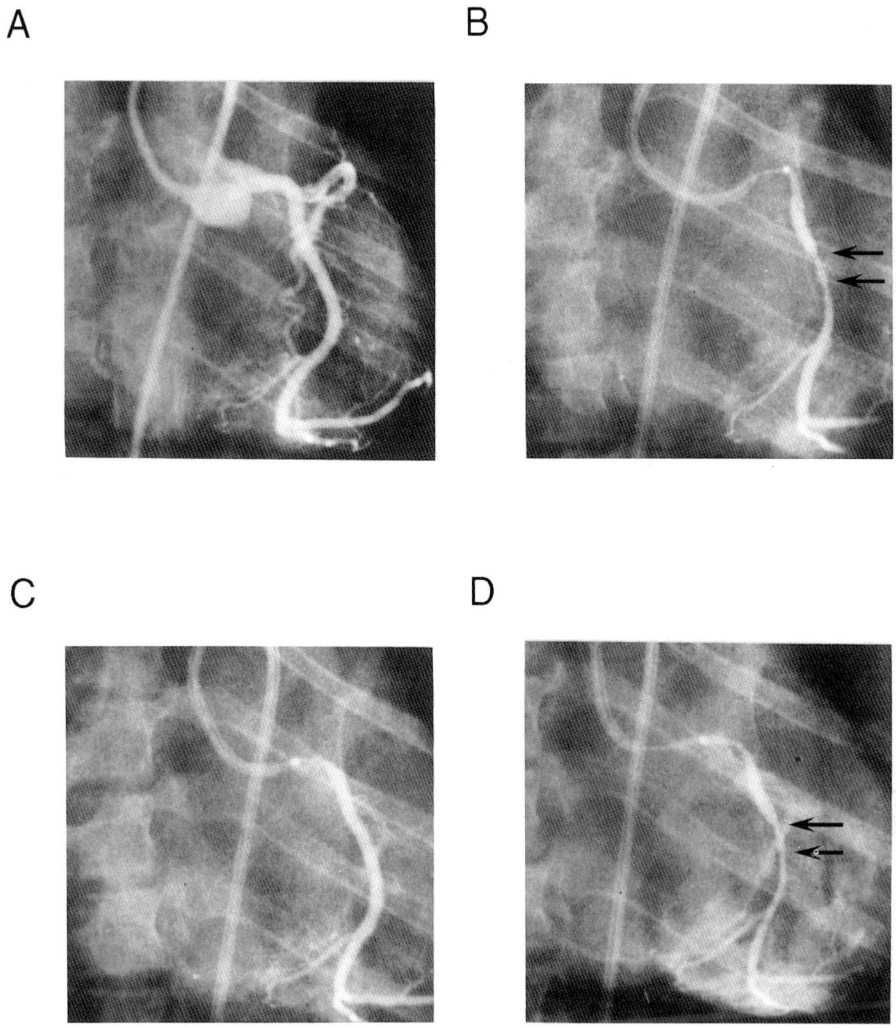


**Figure 17.1.** Characteristic hourglass pattern constrictions that define vasospasm were provoked by catheter injury. The control angiogram (A) is the well dilated reference artery, the coronary artery stretched by a 5F catheter after infusion of 1 ml of 1 mM serotonin over a 30 sec interval (B) showing vasospasms (arrows), the reversal (C) of the vasospasms by injection of 1 μM nitrendipine, and the re-initiation of vasospasms (D) by injection of 1 ml of 1 μM U46619 (thromboxane $A_2$ mimetic). Notice that the vasospasms marked with arrows are several mm downstream from the site of injury just beyond the radio-opaque catheter tip clearly visible in B, and that regions of dilation beyond the focal constrictions easily identify the sites where both constrictor and dilator phenomena consistently appear. Vasospasms redeveloped in D are at exactly the same sites as were found after the serotonin injection in B. The pattern of intense, long duration (5–15 min) constrictions that are usually observed in human vasospasm can be used for comparison with the following 2 figures that avoided any injury to study hyperreactivity based vasospasm. (Reproduced with permission from the *Journal of the American College of Cardiology*.)

A

B

C

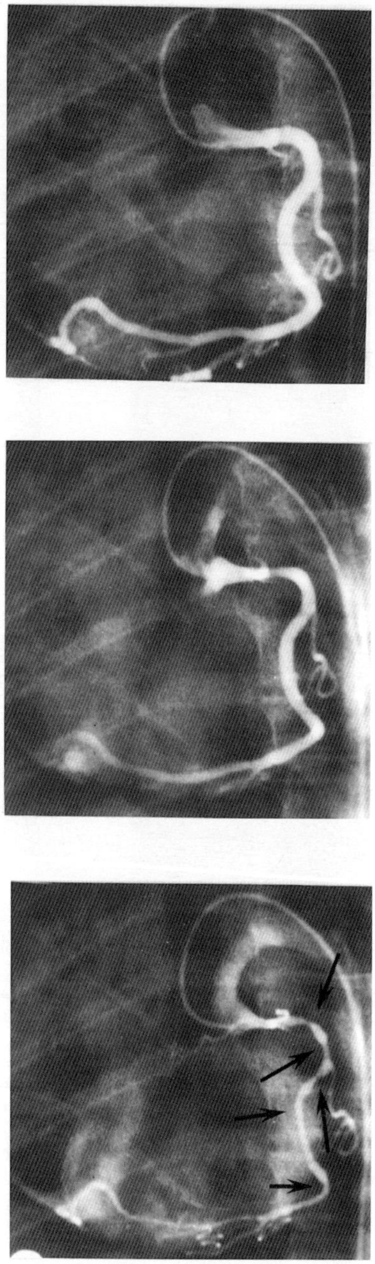

**Figure 17.2.** Drug-induced vasospasm in an uninjured, atherosclerosis-free coronary artery of an ovx monkey. The control normally dilated coronary artery condition (A) is shown. Multiple hour-glass constrictions at 5 min (B) developed in response to injection of the combination of serotonin and U46619. These constrictions were focal and developed into vasospasms shown at 15 min (C) by the arrows. The intense constrictions that nearly completely occluded blood flow persisted for 30 min, continuing at focal sites with reduction by 90% of the original diameter. (Reproduced with permission from the *Journal of the American College of Cardiology*.)

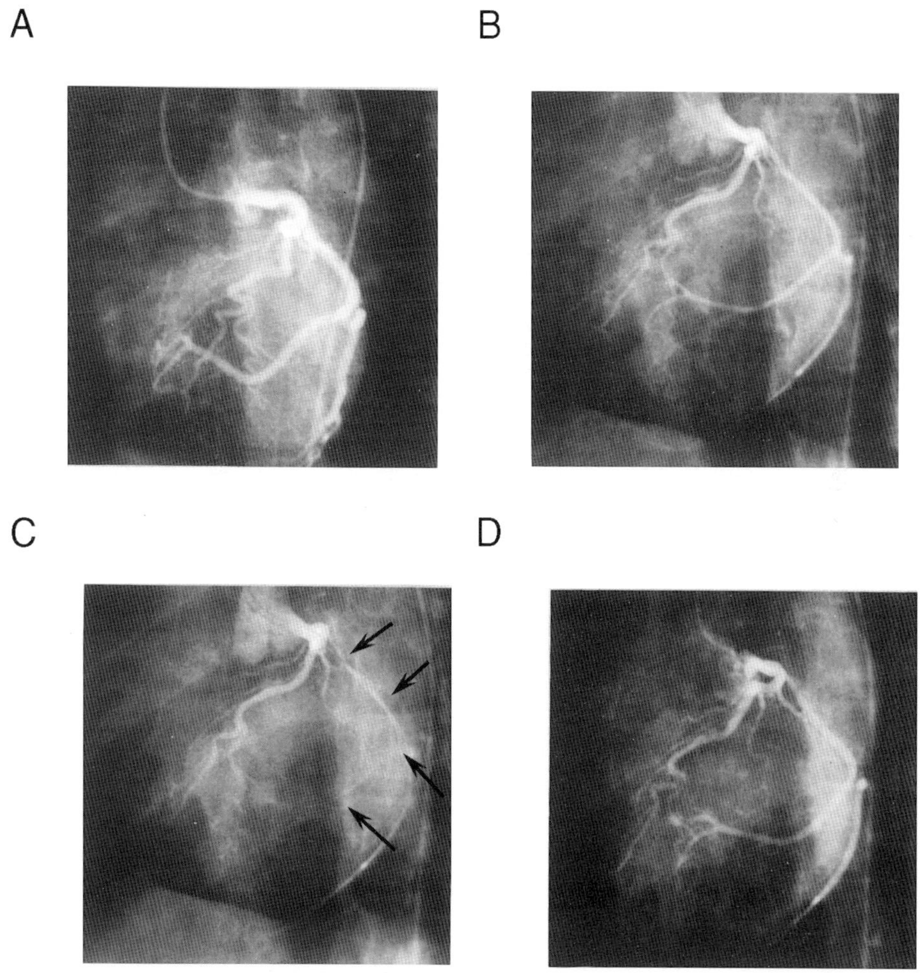

**Figure 17.3.** Drug-induced vasospasm in an uninjured, atherosclerosis-free coronary artery of an ovx monkey and relief by a $Ca^{2+}$ antagonist. In the control condition (A), arteries were well dilated. Strong constrictions in response to serotonin plus U46619 developed within 15 sec (B), and intensified at 3 min (C) as shown by the arrows. Relief of the vasospasm by the novel $Ca^{2+}$ antagonist, mibefradil, returned the diameter to control values (D), even though blood pressure had dropped to cardiogenic shock levels. (Reproduced with permission from the *Journal of the American College of Cardiology*.)

## VMC $Ca^{2+}$, PKC, and Reactivity

Detailed cellular studies of contraction signals support the hyperreactivity hypothesis for vasospasm in the absence of structural occlusion or injury. Coronary artery VMC from monkeys in the W group, which were hypoestrogenic at the time of study, were more reactive than those in the $E_2+P$ group to stimulation with the combination of serotonin and the thromboxane $A_2$ mimetic, U46619. Cellular localizations of $Ca^{2+}$ and PKC signals are shown by digital images in Figure 17.4. In W monkeys,

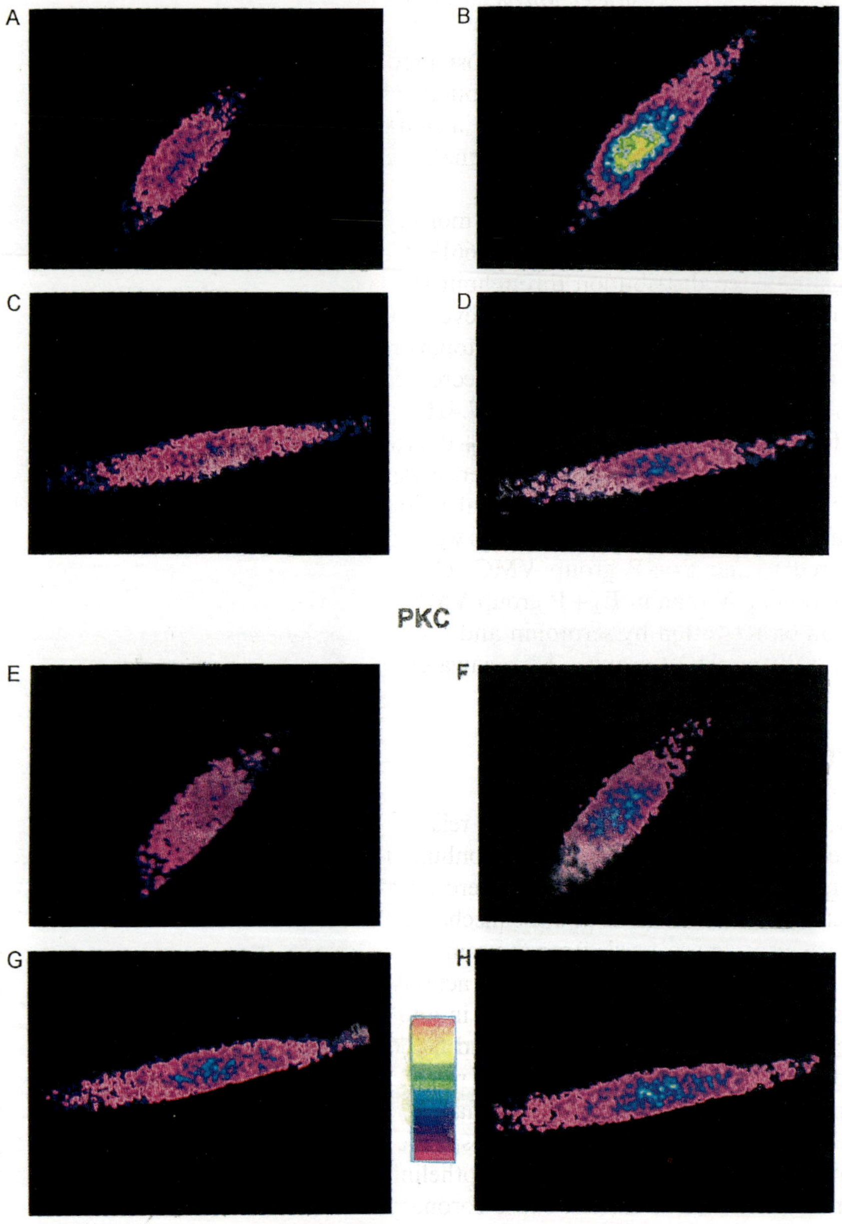


**Figure 17.4.** Effect of steroid treatment *in vivo* on monkey isolated coronary artery muscle cell $Ca^{2+}$ and PKC. Coronary VMC from W monkeys were more reactive to stimulation by 10 μM serotonin plus 100 nM U46619 ($TxA_2$ mimetic) than VMC from $E_2+P$ monkeys, indicated by larger and more prolonged $Ca^{2+}$ and PKC signals. $Ca^{2+}$ levels increased more upon stimulation in W (A versus B) than in $E_2+P$ monkeys (C versus D). PKC levels also increased more in W (E versus F) than in $E_2+P$ monkeys (G versus H). The same VMC from a W monkey in A, B, E, and F is shown at 15 min (B) and 16 min (D) after stimulation, and the same VMC from an $E_2+P$ monkey in C, D, G, and H is shown at 15 (C) and 16 (D) min. The 16 color scale at the bottom of the figure shows increasing intensities as an increase from control with 16 powers of 110% of the control average (100–460%, with black representing 90% or less). (Reproduced with the permission of the American *Journal of Physiology*, 1997.)

$Ca^{2+}$ before stimulation appeared most predominantly in central regions (Figure 17.4A). PKC was more evenly distributed (Figure 17.4E). Stimulation by 10 μM serotonin plus 100 nM U46619 caused a marked increase of $Ca^{2+}$ (Figure 17.4B), most notably in central areas. PKC signals were increased both centrally and at the surface membrane (Figure 17.4F).

However, VMC from $E_2+P$ treated monkeys were less reactive during stimulation with 10 μM serotonin plus 100 nM U46619. $Ca^{2+}$ signals transiently increased from the unstimulated distribution throughout the cytoplasm (Figure 17.4C) during the first minute, and returned to basal levels at 15 min (Figure 17.4D). PKC fluorescence was found throughout the cytoplasm at rest (Figure 17.4G), with only a small membrane translocation and a decrease in intensity at 16 min after stimulation with serotonin plus U46619 (Figure 17.4H).

Differences in W compared to $E_2+P$ groups were found in $Ca^{2+}$, PKC, and contraction. Whole cell $Ca^{2+}$ fluorescence was greater in comparison of W with the $E_2+P$ group, and the peripheral/central ratio was less in W than $E_2+P$ group VMC (Miyagawa *et al.*, 1997b). PKC signals were greater after stimulation in the W group compared to the $E_2+P$ group VMC. Contraction was significantly greater after stimulation in W than in $E_2+P$ group VMC. Based on comparisons of contractions, duration of activation by serotonin and U46619 was between 4 and 9 min in $E_2+P$ versus $>30$ min in W groups (Miyagawa *et al.*, 1997b).

## DISCUSSION

This report summarizes our success in reliably producing coronary vasospasm in the absence of pathology or injury in nonhuman primates based on coronary artery hyperreactivity that is due to ovarian steroid hormone modulation. Coronary reactivity was reliably assessed *in vivo* and the mechanism carefully examined *in vitro* on isolated coronary artery muscle cells from the same monkeys. We believe that this approach is a definitive test of the cardioprotective actions of estrogen and progesterone.

Coronary vasospasms were induced in adult ovariectomized rhesus monkeys with intracoronary injection of serotonin and U46619, a stable thromboxane $A_2$ mimetic. We found that, in our ovariectomized model of the postmenopausal state, injection of this drug combination reliably initiated vasospasms angiographically similar to those observed in humans with vasospastic coronary arteries. Injection of other potent vasoconstrictors, including endothelin and angiotensin II, alone or in combination with serotonin, did not cause coronary vasospasm (Table 17.2; Hermsmeyer *et al.*, 1997). Ovariectomized animals were protected from drug-induced coronary vasospasms when $E_2+P$ were implanted 4 weeks prior to vasospasm challenge.

Similarly, VMC from the left anterior descending, right, and circumflex coronary arteries of the monkeys studied in the catheterization laboratory were hyperreactive, responding to the combined serotonin and thromboxane $A_2$ stimulus with a sustained increase in $Ca^{2+}$ and PKC translocation from the cytosol to the periphery of the cell. Hormone replacement therapy (estrogen in combination with natural progesterone) significantly reduced the fluorescent $Ca^{2+}$ and PKC changes observed, as compared with the hypoestrogenic state.

There is very little information available to explain how estrogen and progesterone directly act to protect against coronary vasospasm and myocardial ischemia. In the baboon, estrogen increases the synthesis of progesterone receptors and alters estrogen receptor distribution in great vessels (Lin *et al.*, 1986). $E_2$ decreases collagen synthesis in cultures of bovine vascular muscle cells (Beldekas *et al.*, 1981) and connective tissue formation in rat aorta (McGill and Stern, 1979). Additionally, estrogen and progesterone limit coronary artery constriction and promote relaxation, perhaps through a $Ca^{2+}$ dependent process (Maseri *et al.*, 1990; Jiang *et al.*, 1992), and estrogens potentiate the release of prostacyclin and nitric oxide (Miller *et al.*, 1995) and inhibit the production of thromboxane, endothelin, and superoxide ions (Stanczyk, 1994). Thus, the ovarian steroid protection phenomenon may be a dominant factor in the risk of coronary vasospasm (Hermsmeyer *et al.*, 1997; Miyagawa *et al.*, 1997a).

In our ovx rhesus, estrogen replacement prevented coronary artery vasospasms. The underlying mechanism by which the hypoestrogenic state is linked to vasospasm appears to be related to the VMC response to vasoactive agents. The sustained increase in intracellular $Ca^{2+}$ needed to maintain the contractile state may be due to altered $Ca^{2+}$ binding or signal proteins. Our data suggests that the $Ca^{2+}$-and phospholipid-dependent protein kinase (PKC) and $Ca^{2+}$ channels are involved. A role of PKC in hyperreactive VMC responses was suggested by Miyagawa *et al.* (1997b). Stimulation of VMC receptors with agonists that activate PLC, transiently elevates intracellular $Ca^{2+}$ by the combination of release from intracellular $Ca^{2+}$ storage sites and influx from extracellular fluid (Missiaen *et al.*, 1995). On the other hand, PKC, which is directly activated by diacylglycerol, provides a slower and longer lasting signal and accelerates or triggers several intracellular functions by phosphorylating key proteins (Gong *et al.*, 1995). The mechanism by which PKC acts to increase excitation–contraction coupling in vascular muscle (for a review, see Khalil and Morgan, 1992) may include enhanced opening of cell membrane and sarcoplasmic reticulum $Ca^{2+}$ channels (Kitamura *et al.*, 1992; Missiaen *et al.*, 1995), allowing $Ca^{2+}$ to reach high, sustained levels in the sarcoplasm (Haller *et al.*, 1995). Although increased cytosolic $Ca^{2+}$ evokes contraction of VMC, high $Ca^{2+}$ levels do not normally persist for more than 2–3 min because processes to reduce intracellular $Ca^{2+}$ are also activated. Vasospasms of coronary arteries are abnormally long-lasting, strong contractions, and PKC activation has the proper time course to play a role in such an abnormally sustained contraction. However, it is not yet clear if enhanced PKC activation is a secondary consequence of a robust $Ca^{2+}$ signal, or whether PKC is the prime target altered during the hypoestrogenic state. Our hypothesis predicts that several VMC proteins involved in $Ca^{2+}$ signaling would be altered after the loss of $E_2$ resulting in hyperreactive coronary arteries.

Additional studies will be needed to clarify the mechanisms by which the hypoestrogenic state increases coronary artery reactivity, the roles estrogen and progesterone play in modulating normal $Ca^{2+}$ and PKC responsiveness, and the optimal plan for hormone replacement therapy from the perspective of reducing the risk of coronary vasospasm.

## ACKNOWLEDGMENTS

This research was supported by the National Institutes of Health grants HL 51723 and HD 18185, and by Bayer Corporation, New Haven, CT.


## REFERENCES


Belchetz, P.E. (1994) Hormonal treatment of postmenopausal women. *N. Engl. J. Med.* **330**: 1062–1071.

Bertrand, M.E., La Blanche, J.M., Tilmant, P.Y. *et al.* (1982) Frequency of provoked coronary arterial spasm in 1089 consecutive patients undergoing coronary arteriography. *Circulation* **65**: 1299–1306.

Bogaty, P., Hackett, D., Davies, G. and Maseri, A. (1994) Vasoreactivity of the culprit lesion in unstable angina. *Circulation* **90**: 5–11.

Egashira, K., Tomoike, H., Hayashi, Y., Yamada, A., Nakamura, M. and Takeshita, A. (1992) Mechanism of ergonovine-induced hyperconstriction of the large epicardial coronary artery in conscious dogs a month after arterial surgery. *Circ. Res.* **71**: 435–442.

Gilligan, D.M., Badar, D.M., Panza, J.A., Quyyumi, A.A. and Cannon, R.O. III. (1994) Acute Vascular effects of estrogen in postmenopausal women. *Circulation* **90**: 786–791.

Golino, P., Piscione, F., Willerson, J.T. *et al.* (1991) Divergent effects of serotonin on coronary-artery dimensions and blood flow in patients with coronary atherosclerosis and control patients. *N. Engl. J. Med.* **324**: 641–648.

Gong, M.C., Kinter, M.T., Somlyo, A.V. and Somlyo, A.P. (1995) Arachidonic acid and diacylglycerol release associated with inhibition of myosin light chain dephosphorylation in rabbit smooth muscle. *J. Physiol.* **486**: 113–122.

Gorodeski, G.I. and Utian, W.H. (1994) Epidemiology and risk factors of cardiovascular disease in postmenopausal women. In: *Treatment of the Postmenopausal Woman: Basic and Cllinical Aspects*, edited by R.A. Lobo, pp. 199–221. New York: Raven.

Gura, T. (1995) Estrogen: key player in heart disease among women. *Science* **269**: 771–773.

Haller, H., Lindschau, C., Quass, P., Distler, A. and Luft, F.C. (1995). Differentiation of vascular smooth muscle cells and the regulation of protein kinase C-α. *Circ. Res.* **76**: 21–29.

Healy, B. (1995) PEPI in perspective, good answers spawn pressing questions. *J. Amer. Med. Assoc.* **273**: 240–241.

Hermsmeyer, K., Miyagawa, K., Kelley, S.T., Rösch, J., Hall, A.S., Axthelm, M.K. and Greenberg, B. (1997) Reactivity based coronary vasospasm independent of atherosclerosis in rhesus monkeys. *J. Am. Coll. Cardiol*, **29**: 671–680.

Ingegno, M.D., Money, S.R., Thelmo, W., Greene, G.L., Davidian, M., Jaffe, B.M. and Pertschuk, L.P. (1988) Progesterone receptors in the human heart and great vessels. *Lab. Invest.* **59**: 353–356.

Jiang, C., Sarrel, P.M., Lindsay, D.C., Poole-Wilson, P.A. and Collins, P. (1991) Endothelium-independent relaxation of rabbit coronary artery by 17b-oestradiol *in vitro*. *Br. J. Pharmacol.* **104**: 1033–1037.

Kalsner, S. (1995) Coronary artery spasm: multiple causes and multiple roles in heart disease. *Biochem. Pharmacol.* **49**: 859–871.

Kaski, J.C., Crea, F., Meran, D., Rodriguez, L., Araujo L., Chierchia, S., Davies, G., and Maseri A. (1986) Local coronary supersensitivity to diverse vasoconstrictive stimuli in patients with variant angina. *Circulation* **74**: 1255–1265.

Kaski, J.C., Maseri, A., Vejar, M., Crea, F. and Hackett, D. (1989) Spontaneous coronary spasm in variant angina is caused by a local hyperreactivity to a generalized constrictor stimulus. *J. Am. Coll. Cardiol.* **14**: 1456–1463.

Khalil, R.A. and Morgan, K.G. (1992) Protein kinase C: A second E-C coupling pathway in vascular smooth muscle? *NIPS* **7**: 10–15.

Kitamura, K., Xiong, A., Teramoto, N. and Kuriyama, H. (1992). Roles of inositol trisphosphate and protein kinase C in the spontaneous outward current modulated by calcium release in rabbit portal vein. *Pflügers Arch* **421**: 539–551.

Lin, A.L., Gonzalez, R. Jr., Carey, K.D. and Shain, S.A. (1986) Estradiol 17-beta affects estrogen receptor distribution and elevates progesterone receptor content in baboon aorta. *Atherosclerosis* **6**: 495–504.

Maseri, A., Davies, G., Hackett, D. and Kaski, J.C. (1990) Coronary artery spasm and vasoconstriction. *Circulation* **81**: 1983–1990.

Maseri, A. (1979) Variant angina and coronary vasospasm: Clues to a broader understanding of angina pectoris. *Cardiovasc. Med.* **4**: 647–667.

McGill, H.C. Jr. and Sheridan, P.J. (1981) Nuclear uptake of sex steroid hormones in the cardiovascular system of the baboon. *Circ. Res.* **48**: 238–244.

McGill, H.C. Jr. and Stern, M.P. (1979) Sex and atherosclerosis. *Atherosclerosis Rev.* **4**: 157.

Miller, V.T. for The PEPI Writing Group. (1995) Effects of estrogen or estrogen/progesterone regimens on heart disease risk factors in postmenopausal women, The Postmenopausal Estrogen/Progestin Interventions (PEPI) Trial. *J. Amer. Med. Assoc.* **273**: 199–208.

Missiaen, L., Parys, J.B., De Smedt, H., Lemaire, F.X., Ksienaert, I., Bootman, M.D. and Casteels, R. (1995) Slow kinetics of $InsP_3$-induced $Ca^{2+}$ release: Differences between uni- and bi-firectional $^{45}Ca^{2+}$ fluxes. *Cell Calcium* **18**: 100–110.

Miyagawa, K., Rösch, J., Stanczyk, F. and Hermsmeyer, K. (1997a) Medroxyprogesterone interferes with ovarian steroid protection against coronary vasospasm. *Nature Med.* **3**: 324–327.

Miyagawa, K., Vidgoff, J. and Hermsmeyer, K. (1997b) $Ca^{2+}$ release mechanism of primate drug-induced coronary vasospasm. *Am. J. Physiol.*, **272**: H2645–H2654.

Ravalli S., Vidali A., Cannon P.J. and Kelly A. (1995) Progesterone receptors are not expressed in normal and atherosclerotic human coronary arteries. *Circulation* **92**: I-107 (abstract).

Rusch, N.J., England, S.K. and Hermsmeyer, K. (1995) Vascular muscle membrane properties in hypertension. In: *Physiology and Pathophysiology of the Heart*, edited by N. Speralakis, pp. 999–1010. Boston: Kluwer Academic Publishers.

Rylance, P.B., Brincat, M., Lafferty, K., De Trafford, J.C., Brincat, S., Parsons, Y. and Studd, J.W.W. (1985) Natural progesterone and antihypertensive action. *Br. Med. J.* **290**: 13–18.

Self, D.A., Bian, K., Mishra, S.K. and Hermsmeyer, K. (1994) Stroke-prone SHR vascular muscle $Ca^{2+}$ current amplitudes correlate with lethal increases in blood pressure. *J. Vasc. Res.* **31**: 359–366.

Stanczyk, F.Z. (1994) Structure-function relationships, potency, and pharmacokinetics of progestogens. In: *Treatment of the Postmenopausal Woman: Basic and Clinical Aspects*, edited by R.A. Lobo, pp. 69–89. New York: Raven.

Sutherland, R.L., Watts, C.K. and Clarke, C.L. (1988) Oestrogen actions. In: *Hormones and Their Actions*, edited by B.A. Cooke, R.J. King and H.J. Van der Molen, pp. 197–215. Amsterdam, The Netherlands: Elsevier.

van den Berg, E.K., Schmits, J.M., Benedict, C.R., Malloy, C.R., Willerson, J.T. and Dehmer, G.J. (1989) Transcardiac serotonin concentration is increased in selected patients with limiting angina and complex coronary lesion morphology. *Circulation* **79**: 116–124.

Wenger, N.K., Speroff, L. and Packard, B. (1993) Cardiovascular health and disease in women. *N. Engl. J. Med.* **329**: 247–256.

Willerson, J.T. (1995) Conversion from chronic to acute coronary heart disease syndromes. Role of platelets and platelet products. *Texas Heart Inst. J.* 22: 13–19.

Williams, J.K., Honore, E.K., Washburn, S.A. and Clarkson, T.B. (1994) Effects of hormone replacement therapy on reactivity of atherosclerotic coronary arteries in cynomolgus monkeys. *J. Amer. Coll. Cardiol.* **24**: 1757–1761.

# 18 Estrogen's Effects on Human Coronary and Peripheral Vasoreactivity

**Scott Kinlay, Marie Gerhard, Andrew P. Selwyn, Mark A. Creager * and Peter Ganz**

*Cardiovascular Division, Brigham and Women's Hospital, Harvard Medical School, 75 Francis Street, Boston MA 02115, USA.*
**Corresponding author, Tel.: 001 617 732 5267, Fax: 001 617 232 2749.*

Observational studies have shown that post-menopausal women who take estrogen, with or without progesterone, have a reduced risk of coronary artery disease, compared to women who have never used hormone replacement therapy. Estrogen may lower the risk of heart disease in post-menopausal women by favorably altering serum lipid levels, by reducing the oxidation of low density lipoprotein cholesterol, by decreasing fibrinolytic activity, or by improving vasomotor tone. The effect of estrogen on vasomotor function is dependent on the length of exposure, preparation and dose used, addition of progesterone, and the lipid altering effects of hormone replacement therapy. Studies of the acute administration of intracoronary or intravenous 17$\beta$-estradiol have shown improvement in endothelium-dependent coronary and peripheral vasomotor responses. These rapid changes suggest mechanisms other than those involving the classic intracellular estrogen receptor. Importantly, long-term estrogen administration has also improved endothelium-dependent vasodilation. It is postulated that hormone replacement therapy will prevent coronary events, in part, by improving endothelium-dependent vasomotor function. Improved vasomotor tone may reduce myocardial ischemia. Also, enhancement of endothelium-dependent vasodilation may be a surrogate marker indicating beneficial changes in other endothelial functions that determine thrombogenicity and the vulnerability of the atherosclerotic plaque.

KEYWORDS: Women, hormone replacement therapy, estrogen, coronary artery disease, endothelial function.

Each year in the United States, nearly half a million women die from heart attacks and stroke (National Center for Health Statistics, 1993; American Heart Association, 1992). Many more women survive these events but with significant morbidity. The risk of myocardial infarction and stroke increases after menopause, and the loss of endogenous estrogen likely contributes to this increased rate of cardiovascular events.

Data supporting the concept that estrogen therapy prevents heart disease is derived largely from observational cohort studies (Grodstein, 1995; Stampfer, 1991a,b; Bush, 1987). These studies have shown that post-menopausal women using estrogen replacement had less overt heart disease than non-users. Currently, recommended estrogen therapy is supplemented with a progesterone to oppose the proliferative effect, and possible neoplastic risk, on the uterus in women who have not undergone hysterectomy (The Writing Group for the PEPI Trial, 1996; Grady, 1992). Until recently, there were concerns that progesterone may reduce some of the potential cardiovascular benefits of estrogen therapy. However, in the Nurses' Health Study, women taking progesterone with estrogen had a similar reduction in cardiovascular risk as those taking unopposed estrogens (Grodstein, 1996a). The effect of estrogen and progesterone therapy on heart disease is currently being tested in prospective randomized controlled trials.

This review will discuss the possible reasons why estrogen therapy may have a beneficial effect on cardiovascular disease. These include the effects of estrogen on the lipid profile, vascular reactivity, and other mechanisms which are now thought to play an important part in preventing the vascular events that cause clinically evident cardiovascular disease.

## EFFECTS ON THE LIPID PROFILE

Prior to menopause, women have lower serum low-density lipoprotein (LDL) cholesterol and higher high-density lipoprotein (HDL) levels compared to men. After menopause, LDL cholesterol increases and HDL cholesterol falls (Stevenson, 1993; Matthews, 1989). Oral estrogen replacement therapy restores the lipid profile toward premenopausal levels (The Writing Group of the PEPI Trial, 1995; Hong, 1992; Walsh, 1991). Reduction in LDL cholesterol and increase in HDL cholesterol with estrogen therapy has been associated with a reduced risk of coronary artery disease in women in observational studies (Lafferty, 1994). Estrogen therapy also reverses the rise in lipoprotein(a) after menopause (Sacks, 1994a; Soma, 1993). However, the clinical benefit of this change is uncertain, as lipoprotein(a) has not been consistently linked to the risk of cardiovascular events (Sunayama, 1996; Kinlay, 1996; Ridker, 1993), and the effect of lowering lipoprotein(a) is unknown. It is of concern that very low density lipoprotein is increased with estrogen replacement (The Writing Group of the PEPI Trial, 1995; Walsh, 1991), although this change does not have a proven effect on atheroma or risk of coronary events (Sacks, 1994b).

Two recent randomized controlled trials (Sacks, 1996; Kjeskshus, 1995) have conclusively proven the concept that lipid lowering can markedly reduce coronary events in women with coronary artery disease. Compared to men, women in these studies treated with HMG CoA reductase inhibitors had similar or greater reductions in the incidence of coronary events (Sacks, 1996; Kjeskshus, 1995). Although estrogen replacement likely reduces the risk of cardiovascular events by its favorable impact on lipid profiles, this mechanism does not fully account for the reduction in risk noted in the observational studies. In the Lipid Research Clinics Follow-up study, only about half of the reduced risk associated with estrogen therapy could be ascribed to the changes in the lipid profile (Bush, 1987). Furthermore, in patients with heterozygous familial hyperlipidemia, the more favorable course in women compared to men cannot be adequately explained by the differences in lipid profiles.

Despite higher levels of total and LDL cholesterol compared to men, women have less frequent, and more delayed onset of coronary disease (Hill, 1991).

## OTHER VASCULOPROTECTIVE MECHANISMS

Estrogen may prevent the oxidation of LDL within the arterial wall (Sack, 1994; Keaney, 1994; Rifici, 1992; Maziere, 1991; Yagi, 1986). Oxidized LDL plays a pivotal role in the activation of cellular components that promote the growth of an atherosclerotic plaque and make it vulnerable to rupture (Selwyn, 1996). 17$\beta$-estradiol shares the hydroxyphenol ring of other anti-oxidants such as vitamin E and probucol, and is capable of donating an electron to act as a reducing agent and lower "oxidative stress" within a plaque (Keaney, 1994; Sugioda, 1987). However, *in vitro* studies require supraphysiological concentrations of 17$\beta$-estradiol to achieve an anti-oxidant effect (Rifici, 1992; Maziere, 1991; Yagi, 1986). *In vivo* experimental studies have found that lower physiological concentrations of 17$\beta$-estradiol ($10^{-9}$ M) are able to exert an anti-oxidant effect (Sack, 1994). These data suggest that this effect of 17$\beta$-estradiol is due to generation of another antioxidant from the vessel wall (Guetta, 1996).

Estrogen exerts antithrombotic effects that may reduce cardiovascular risk. In the past estrogen was shown to have prothrombotic properties, partly based on the increase in the incidence of myocardial infarction in those men participating in the Coronary Drug Project who were randomized to receive 5 mg of conjugated estrogen per day (The Coronary Drug Project Research Group, 1970). However, this is a much higher dose than commonly used in post-menopausal women (0.625 mg per day). In the Postmenopausal Estrogen/Progestin Interventions trial (The Writing Group for the PEPI Trial, 1995), 0.625mg per day of conjugated equine estrogen reduced fibrinogen concentrations in post-menopausal women. In the Framingham Offspring Study (Gebara, 1995), and the Atherosclerosis Risk In Communities (ARIC) Study (Shahar, 1996), post-menopausal women receiving estrogen had lower levels of plasminogen activator inhibitor-1 antigen and tissue-type plasminogen activator antigen than similar women who were not taking estrogen replacement therapy. The levels of these endothelial factors that reflect fibrinolytic activity, were similar to those in women before menopause (Gebara, 1995), and may in part, reflect a reduction in oxidized LDL levels within the endothelium (Kugiyama, 1993). Experimental studies have shown that estrogen can also reduce platelet aggregation (Mitchell, 1974), although the effect has not been investigated *in vivo*.

These effects appear to conflict with recent observational studies describing a small, but statistically significant increase in risk of venous thromboembolism with estrogen replacement therapy (Daly, 1996; Grodstein, 1996b; Jick, 1996). In the observational studies estrogen replacement therapy doubles the risk of deep venous thrombosis and pulmonary embolism in otherwise healthy post-menopausal women. The absolute risk of venous thromboembolism was estimated to be very small (annual extra risk of 1–2 cases per 10,000 women treated) (Grodstein, 1996b; Jick, 1996), and much smaller than the potential benefits on cardiovascular risk. The increased risk of venous thromboembolism could be an artifact of a diagnostic bias, where women on hormone replacement therapy are more likely to be investigated, or it could relate to mechanisms that affect venous, but not arterial thrombosis. This uncertainty about the thromboembolic risk of estrogen replacement add further importance to the current randomized trials that will be able to assess the risks and benefits of this therapy.

Studies of atherosclerosis models in rabbits have demonstrated that estradiol can change the proportion of collagen and elastin and the ratio of procollagen Type I and Type III in the blood vessel wall (Fischer, 1985; 1972; Beldekas, 1981). If these changes occur in human coronary arteries, they could have an important impact on vascular distensibility or vulnerability of atherosclerotic plaques to rupture, and thereby, on the risk of clinical events.

## EFFECTS ON VASOMOTOR FUNCTION

Perhaps the most intriguing recent findings regarding estrogens relate to their ability to modulate coronary and systemic vasomotor tone. The effect of estrogen is highly dependent on the length of exposure, the preparation and dose used, addition of progesterone, and its lipid altering effects. Although most studies of estrogen have focused on endothelium-dependent vasodilator function (see below), estrogens also elicit a direct relaxation of vascular smooth muscle. However, this direct effect occurs largely at supraphysiological estrogen concentrations ($>10^{-9}$ M) (Chester, 1995).

### Acute Effects of Estrogen on Vasomotor Function

Four studies have examined the effect of the acute administration of intracoronary or intravenous estrogens on coronary vasomotion. In one study, physiological concentrations of intracoronary 17$\beta$-estradiol were administered to post-menopausal women undergoing cardiac catheterization (Gilligan, 1994a). Estradiol had no discernable direct effect on vascular tone, but improved the endothelium-dependent epicardial coronary vasomotor response to acetylcholine (Gilligan, 1994a). Enhancement of endothelium-dependent vasodilation was greater in those women with more impaired baseline endothelium-dependent responses. In addition, 17$\beta$-estradiol augmented the fall in coronary vascular resistance to acetylcholine reflecting improved vasodilation of the coronary microcirculation (Gilligan, 1994a). A study using supraphysiological concentration of estrogens also has shown improved endothelium-dependent vasomotion of the epicardial coronary arteries of post-menopausal women (Reis, 1994). In this study, the synthetic estrogen, ethinyl estradiol, also had a direct vasodilating effect on the coronary conduit and resistance vessels (Reis, 1994). This study suggests that a high dose of estrogen, and/or the particular synthetic estrogen preparation, may cause endothelium-independent vasodilation. If this non-specific dilation of resistance vessels occurred in a coronary bed with stenoses, it might lead to an adverse myocardial steal phenomenon that could increase ischemia (a persantine-like effect). We have studied the effect of Premarin™, the estrogen used most often in observational studies of cardiovascular risk, on coronary vasoreactivity of post-menopausal women. Intravenous Premarin™ (10 mg), improved acetylcholine-induced endothelium-dependent vasomotion of atherosclerotic epicardial arteries, but had no direct effect on resting vascular tone (Lieberman, 1993). In one study, endothelium-dependent responses to intracoronary 17$\beta$-estradiol improved in post-menopausal women, but not in men with coronary atherosclerosis (Collins, 1995). Thus, acute estrogen administration appears to benefit the cardiovascular system of post-menopausal women only.

Similar results have been observed in forearm resistance vessels following the acute administration of estrogens into the brachial artery. Changes in forearm blood flow reflect the changes in vascular tone of the microvascular circulation, and can be mediated by endothelium-dependent or independent mechanisms. In post-menopausal women, physiological concentrations of 17$\beta$-estradiol potentiated the forearm blood flow response to the endothelium-dependent vasodilator, acetylcholine (Gilligan, 1994b). 17$\beta$-estradiol also augmented the vasodilation with nitroprusside infusions in women with risk factors for coronary artery disease, but had no incremental effect on this endothelium-independent vasodilation in women without risk factors (Gilligan, 1994b). Premarin™, when administered intravenously in pharmacological doses, improved endothelium dependent vasodilation in forearm resistance vessels of healthy post-menopausal women (Gerhard, 1994). Supraphysiological doses of 17$\beta$-estradiol, when administered sublingually have been found to have a vasodilator effect in post-menopausal women (Volterrani, 1995). These studies are consistent with the results in the coronary circulation, and suggest that estrogens affect vasomotion through more than one mechanism.

The results of the acute administration of estrogens cannot be explained by an action on the classic intracellular estrogen receptor, or by changes in the lipid profile, as these would be affected over a much longer time course than that observed in the above studies.

Recently, 17$\beta$-estradiol has been found to modulate gene expression through DNA response elements that do not require the DNA binding domain of the classic estrogen receptor (Yang, 1996). A second nuclear estrogen receptor has also been detected in the rat (estrogen receptor $\beta$) (Kuiper, 1996). Multiple pathways and receptors that mediate estrogen effects could explain the different results on vasoactivity with different estrogens, doses or lengths of exposure.

### Long-term Effects of Estrogen on Vasomotor Function

The effect of long-term estrogen administration on vasomotor function has been evaluated in the brachial artery using ultrasound. This method can assess endothelial function by measuring the vasodilator response at a conduit artery, such as the brachial artery, to increases in shear stress induced by a rise in blood flow (flow mediated dilation). Flow induced dilation of the human brachial artery is mediated by nitric oxide, and it is inhibited by the nitric oxide synthase antagonist $N^6$monomethyl-L-arginine (Lieberman, 1996). As in the coronary studies, changes in the diameter of the brachial artery due to flow mediated dilation, can be compared to dilation from endothelium independent mechanisms, such as by exogenous nitrates.

Flow mediated dilation of the brachial artery closely reflects the coronary endothelium-dependent vasomotor responses to acetylcholine. Both are impaired by the major risk factors for coronary artery disease, such as smoking, hyperlipidemia and diabetes (Zeiher, 1995; Celermajer, 1994a; Lieberman, 1996; Celermajer, 1993; Egashira, 1993; Zeiher, 1991; Vita, 1990). In addition, there is a strong correlation between brachial artery flow mediated dilation, and vascular responses of the coronary artery to acetylcholine in individuals who have had both tests (Anderson, 1995a,b).

During the ovarian cycle, flow mediated dilation of the brachial artery increases with the rise in estrogen during the mid-cycle follicular stage, and decreases only modestly if at all, in the late-cycle luteal stage when progesterone and estrogen are both elevated (Hashimoto, 1995). Studies of brachial artery endothelium-dependent flow mediated dilation have revealed interesting differences between men and women. In one study, the decline in flow mediated dilation with age was delayed by approximately 10 years in women compared to men, occurring at the time of menopause (Celermajer, 1994b). As with acute infusion, longer term administration of endogenous estrogen may protect the endothelium from vasodilator dysfunction. Lieberman *et al.*, compared the effect of daily oral doses of 1 mg 17$\beta$-estradiol or 2 mg 17$\beta$-estradiol, to placebo on brachial artery flow mediated dilation of post-menopausal women (Lieberman, 1994). After 9 weeks of treatment, both doses of estrogen increased flow mediated dilation, compared to placebo, but had no effect on endothelium-independent dilation to oral nitroglycerin (Leiberman, 1994).

Studies in oophorectomized cholesterol fed primates have suggested that androgenic progesterones abolish most or all of the estrogen effects on endothelium-dependent vasodilation (Williams, 1994). This may not be the case with naturally occurring progesterones. Preliminary results from a recent study suggest that progesterone does not attenuate the vasomotor effects of estrogens in post-menopausal women (Gerhard, 1996).

Thus, the studies of the effect of longer term therapy of estrogens on vasomotor function suggest that replacement with physiological doses of estrogen and progesterone improves endothelial function. The effect is unlikely to depend on the changes in the lipid profile, which were small in each study, and may provide a mechanism to explain some of the direct benefit of estrogen replacement therapy on cardiovascular risk in post-menopausal women.

## THE EFFECT OF ESTROGENS ON MYOCARDIAL ISCHEMIA

The acute effect of estrogen on myocardial ischemia was studied in post-menopausal women with coronary artery disease (Rosano, 1993), who were given 1mg of 17$\beta$-estradiol sublingually or placebo prior to performing an exercise tolerance test. The acute administration of 17$\beta$-estradiol increased the time to electrocardiographic manifestation of ischemia, and increased the total exercise time compared to placebo. At peak exercise, there were no significant differences in the heart rate, systolic blood pressure, or rate pressure product, suggesting the reduction in ischemia may be due to an improvement in coronary blood flow. Preliminary results from another study suggest that medroxyprogesterone may attenuate the beneficial effects of 17$\beta$-estradiol on exercise induced myocardial ischemia more than natural progesterones (Rosano, 1996). The evidence from the studies of estrogen and vasomotion discussed earlier, suggest that increased coronary blood flow could be a result of augmented endothelium-dependent vasodilation in response to increased myocardial demand, direct vasodilation by estrogen, or a combination of both mechanisms.

## CONCLUSIONS

The risk of coronary artery disease in women increases at the time of menopause, corresponding to a time of decreasing estrogen production. Observational data are very consistent in demonstrating that post-menopausal women who receive estrogen therapy have fewer coronary events than those who are not on estrogen therapy. The reduced risk of coronary artery disease may be partly explained by oral hormone replacement therapy which alters lipids to a less atherogenic profile. In addition, the physiological replacement of estrogens may play an important role in the vascular biology of atherosclerosis, and help preserve endothelium-dependent vasomotor function. Improved vasomotor function helps to prevent constriction at the site of coronary stenoses that cause myocardial ischemia, and may reflect beneficial changes in other endothelial functions that determine thrombogenicity and the vulnerability of the atherosclerotic plaque. The vasomotor changes that are associated with estrogen replacement are not abolished by physiological replacement with progesterone. Adequate replacement of estrogen to physiological levels, with or without progesterone may contribute to the reduced risk of coronary artery disease observed in post-menopausal women who are taking hormone replacement therapy.

## REFERENCES

American Heart Association. (1992) Heart and Stroke Facts. Dallas, Tex.

Anderson TJ, Gerhard MD, Meredith IT *et al.* (1995a) Systemic nature of endothelial dysfunction in atherosclerosis. *Am. J. Cardiol.* **75**: 71B–74B.

Anderson TJ, Uehata A, Gerhard MD, Meredith IT, Knab S, Delagrange D, Lieberman EH, Ganz P, Creager MA, Yeung AC, Selwyn AP. (1995b) Close relationship of endothelial function in the human coronary and peripheral circulation. *J. Am. Coll. Cardiol.* **26**: 1235–1241.

Beldekas JC, Smith B, Gerstenfeld LC, Sonenshein GE, Franzblau C. (1981) Effects of 17$\beta$ estradiol on the biosynthesis of collagen in cultured bovine aortic smooth muscle cells. *Biochemistry* **20**: 2162–2167.

Bush TL, Barrett-Connor E, Cowan LD *et al.* (1987) Cardiovascular mortality and non-contraceptive estrogen use in women: results from the lipid research clinics program follow-up study. *Circulation* **75**: 1002–1009.

Celermajer DS, Sorensen KE, Gooch VM *et al.* (1993) Cigarette smoking is associated with dose-related and potentially reversible impairment of endothelium-dependent dilation in healthy young adults. *Circulation* **88**: 2149–2155.

Celermajer DS, Sorensen KE, Bull C, Robinson J, Deanfield JE. (1994a) Endothelium-dependent dilation in the systemic arteries of asymptomatic subjects relates to coronary risk factors and their interaction. *J. Am. Coll. Cardiol.* **24**: 1468–1474.

Celermajer DS, Sorensen KE, Spiegelhalter DJ, Georgakopoulos D, Robinson J, Deanfield JE. (1994b) Aging is associated with endothelial dysfunction in healthy men years before the age-related decline in women. *J. Am. Coll. Cardiol.* **24**: 471–476.

Chester AH, Jiang C, Borland JA, Yacoub MH, Collins P. (1995) Oestrogen relaxes human epicardial coronary arteries through non-endothelium-dependent mechanisms. *Coronary Artery Disease* **6**: 417–422.

Collins P, Rosano GMC, Sarrel PM *et al.* (1995) 17$\beta$-estradiol attenuates acetylcholine-induced coronary arterial constriction in women but not men with coronary heart disease. *Circulation* **92**: 24–30.

Cox DA, Cohen ML. (1996) Effects of oxidized low-density lipoprotein on vascular contraction and relaxation: Clinical and pharmacological implications in atherosclerosis. *Pharmacological Reviews* **48**: 3–19.

Daly E, Vessey MP, Hawkins MM, Carson JL, Gough P, Marsh S. (1996) Risk of venous thromboembolism in users of hormone replacement therapy. *Lancet* **348**: 977–980.

Egashira K, Inou T, Hirooka Y *et al.* (1993) Impaired coronary blood flow response to acetylcholine in patients with coronary risk factors and proximal atherosclerotic lesions. *J. Clin. Invest.* **91**: 29–37.

Fischer GM, Swain ML. (1985) Effects of estradiol and progesterone on the increased synthesis of collagen in atherosclerotic rabbit aortas. *Atherosclerosis* **54**: 1770–1785.

Fischer GM. (1972) *In vivo* effects of estradiol on collagen and elastin dynamics in rat aorta. *Endocrinology* **9**: 1227–1232.

Gebara OCE, Mittleman MA, Sutherland P *et al.* (1995) Association between increased estrogen status and increased fibrinolytic potential in the Framingham offspring study. *Circulation* **91**: 1952–1958.

Gerhard M, Tawakol A, Haley E, Creager SJ, Walsh BW, Ganz P, Creager MA. (1996) Long-term estradiol therapy with or without progesterone improves endothelium-dependent vasodilation in post-menopausal women. *Circulation* **94** (Suppl): I-279.

Gerhard MD, Roddy M-A, Knab ST, Creager SJ, Creager MA. (1994) Acute estrogen administration improves endothelium-dependent vasodilation in postmenopausal women. *Circulation* **90** (Suppl): I-86.

Gilligan DM, Quyyumi AA, Cannon RO. (1994a) Effects of physiological levels of estrogen on coronary vasomotor function in postmenopausal women. *Circulation* **89**: 2545–2551.

Gilligan DM, Badar DM, Panza JA, Quyyumi AA, Cannon RO. (1994b) Acute vascular effects of estrogen in postmenopausal women. *Circulation* **90**: 786–791.

Grady D, Rubin SM, Petitti DB, Fox CS *et al.* (1992) Hormone therapy to prevent disease and prolong life in postmenopausal women. *Ann. Intern. Med.* **117**: 1016–1037.

Grodstein F, Stampfer MJ, Manson JE *et al.* (1996a) Postmenopausal estrogen and progestin use and the risk of cardiovascular disease. *N. Engl. J. Med.* **335**: 453–461.

Grodstein F, Stampfer MJ, Goldhaber SZ *et al.* (1996b) Prospective study of exogenous hormones and risk of pulmonary embolism in women. *Lancet* **348**: 983–987.

Grodstein F, Stampfer M. (1995) The epidemiology of coronary heart disease and estrogen replacement in postmenopausal women. *Progress in Cardiovascular Diseases* **38**: 199–210.

Guetta V, Cannon RO III. (1996) Cardiovascular effects of estrogen and lipid-lowering therapies in postmenopausal women. *Circulation* **93**: 1928–1937.

Hashimoto M, Akishita M, Eto M *et al.* (1995) Modulation of endothelium-dependent flow-mediated dilation of the brachial artery by gender and menstrual cycle. *Circulation* **92**: 3431–3435.

Hill JS, Hayden MR, Frolich J, Pritchard OH. (1991) Genetic and environmental factors affecting the incidence of coronary artery disease in heterozygous familial hypercholesterolemia. *Arterioscler. Thromb.* **11**: 290–297.

Hong MK, Romm PA, Reagan K, Green CE, Rackley CE. (1992) Effects of estrogen replacement therapy on serum lipid values and angiographically defined coronary artery disease in post-menopausal women. *Am. J. Cardiol.* **69**: 176–178.

Jick H, Derby LE, Myers MW, Vasilakis C, Newton K. (1996) Risk of hospital admission for idiopathic venous thromboembolism among users of postmenopausal oestrogens. *Lancet* **348**: 981–983.

Keaney JF, Shwaery GT, Xu A, Nicolosi RJ, Loscalzo J, Foxall TL, Vita JA. (1994) 17$\beta$-estradiol preserves endothelial vasodilator function and limits low-density lipoprotein oxidation in hypercholesterolemic swine. *Circulation* **89**: 2251–2259.

Kinlay S, Dobson AJ, Heller RF, McElduff P, Alexander H, Dickeson J. (1996) Risk of primary and recurrent acute myocardial infarction from lipoprotein(a) in men and women. *J. Am. Coll. Cardiol.* **28**: 870–875.

Kjeskshus J, Pedersen TR. (1995) Reducing the risk of coronary events: evidence from the Scandinavian Simvastatin Survival Study (4S). *Am. J. Cardiol.* **76**: 64C–68C.

Kugiyama K, Sakamoto T, Misumi I *et al.* (1993) Transferable lipids in oxidized low-density lipoprotein stimulate plasminogen activator release from endothelial cells. *Circ. Res.* **73**: 335–343.

Kuiper GG, Enmark E, Pelto-Huikko M, Nilsson S, Gustafsson JA. (1996) Cloning of a novel receptor expressed in rat prostate and ovary. *Proc. Natl. Acad. Sci. USA* **93**: 5925–5930.

Lafferty FW, Fiske ME. (1994) Postmenopausal estrogen replacement: a long-term cohort study. *Am. J. Med.* **97**: 66–76.

Lieberman EH, Gerhard MD, Uheta A, Selwyn AP, Ganz P, Yeung AC, Creager MA. (1996) Flow-induced vasodilation of the human brachial artery is impaired in patients <40 years of age with coronary artery disease. *Am. J. Cardiol.* **78**: 1210–1214.

Lieberman EH, Gerhard MD, Uehata A, Walsh BW, Selwyn AP, Ganz P, Yeung AC, Creager MA. (1994) Estrogen improves endothelium-dependent flow-mediated vasodilation in postmenopausal women. *Ann. Intern. Med.* **121**: 936–941.

Lieberman EH, Gerhard M, Yeung AC, Anderson TJ, Meredith IT, Ganz P, Selwyn AP. (1993) Estrogen improves coronary vasomotor responses to acetylcholine in post-menopausal women. *Circulation* **88** (Suppl.): I-79.

Matthews KA, Meilahn E, Kuller LH, Kelsey ST, Caggiula AW, Wing RR. (1989) Menopause and risk factors for coronary artery disease. *N. Engl. J. Med.* **321**: 641–646.

Maziere C, Auclair M, Ronveaux M-F, Salmon S, Santus R, Maziere J-C. (1991) Estrogens inhibit copper and cell-mediated modification of low density lipoprotein. *Atherosclerosis* **89**: 175–182.

Mitchell HC. (1974) Effect of estrogens and a progestogen on platelet adhesiveness and aggregation in rabbits. *J. Lab. Clin. Med.* **83**: 79–89.

National Center for Health Statistics. Vital statistics of the United States. (1989) Vol. II. Mortality. Part A. Washington, D.C.: Government Printing Office, 1993. (DHHS publication no. (PHS) 93–1101).

Reis SE, Gloth ST, Blumenthal RS, Resar JR, Zacur HA, Gerstenblith G, Brinker JA. (1994) Ethinyl estradiol acutely attenuates abnormal coronary vasomotor responses to acetylcholine in post-menopausal women. *Circulation* **89**: 52–60.

Ridker PM, Hennekens CH, Stampfer MJ. (1993) A prospective study of Lipoprotein(a) and the risk of myocardial infarction. *JAMA* **270**: 2195–2199.

Rifici VA, Khachadurian AK. (1992) The inhibition of low-density lipoprotein oxidation by 17$\beta$-estradiol. *Metabolism* **41**: 1110–1114.

Rosano GMC, Sarrel PM, Cheirchia SL *et al.* (1996) Medroxyprogesterone but not natural progesterone, reverses the beneficial effect of estradiol 17 beta upon exercise-induced myocardial ischemia. *Circulation* **94** (Suppl): I-18.

Rosano GMC, Sarrel PM, Poole-Wilson PA, Collins P. (1993) Beneficial effect of estrogen on exercise-induced myocardial ischaemia in women with coronary artery disease. *Lancet* **342**: 133–136.

Sack MN, Rader DJ, Cannon RO. (1994) Oestrogen and inhibition of oxidation of low-density lipoproteins in postmenopausal women. *Lancet* **343**: 269–270.

Sacks FM, Pfeffer MA, Moye LA *et al.* (1996) The effect of Pravastatin on recurrent coronary events after myocardial infarction in patients with average cholesterol levels. *N. Engl. J. Med.* **335**: 1001–1009.

Sacks FM, McPherson R, Walsh BW. (1994a) Effect of postmenopausal estrogen replacement on plasma Lp(a) lipoprotein concentrations. *Arch. Intern. Med.* **154**: 1106–1110.

Sacks FM, Walsh BW. (1994b) Sex hormones and lipoprotein metabolism. *Curr. Opin. Lipidol.* **5**: 236–240.

Selwyn AP, Kinlay S, Libby P, Ganz P. (1997) Atherogenic lipids, vascular dysfunction, and clinical signs of ischemic heart disease. *Circulation* **95**: 5–7.

Shahar E, Folsom AR, Salomaa VV *et al.* (1996) For the Atherosclerosis Risk in Communities (ARIC) Study Investigators. Relation of hormone-replacement therapy to measures of plasma fibrinolytic activity. *Circulation* **93**: 1970–1975.

Soma MR, Osnago-Gadda I, Paoletti R, Fumagalli R, Morrisett JD, Meschia M, Crosignani P. (1993) The lowering of lipoprotein(a) induced by estrogen plus progesterone replacement therapy in postmenopausal women. *Arch. Intern. Med.* **153**: 1462–1468.

Stampfer MJ, Colditz GA, Willett WC, Manson JE, Rosner B, Speizer FE, Hennekens CH. (1991a) Postmenopausal estrogen therapy and cardiovascular disease: Ten-year follow-up from the Nurses' Health Study. *N. Engl. J. Med.* **325**: 756–762.

Stampfer MJ, Colditz GA. (1991b) Estrogen replacement therapy and coronary heart disease: A quantitative assessment of the epidemiologic evidence. *Prev. Med.* **20**: 47–63.

Stevenson JC, Crook D, Godsland IF. (1993) Influence of age and menopause on serum lipids and lipoproteins in healthy women. *Atherosclerosis* **98**: 83–90.

Sugioka JM, Shimosegawa Y, Nakano MM. (1987) Estrogens as natural antioxidants of membrane lipid peroxidation. *FEBS Lett.* **210**: 37–39.

Sunayama S, Daida H, Mokuno H *et al.* (1996) Lack of increased coronary atherosclerotic risk due to elevated lipoprotein(a) in women ⩾55 years of age. *Circulation* **94**: 1263–1268.

The Coronary Drug Project Research Group. (1970) The Coronary Drug Project: initial findings leading to modifications of its research protocol. *JAMA* **214**: 1303–1313.

The Writing Group for the PEPI Trial. (1996) Effects of hormone replacement therapy on endometrial histology in postmenopausal women. The Postmenopausal Estrogen/Progestin Interventions (PEPI) Trial. *JAMA* **275**: 370–375.

The Writing Group for the PEPI Trial. (1995) Effects of estrogen or estrogen/progestin regimens on heart disease risk factors in postmenopausal women: The Postmenopausal Estrogen/Progestin Interventions (PEPI) Trial. *JAMA* **273**: 199–208.

Vita JA, Treasure CB, Nabel EG, McLenachan JM, Fish RD, Yeung AC, Vekshtein VI, Selwyn AP, Ganz P. (1990) The coronary vasomotor response to acetylcholine relates to risk factors for coronary artery disease. *Circulation* **81**: 491–497.

Volterrani M, Rosano G, Coats A, Beale C, Collins P. (1995) Estrogen acutely increases peripheral blood flow in postmenopausal women. *Am. J. Med.* **99**: 119–122.

Walsh BW, Schiff I, Rosner B, Greenberg L, Ravnikar V, Sacks FM. (1991) Effects of postmenopausal estrogen replacement on the concentrations and metabolism of plasma lipoproteins. *N. Engl. J Med.* 1196–1204.

Williams JK, Honore EK, Washburn SA, Clarkson TB. (1994) Effects of hormone replacement therapy on reactivity of atherosclerotic coronary arteries in cynomolgus monkeys. *J. Am. Coll. Cardiol.* **24**: 1757–1761.

Yagi K, Komura S. (1986) Inhibitory effect of female hormones on lipid peroxidation. *Biochem. Int.* **13**: 1051–1055.

Yang NN, Venugopalan M, Hardikar S, Glasebrook A. (1996) Identification of an estrogen response element activated by metabolites of 17$\beta$-estradiol and raloxifene. *Science* **273**: 1222–1225.

Zeiher AM, Schachinger V, Minners J. (1995) Long-term cigarette smoking impairs endothelium-dependent coronary arterial vasodilator function. *Circulation* **92**: 1094–1100.

Zeiher AM, Drexler H, Wollschläger H, Just H. (1991) Modulation of coronary vasomotor tone in humans: Progressive endothelial dysfunction with different early stages of coronary atherosclerosis. *Circulation* **83**: 391–401.

# 19 Prevention of Cardiovascular Disease by Estrogen

**David M. Herrington**

*Division of Cardiology, The Bowman Gray School of Medicine, Medical Center Blvd., Winston-Salem, NC 27157-1045, USA, Tel.: (910) 716-4950, Fax: (910) 716-9188, E-mail: dmh@osler.card.wfu.edu.*

There are numerous observational studies suggesting that postmenopausal estrogen replacement is associated with lower risk of clinical manifestations of coronary heart disease. These data are consistent and supported by several biologically plausible mechanisms through which estrogen could influence the atherosclerotic process. Until recently, however, there have been few data on the relationship between estrogen use and directly measured atherosclerosis in humans. These new data, along with the plans for several new angiographic endpoint clinical trials of estrogen, are presented in detail. These trials will complement the primary and secondary prevention trials already underway to provide a comprehensive assessment of the role of estrogen in the prevention of cardiovascular disease.

KEYWORDS: Coronary heart disease, angiography, estrogen.

## INTRODUCTION

Much has been written about the potential role of estrogen in the prevention of cardiovascular disease. Indeed, many investigators and clinicians appear to have accepted as fact that estrogen reduces the risk for cardiovascular events by 50% (Wren, 1992). This conclusion is based largely on the volume of observational studies of clinical cardiovascular events in postmenopausal women coupled with the well-established lipid effects of estrogen (The writing Group for the PEPI Trial, 1995). These lipid effects, combined with the increasingly dramatic effects of estrogen on the vessel wall, (Herrington, 1996) discussed in other chapters of this book, serve as compelling biologically plausible mechanisms through which estrogen could act to lower risk of heart disease.

Nonetheless, the body of evidence concerning the cardiovascular benefits of estrogen is far from complete. Recently, several studies have examined the relationship between estrogen and directly measured manifestations of atherosclerosis in humans, rather than the clinical events that typically ensue. These studies provide more direct insight into the effects of estrogen on the pathogenesis and prevention of atherosclerosis. Several angiographic and clinical endpoint trials are now underway which will provide the most conclusive evidence to date of the role of estrogen in prevention of cardiovascular disease. This chapter will briefly review the observational studies of clinical endpoints and then focus on the relatively new data and new clinical trials concerning estrogen and directly measured atherosclerosis in humans. A discussion of data concerning the effects of estrogen on the pathogenesis of atherosclerosis in animal models and the effects of estrogen on endothelial function are the subjects of other chapters.

## POSTMENOPAUSAL ESTROGEN REPLACEMENT AND RISK FOR CLINICAL CARDIOVASCULAR EVENTS

There have been over seventy observational studies published since 1966 examining the relationship between postmenopausal estrogen use and risk of cardiovascular disease. In general, these studies have been amazingly consistent in demonstrating an inverse association between postmenopausal estrogen use and risk of coronary heart disease. This consistency is best illustrated in three independently conducted meta-analyses of the available published data. (Psaty *et al.*, 1993; Grady *et al.*, 1992; Stampfer and Colditz, 1991) Among the three meta-analyses, the overall risk of coronary heart disease ranged from 0.50–0.65 depending on the types of studies included. The results of the meta-analysis by Stampfer *et al.* are found in Figure 19.1.

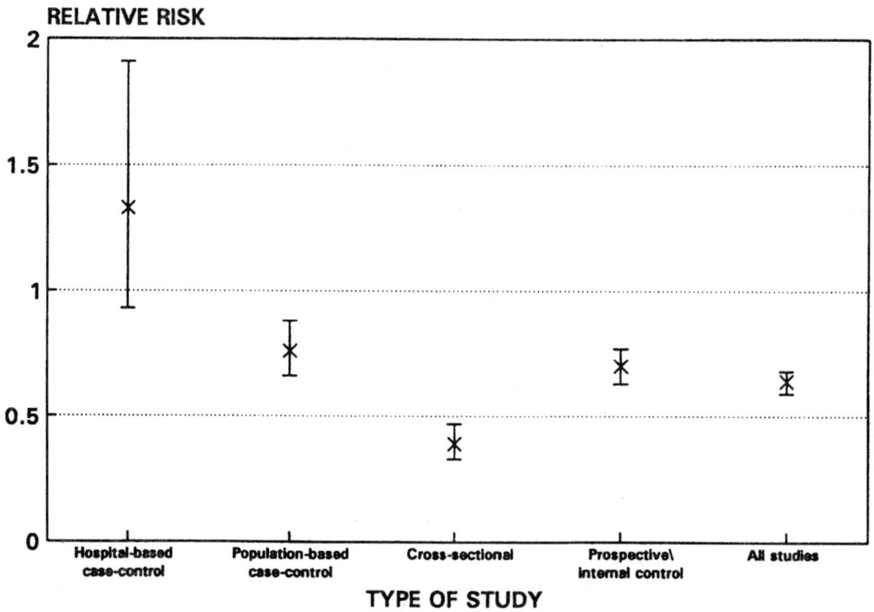


**Figure 19.1.** Summary of relative risks and 95% confidence intervals for postmenopausal estrogen use and risk of coronary heart disease. The point estimates are stratified according to study design. (Adapted with permission from Grodstein and Stampfer, *Progress in Cardiovascular Diseases*, Volume 38, pp. 199–210, 1995.

The data on estrogen use and other manifestations of cardiovascular disease are less clear. Among the fifteen published studies of estrogen use and risk of stroke, the pooled estimate of relative risk was 0.96 (Grady *et al.*, 1992). This author is unaware of any published studies of the relationship between estrogen use and peripheral arterial disease.

The strength and consistency of the findings regarding estrogen use and reduced risk of coronary events is remarkable. Indeed, they fulfill two of the criteria commonly accepted in support of a causal inference. Exposures may be causally related to outcomes of interest if there is a strong association between the exposure and the outcome, and if the association is reproducible in different studies from different populations (Gordis, 1996). However, there remain lingering questions about potential bias in these observational studies. In general, users of post-menopausal estrogen have been healthier than similar women not on estrogen. Derby *et al.* (1995) reported 842 women participating in a random survey of women from communities in New England. Users of estrogen were less likely to be smokers and more likely to have had their cholesterol checked and exercise regularly ($p<0.05$). Similar observations were made from the Rancho Bernardo cohort by Barrett-Connor and Bush (1991). While the possibility of bias is clear, it is highly unlikely that this bias could account for all of the apparent reduction in risk from the observational studies. It does, however, suggest that our current estimates of the size of the effect remain uncertain. Controlled clinical trials are required to clarify this point.

## CROSS-SECTIONAL STUDIES OF ESTROGEN REPLACEMENT AND VASCULAR ATHEROSCLEROSIS

Despite the large number of studies of estrogen replacement therapy and clinical cardiovascular events, there are relatively few studies examining the relationship between estrogen use and extent or progression of directly measured vascular atherosclerosis. Nonetheless, the currently available data are useful to review. There are four cross-sectional studies examining the relationship between estrogen use and the extent and severity of angiographically-defined coronary disease (Table 19.1) (Hong *et al.*, 1992; McFarland *et al.*, 1989; Gruchow *et al.*, 1988;

**Table 19.1** Cross-sectional studies of estrogen use and directly measured atherosclerosis

| *Angiographic studies* | *Relative risk* |
|---|---|
| ◆ Gruchow | 0.4–0.6 |
| ◆ Sullivan | 0.6 |
| ◆ McFarland | 0.5 |
| ◆ Hong | 0.1 |
| *Carotid ultrasound studies* | |
| ◆ Punnonen | 0.1–0.2 |
| ◆ Jonas | 0.6–0.8 |
| ◆ Nabulsi | N.S. |

Sullivan *et al.*, 1988). In a study of 2188 women undergoing clinically-indicated coronary angiography, Sullivan *et al.* (1988) reported that women with normal coronaries were far more likely to be current users of estrogen than women with at least one ≥70% stenosis ($p<0.05$). Among the 744 women with angiographic evidence of significant coronary disease, only 2.7% were current users of estrogen compared with 7.7% of the 1444 women with normal coronaries. In a similar study, McFarland *et al.* (1989) reported on 345 women aged 35–59 years with angiographic assessment of their coronary anatomy. The 137 women with at least one ≥70% coronary stenosis were half as likely to be using estrogen replacement as the 208 women with normal coronaries (O.R.=0.50; $p<0.01$).

Similar results were reported by Gruchow *et al.* (1988) in a study of 933 women examined by angiography. In that study, use of postmenopausal estrogen was inversely related to occlusion scores after adjustment for age, type of menopause and most cardiovascular risk factors ($p<0.05$). This inverse relationship was strongest among women with the most severe occlusion scores. The one risk factor that seemed to account for some of the association between estrogen and angiographic disease was plasma HDL-cholesterol level. In a much smaller study, Hong *et al.* (1992) examined 90 consecutive postmenopausal women undergoing diagnostic angiography. Among the eighteen women on estrogen replacement, only four (22%) had any lesion greater than or equal to 25% diameter stenosis. In contrast, 49 of the 72 women (68%) not on estrogen had angiographic evidence of atherosclerosis (OR=0.13, $p<0.001$). In multiple logistic regression models including other predictors of CAD, absence of estrogen replacement was the most powerful predictor of atherosclerosis ($p<0.001$). As in the Gruchow study, plasma HDL was also a significant predictor of disease, even after adjusting for estrogen use ($p<0.01$).

Cross-sectional data on estrogen and vascular atherosclerosis are not limited to coronary arteries. There are several studies examining the relationship between estrogen use and carotid, femoral, and aortic atherosclerosis as measured by vascular ultrasound (Table 19.1). In some cases, the data come from large cohort studies of cardiovascular disease and its risk factors. These studies avoid the problem of selection biases inherent in studies of cath lab cohorts. Ultrasound studies of carotid atherosclerosis also provide a means to examine the relationship between estrogen use and atherosclerosis at an earlier stage in the disease process, even before any clinical or angiographic evidence of disease is present.

Punnonen *et al.* (1995) performed vascular ultrasound on a small group of women on estradiol valerate with and without levonorgestrel and compared the findings to similar studies in postmenopausal women who had never used hormone replacement. The prevalence of abdominal, carotid and femoral fibrous and calcific plaques was significantly lower in the women using hormone replacement ($p<0.01$). In the Cardiovascular Health Study (CHS), the relationship between carotid wall thickness and postmenopausal estrogen replacement was examined in 2962 women over the age of 65 years (Jonas *et al.*, 1996; Manolio *et al.*, 1993). In this cohort, the risk of any abnormal thickening (≥1% stenosis) in the internal and common carotid artery was significantly lower in current users of estrogen replacement when compared with non-users ($p<0.05$). The average wall thicknesses of the internal and common carotid arteries were significantly greater than in non-users of estrogen even after adjustment for several important covariates. Despite these observations, an inverse association between estrogen and vascular atherosclerosis measured by ultrasound has not been consistently reported. Nabulsi *et al.* (1996) found no difference in wall thickness between users and non-users of postmenopausal hormone replacement among 5436 female participants in the Atherosclerosis Risk in Communities (ARIC) Study.

## PROSPECTIVE DATA ON ESTROGEN USE AND PROGRESSION OF ATHEROSCLEROSIS

One of the limitations of cross-sectional data is that the temporal relationship between exposure to estrogen and progression of vascular disease cannot be established. In these studies, it is impossible to tell whether the estrogen deprivation was a cause or a consequence of atherosclerosis. For many years the Physicians Desk Reference has stated that estrogens "should be used with caution in patients with cerebrovascular and cardiovascular disease..." (Anonymous, 1996). Thus, in the past many physicians may have withheld estrogen in their female patients once they developed cardiovascular disease. To overcome the question of the temporal relationship between estrogen use and subsequent disease progression, prospective data are required. Unfortunately, there are very few prospective data on estrogen use and progression of atherosclerosis.

The best prospective data come from secondary analyses of cardiovascular clinical trials of other types of cardiovascular interventions. In an analysis of data from the Asymptomatic Carotid Atherosclerosis Progression Study (ACAPS), Espeland *et al.* (1995) compared women who were users and non-users of postmenopausal estrogen at the time of randomization to lovastatin or placebo (Figure 19.2). Carotid intimal-medial wall thickness was measured at baseline and serially over three years of follow-up. Among the 186 postmenopausal women, those who were on estrogen at baseline ($N=63$) had significant regression of their carotid wall thickness regardless of their assignment to lovastatin or placebo ($p=0.05$). In contrast, the women on placebo who were not taking estrogen had significant progression of their disease. Similarly, O'Brien *et al.* (1996) described the progression of coronary luminal narrowing six months following percutaneous coronary interventions in 204 women enrolled in the Coronary Angioplasty vs. Excisional Atherectomy Trial I (CAVEAT I). The 47 postmenopausal women on estrogen at the time of the intervention had significantly less restenosis than similar women not taking estrogen at baseline ($p=0.055$). Specifically, the women on estrogen who underwent directional coronary atherectomy had only a 2790 restenosis rate compared to a 5790 restenosis rate in women not on estrogen. The dramatic prospective relationship between estrogen use and restenosis is consistent with studies of estrogen in animal models of restenosis and supports the concept that estrogen may play a fundamental role in the response to injury that is a central element in the pathogenesis of atherosclerosis (Espinosa, Oemar and Luscher, 1996; Chen *et al.*, 1996).

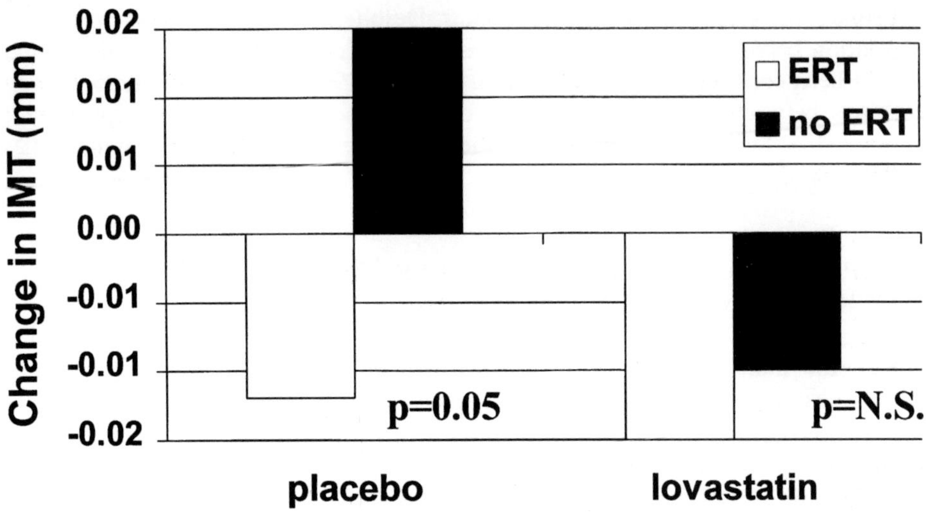


**Figure 19.2.** Progression/regression of carotid intimal-medial wall thickness (IMT) according to estrogen use in the Asymptomatic Carotid Artery Study. ERT = estrogen replacement therapy. (Adapted with permission from Espeland *et al.*, *American Journal of Epidemiology*, Volume 142, pp. 1011–1019, 1995.)

## CLINICAL TRIALS OF ESTROGEN REPLACEMENT TO SLOW PROGRESSION OR INDUCE REGRESSION OF CORONARY ATHEROSCLEROSIS

Definite proof of the effects of estrogen replacement on the pathogenesis, progression and regression of coronary atherosclerosis will depend on well-designed randomized clinical trials in postmenopausal women. Clinical trials with sufficient numbers of women are the only way to assure, with a high degree of confidence, that a difference between treatment groups is temporally related to exposure to estrogen and unlikely to be due to unequal distribution of other factors that influence the outcomes of interest. While no such clinical trials have been completed, there are four underway, all sponsored by the National Institutes of Health, that will provide answers to many of the important questions regarding estrogen and progression and regression of coronary atherosclerosis (Table 19.2).

The first angiographic endpoint trial of estrogen replacement therapy is the Estrogen Replacement and Atherosclerosis (ERA) Trial (PI: D. Herrington). Beginning in December, 1994, 300 women with angiographically proven coronary atherosclerosis were recruited. After baseline coronary angiography, participants were randomized to receive conjugated equine estrogen 0.625 mg daily, conjugated equine estrogen 0.625 mg plus medroxyprogesterone acetate (MPA) 2.5 mg daily or placebo. The women are followed for approximately three years after which they undergo repeat coronary angiography. Quantitative coronary angiography (QCA) will be used to measure changes in the severity and extent of angiographically evident luminal encroachment. The primary outcome will be based on the minimum diameter of each of fifteen standardized segments of the coronary tree. The incidence

**Table 19.2** Angiographic endpoint trials of estrogen replacement therapy

| *Trial name* | *N=* | *Intervention(s)* | *Design* |
|---|---|---|---|
| ERA (PI: D. Herrington) | 300 | CEE 0.625 mg<br>CEE 0.625 mg + MPA 2.5 mg† | 3-arm placebo controlled |
| WELL-HART (PI: H. Hodis) | 214 | LL & $E_2$ 1.0 mg<br>LL & $E_2$ 1.0 mg + MPA 5.0 mg‡ | 3-arm placebo controlled |
| EAGAR (PI: P. Ouyang) | 200 | $E_2$ 1.0 mg + MPA 2.5 mg | 2-arm placebo controlled |
| WAVE (PI: D. Waters *) | 450 | CEE 0.625 mg +/− MPA 2.5 mg†<br>+/− Vitamin E & C | 2 × 2 factorial placebo controlled |

CEE = Conjugated equine estrogen; LL = Lipid lowering therapy; $E_2$ = Estradiol; MPA = Medroxyprogesterone acetate; *—Chair of Steering Committee; † = continuous; ‡ = cyclic.

of clinical cardiovascular events will also be compared among the three treatment arms. An important feature of ERA is the measurement of plasma factors that may help clarify the mechanisms through which estrogen exerts its effects. In addition to plasma lipids and lipoproteins, measures of oxidizability of LDL, insulin resistance, and certain components of the renin/angiotensin cascade will offer helpful opportunities to clarify which of the metabolic effects of estrogen are most important in modulating the progression of coronary atherosclerosis.

A second angiographic trial underway is called the Womens Estrogen Progestin Lipid Lowering Hormone Atherosclerosis Regression Trial (WELL-HART) (PI: H. Hodis). In this trial, 214 women are intended for randomization to estradiol 1 mg, estradiol 1 mg plus cyclic medroxyprogesterone acetate (5 mg, days 1–10) or placebo. All women will receive pravastatin as required to lower LDL cholesterol to <100 mg/dl. As in the ERA trial, QCA evaluation of baseline and 3-year follow-up coronary angiography will be used to measure the effects of hormone replacement on progression of disease. The notable difference between ERA and WELL-HART rests in the management of plasma lipids during the trials. Since ERA will leave lipid management to the participants' usual source of medical care, it will be testing the effects of adding estrogen replacement to the care normally offered postmenopausal women with coronary artery disease. In contrast, WELL-HART will determine the marginal benefits of estrogen replacement over and above what can be achieved by aggressive lipid lowering therapy. Differences in the formulation of estrogen used (conjugated equine estrogen vs. estradiol) and the dosing regimen of progestin (continuous vs. cyclic) will also provide important additional contrasts between the two trials.

There are two other angiographic endpoint trials of estrogen replacement. One is Estrogen and Bypass Graft Atherosclerosis Regression (EAGAR) (PI: P. Ouyang). This trial is designed to determine if estrogen replacement will slow the development of native artery and vein graft atherosclerosis in postmenopausal women who have undergone coronary artery bypass graft (CABG) surgery. In this trial, 200 women will be randomized to oral estradiol (1 mg) plus continuous low dose MPA (2.5 mg) or placebo within four months of CABG surgery. Six months after randomization a baseline angiogram and intracoronary ultrasound will be performed. The primary outcome will be a visual assessment of global change in both native and vein graft atherosclerosis. However, QCA assessment of progression of disease and intracoronary ultrasound assessment of progression of vein graft atherosclerosis will be important secondary outcomes.

The final angiographic endpoint trial is the Womens Angiographic Vitamin/ Estrogen Trial (WAVE) (Chair-Steering Committee: D. Waters). This trial is intended to be a $2 \times 2$ factorial design to test the independent and joint effects of estrogen replacement therapy and antioxidant therapy to slow progression of coronary atherosclerosis. Starting in the spring of 1997, 450 women undergoing clinically-indicated coronary angiography were randomly assigned to estrogen replacement or placebo and antioxidant vitamins or placebo. The estrogen replacement will be conjugated equine estrogen (0.625 mg + MPA 2.5 mg). As in the other three trials, the follow-up period will be three years.

In the aggregate, these trials will provide a comprehensive assessment of the effects of estrogen on progression of coronary atherosclerosis. The marginal effects of estrogen over and above other forms of therapy including lipid lowering therapy and anti-oxidant therapy will be examined. Important information concerning the likely mechanisms through which estrogen may exert its effects will also be available. This information will be important in order to develop more mechanistically specific forms of therapy, and perhaps new therapeutic options that will be applicable for men.

## CARDIOVASCULAR CLINICAL ENDPOINT TRIALS OF ESTROGEN REPLACEMENT

Despite the important contribution of these angiographic trials to the understanding of estrogen and the pathogenesis and prevention of atherosclerosis, none of them will be sufficiently large to test hypotheses concerning the ability of postmenopausal estrogen to reduce clinical cardiovascular events. However, there are two large clinical endpoint trials currently underway that will address this issue. The Heart and Estrogen/Progestin Replacement Study (HERS) is a randomized double-blinded, placebo-controlled secondary prevention trial of estrogen plus progestin for the prevention of recurrent cardiovascular events in women. Starting in 1993, 2763 women with established coronary artery disease were recruited and randomized to active treatment (conjugated equine estrogen 0.625 mg plus medroxyprogesterone acetate 2.5 mg daily) or placebo. They will be followed for an average of 4.2 years. The primary outcome will be fatal and non-fatal coronary heart disease events (fatal and non-fatal MI and sudden death).

The Womens Health Initiative (WHI) includes a large primary prevention trial of estrogen replacement for prevention of CAD events in healthy women. Approximately 25000 women will be randomized to estrogen replacement or placebo and followed for nine years. Women with an intact uterus will receive CEE 0.625 mg plus MPA 2.5 mg daily or placebo while those without a uterus will receive unopposed estrogen or placebo. The results of this trial are not anticipated before 2005.

HERS and WHI will provide clinical trial evidence of the safety and efficacy of estrogen replacement for the prevention of recurrent or first CHD events. Although randomized, double-blinded clinical endpoint trials are considered the gold standard for determining the benefits of cardiovascular interventions, it is important to realize that even clinical trials have their limitations. For example, clinical trials invariably recruit a healthier cohort of participants than the actual intended target population for the intervention under study. However, interventions that are safe and effective in generally healthier subjects might result in greater adverse events in sicker patients. In this case, relative benefits of the intervention under study could be overestimated. Conversely, if the intervention is more effective in more severely diseased individuals, then a trial in mostly healthier subjects might miss an important treatment benefit. The comparability of the participants in the estrogen angiographic and clinical endpoint trials to the broader population of women with or at risk for coronary disease has not yet been established. Another limitation of clinical trials is the extensive and unrealistic ways in which compliance is encouraged. If a drug, like estrogen, suffers from poor compliance in real life, a benefit observed in a clinical trial may not be achievable under the ordinary circumstances of routine clinical medicine. Thus, it is important to remember that it is the totality of evidence, not simply the results of clinical trials, that should guide the use of new interventions for the prevention and treatment of coronary disease. In the case of estrogen replacement, the combination of the observational data, biologically plausible mechanisms, and experimental data in animals and humans (pending) should all be considered when seeking to understand the role of estrogen in the prevention of CAD.

## SUMMARY

Much attention has been paid to the numerous observational studies of postmenopausal estrogen use and risk of clinical cardiovascular events. Recently, these data have been extended by studies of estrogen and directly measured atherosclerosis in humans. These data, in conjunction with the newly initiated angiographic endpoint trials, will provide substantial additional information concerning the role of estrogen in the pathogenesis and prevention of atherosclerosis. They will also help clarify the mechanisms through which estrogen may exert its effects. Nonetheless, there will still remain numerous other pressing clinical questions regarding estrogen and coronary disease. For example, are there other formulations of estrogen replacement that are better tolerated or more effective than simple conjugated equine estrogen; how does the dose, type and route of administration of progestins modify the effects of estrogen; and is there a role for other estrogenic compounds (i.e. soy protein, tamoxifen analogues, etc.) in the prevention of atherosclerosis in women and men? These and other questions will undoubtedly be the subjects of future investigation.

## REFERENCES

Anonymous. (1996) *Physicians Desk Reference*, 50th edn. Montvale, NJ: Medical Economics Company.

Barrett-Connor, E. and Bush, T.L. (1991) Estrogen and coronary heart disease in women. *JAMA*, **265**, 1861–1867.

Chen, S.J., Li, H., Durand, J., Oparil, S. and Chen, Y.F. (1996) Estrogen reduces myointimal proliferation after balloon injury of rat carotid artery. *Circulation*, **93**, 577–584.

Derby, C.A., Hume, A.L., McPhillips, J.B., Barbour, M.M. and Carleton, R.A. (1995) Prior and current health characteristics of postmenopausal estrogen replacement therapy users compared with nonusers. *American Journal of Obstetrics & Gynecology*, **173**, 544–550.

Espeland, M.A., Applegate, W., Furberg, C., Lefkowitz, D., Rice, L. and Hunninghake, D. (1995) Estrogen replacement therapy and progression of intimal-medial thickness in the carotid arteries of postmenopausal women. *American Journal of Epidemiology*, **142**, 1011–1019.

Espinosa, E., Oemar, B.S. and Luscher, T.F. (1996) 17 beta-Estradiol and smooth muscle cell proliferation in aortic cells of male and female rats. *Biochemical & Biophysical Research Communications*, **221**, 8–14.

Gordis, L. (1996) From association to causation: deriving differences from epidemiologic studies. In: *Epidemiology*, edited by L. McGrew, pp. 167–182. Philadelphia, Pennsylvania: W.B. Saunders Company.

Grady, D., Rubin, S.M., Petitti, D.B., Fox, C.S., Black, D., Ettinger, B. *et al.* (1992) Hormone therapy to prevent disease and prolong life in postmenopausal women. *Annals of Internal Medicine*, **117**, 1016–1037.

Gruchow, H.W., Anderson, A.J., Barboriak, J.J. and Sobocinski, K.A. (1988) Postmenopausal use of estrogen and occlusion of coronary arteries. *American Heart Journal*, **115**, 954–963.

Herrington, D.M. (1996) Effects of estrogen on the anatomic and functional sequelae of coronary atherosclerosis. In: *Hormonal, Metabolic, and Cellular Influences on Cardiovascular Diseases in Women*, edited by T.M. Forte, Armonk, New York: Futura Publishing Company, Inc. (in press).

Hong, M.K., Romm, P.A., Reagan, K., Green, C.E. and Rackley, C.E. (1992) Effects of estrogen replacement therapy on serum lipid values and angiographically defined coronary artery disease in postmenopausal women. *American Journal of Cardiology*, **69**, 176–178.

Jonas, H.A., Kronmal, R.A., Psaty, B.M., Manolio, T.A., Meilahn, E.N., Tell, G.S. *et al.* (1996) Current estrogen-progestin and estrogen replacement therapy in elderly women: association with carotid atherosclerosis. *Annals of Epidemiology*, **6**, 314–323.

Manolio, T.A., Furberg, C.D., Shemanski, L., Psaty, B.M., O'Leary, D.H., Tracy, R.P. *et al.* (1993) Associations of postmenopausal estrogen use with cardiovascular disease and its risk factors in older women. The CHS Collaborative Research Group. *Circulation*, **88**, 2163–2171.

McFarland, K.F., Boniface, M.E., Hornung, C.A., Earnhardt, W. and Humphries, J.O. (1989) Risk factors and noncontraceptive estrogen use in women with and without coronary disease. *American Heart Journal*, **117**, 1209–1214.

Nabulsi, A.A., Folsom, A.R., Szklo, M., White, A., Higgins, M. and Heiss, G. (1996) No association of menopause and hormone replacement therapy with carotid artery intima-media thickness. *Circulation*, **94**, 1857–1863.

O'Brien, J.E., Peterson, E.D., Keeler, G.P., Berdan, L.G., Ohman, M., Faxon, D.P., Jacobs, A.K., Topol, E.J. and Califf, R.M. (1996) Relation between estrogen replacement therapy and restenosis after percutaneous coronary interventions. *Journal of the American College of Cardiology*, **28**, 1111–1118.

Psaty, B.M., Heckbert, S.R., Atkins, D., Siscovick, D.S., Koepsell, T.D., Wahl, P.W. *et al.* (1993) A review of the association of estrogens and progestins with cardiovascular disease in postmenopausal women. *Archives of Internal Medicine*, **153**, 1421–1427.

Punnonen, R.H., Jokela, H.A., Dastidar, P.S., Nevala, M. and Laippala, P.J. (1995) Combined oestrogen-progestin replacement therapy prevents atherosclerosis in postmenopausal women. *Maturitas*, **21**, 179–187.

Stampfer, M.J. and Colditz, G.A. (1991) Estrogen replacement therapy and coronary heart disease: a quantitative assessment of the epidemiologic evidence. *Preventive Medicine*, **20**, 47–63.

Sullivan, J.M., Vander Zwaag, R., Lemp, G.F., Hughes, J.P., Maddock, V., Kroetz, K.W. *et al.* (1988) Postmenopausal estrogen use and coronary atherosclerosis. *Annals of Internal Medicine*, **108**, 358–363.

The Writing Group for the PEPI Trial (1995) Effects of estrogen or estrogen/progestin regimens on heart disease risk factors in postmenopausal women. The Postmenopausal Estrogen/Progestin Interventions (PEPI) Trial. *JAMA*, **273**, 199–208.

Wren, B.G. (1992) The effect of oestrogen on the female cardiovascular system. *Medical Journal of Australia*, **157**, 204–208.

# INDEX